METHODS IN MOLECULAR BIOLOGY™

Series Editor
John M. Walker
School of Life Sciences
University of Hertfordshire
Hatfield, Hertfordshire, AL10 9AB, UK

For further volumes:
http://www.springer.com/series/7651

Rho GTPases

Methods and Protocols

Edited by

Francisco Rivero

Centre for Cardiovascular and Metabolic Research, The Hull York Medical School, University of Hull, Hull, UK

 Humana Press

Editor
Francisco Rivero
Centre for Cardiovascular and Metabolic Research
The Hull York Medical School, University of Hull
Hull, UK
francisco.rivero@hyms.ac.uk

ISSN 1064-3745   e-ISSN 1940-6029
ISBN 978-1-61779-441-4   e-ISBN 978-1-61779-442-1
DOI 10.1007/978-1-61779-442-1
Springer New York Dordrecht Heidelberg London

Library of Congress Control Number: 2011942754

Printed on acid-free paper

Humana Press is part of Springer Science+Business Media (www.springer.com)

Preface

The function of Rho GTPases in reorganizing the actin cytoskeleton was first demonstrated in yeast and mammalian cells in the early 1990s. From that time dates the paradigm of Rho, Rac, and Cdc42 being involved in the formation of stress fibers, lamellipodia, and membrane ruffles, respectively. The identification and characterization of effectors and regulators followed soon, and since then the field has been experiencing steady expansion, to the point that the key cellular processes where Rho GTPases do not participate in some way are becoming less and less. Thus, although initially described as major regulators of cytoskeletal remodeling, a "title" that they still conserve, Rho GTPases have been implicated in the establishment of polarity, endocytosis, vesicle trafficking, morphogenesis, cytokinesis, transcriptional activation, cell cycle progression, and apoptosis, to mention a few. Rho GTPases have acquired medical relevance because of their participation in tumorigenesis and metastasis, in cardiovascular conditions and as targets of infectious agents. With the classical triad of Rho/Rac/Cdc42 having been extensively (though not yet exhaustively) studied, more recently the so-called atypical Rho GTPases, those that follow patterns of regulation that deviate from the classical switching between the GTP-bound and the GDP-bound form, are starting to attract attention. The field of Rho GTPases has broadened even more with the contribution of studies in model organisms (plants, amoebas, fungi, invertebrates), each adding a particular view to the complexity of the family and vastly enriching our perception of these important signaling components.

Reflecting the importance of the Rho family, numerous methodological approaches have been developed over the years, covering a wide spectrum of disciplines, from biochemistry and structural biology to cell biology and physiology. This book presents a collection of techniques that are standard for researchers in the field. It should prove useful to those whose work brings them in the need of touching or entering the Rho GTPase field but also to those who, being already familiar with some of the techniques, wish or need to explore additional aspects. Each chapter provides the general principles that support the technique, followed by detailed step-by-step instructions and detailed notes that provide troubleshooting and help avoiding pitfalls.

This book is divided into five major parts. The first part provides an historical overview of the field and an account of the phylogenetics of the Rho family. The second part describes general biochemical methods, while the third part is dedicated to functional assays that allow monitoring the consequences of manipulating Rho GTPases in a variety of contexts. The fourth part is devoted to advanced imaging methods and to recently developed high-throughput methods. The last part includes techniques specifically designed for studies in selected nonmammalian model organisms. It is hoped that *RhoGTPase Protocols* will constitute an invaluable tool for all those with an interest in this remarkable family of signaling proteins.

Hull, UK   *Francisco Rivero*

Contents

Contributors

MOHAMMAD R. AHMADIAN • *Medical Faculty, Institute of Biochemistry and Molecular Biology II, Heinrich-Heine University, Düsseldorf, Germany*

JAMIE K. ALAN • *Department of Molecular Biosciences, University of Kansas, Lawrence, KS, USA*

ASTRID BASSE • *Molecular Pathology Section, Department of Biomedical Sciences, BRIC, University of Copenhagen, Copenhagen, Denmark*

DAMARIS BAUSCH-FLUCK • *Institute of Molecular Systems Biology, NCCR Neuro Center for Proteomics, ETH Zürich, Zürich, Switzerland*

HEIKE BIELEK • *Institute for Experimental and Clinical Pharmacology and Toxicology, University of Freiburg, Freiburg, Germany*

ETIENNE BOULTER • *Institut National de la Santé et de la Recherche Médicale Avenir Team, U634, Sophia-Antipolis University, Nice, France*

VANIA M.M. BRAGA • *Molecular Medicine Section, National Heart and Lung Institute, Faculty of Medicine, Imperial College London, London, UK*

CORD BRAKEBUSCH • *Molecular Pathology Section, Department of Biomedical Sciences, BRIC, University of Copenhagen, Copenhagen, Denmark*

RICHARD A. CERIONE • *Department of Chemistry and Chemical Biology, Cornell University, Ithaca, NY, USA; Department of Molecular Medicine, Cornell University, Ithaca, NY, USA*

EMILY J. CHENETTE • *Lineberger Comprehensive Cancer Center, University of North Carolina, Chapel Hill, NC, USA*

PATRICK CHINESTRA • *INSERM UMR 1037, Cancer Research Centre of Toulouse, Claudius Regaud Cancer Institute, University of Toulouse, Toulouse, France*

CHANNING J. DER • *Lineberger Comprehensive Cancer Center and Department of Pharmacology, University of North Carolina, Chapel Hill, NC, USA*

ANA FILIPA DOMINGUES • *Instituto de Biologia Molecular e Celular, Universidade do Porto, Porto, Portugal*

ANNE DOYE • *Faculté de Médecine, INSERM, U895, Centre Méditerranéen de Médecine Moléculaire, C3M, Institut Signalisation et Pathologie, IFR50, Université de Nice-Sophia-Antipolis, Nice, France*

BADRI N. DUBEY • *Medical Faculty, Institute of Biochemistry and Molecular Biology II, Heinrich-Heine University, Düsseldorf, Germany*

GARY EITZEN • *Department of Cell Biology, University of Alberta, Edmonton, AB, Canada*

MAREK ELIÁŠ • *Faculty of Science, University of Ostrava, Ostrava, Czech Republic; Faculty of Science, Charles University in Prague, Prague, Czech Republic*

JENNIFER C. ERASMUS • *Faculty of Medicine, Molecular Medicine Section, National Heart and Lung Institute, Imperial College London, London, UK*

JON W. ERICKSON • *Department of Chemistry and Chemical Biology, Cornell University, Ithaca, NY, USA*

GILLES FAVRE • *INSERM UMR 1037, Cancer Research Centre of Toulouse, Claudius Regaud Cancer Institute, University of Toulouse, Toulouse, France*

JEAN-CHARLES FAYE • *INSERM UMR 1037, Cancer Research Centre of Toulouse, Claudius Regaud Cancer Institute, University of Toulouse, Toulouse, France*

DANIEL FELTRIN • *Department of Biomedicine, Institute of Biochemistry and Genetics, University of Basel, Basel, Switzerland*

RAFAEL GARCIA-MATA • *Department of Cell and Developmental Biology, University of North Carolina at Chapel Hill, Chapel Hill, NC, USA*

PETER GIERSCHIK • *Institute of Pharmacology and Toxicology, University of Ulm Medical Center, Ulm, Germany*

MICHAEL GLOGAUER • *Faculty of Dentistry, Matrix Dynamics Group, University of Toronto, Toronto, ON, Canada*

LOTHAR GREMER • *Medical Faculty, Institute of Biochemistry and Molecular Biology II, Heinrich-Heine University, Düsseldorf, Germany*

LOUIS HODGSON • *Department of Anatomy and Structural Biology and Gruss-Lipper Biophotonics Center, Albert Einstein College of Medicine of Yeshiva University, Bronx, NY, USA*

ADAM D. HOPPE • *Department of Chemistry and Biochemistry, South Dakota State University, Brookings, SD, USA*

MAMTA JAISWAL • *Medical Faculty, Institute of Biochemistry and Molecular Biology II, Heinrich-Heine University, Düsseldorf, Germany*

JARED JOHNSON • *Department of Chemistry and Chemical Biology, Cornell University, Ithaca, NY, USA*

THOMAS O. JOOS • *Department of Biochemistry, NMI Natural and Medical Sciences Institute at the University of Tübingen, Reutlingen, Germany*

VLADIMÍR KLIMEŠ • *Faculty of Science, Charles University in Prague, Prague, Czech Republic*

KATJA T. KOESSMEIER • *Medical Faculty, Institute of Biochemistry and Molecular Biology II, Heinrich-Heine University, Düsseldorf, Germany*

ISABELLE LAJOIE-MAZENC • *INSERM UMR 1037, Cancer Research Centre of Toulouse, Claudius Regaud Cancer Institute, University of Toulouse, Toulouse, France*

TINE LEFEVER • *Molecular Pathology Section, Department of Biomedical Sciences, BRIC, University of Copenhagen, Copenhagen, Denmark*

EMMANUEL LEMICHEZ • *Faculté de Médecine, INSERM, U895, Centre Méditerranéen de Médecine Moléculaire, C3M, Institut Signalisation et Pathologie, IFR50, Université de Nice-Sophia-Antipolis, Nice, France*

ROLAND LEUNG • *Matrix Dynamics Group, University of Toronto, Toronto, ON, Canada*

MICHAEL R. LOGAN • *Department of Cell Biology, University of Alberta, Edmonton, AB, Canada*

BOON CHUAN LOW • *Cell Signaling and Developmental Biology Laboratory, Department of Biological Sciences, Mechanobiology Institute, National University of Singapore, Singapore, Singapore*

ERIK A. LUNDQUIST • *Department of Molecular Biosciences, University of Kansas, Lawrence, KS, USA*

AMEL METTOUCHI • *INSERM, U634, UNSA, UFR Médecine, IFR50, Université de Nice-Sophia-Antipolis, Nice, France*

NATALIA MITIN • *Lineberger Comprehensive Cancer Center, University of North Carolina, Chapel Hill, NC, USA*

KEI MIYANO • *Department of Biochemistry, Kyushu University Graduate School of Medical Sciences, Fukuoka, Japan*

LAURA MONTANI • *Instituto de Biologia Molecular e Celular, Universidade do Porto, Porto, Portugal*

SÉBASTIEN NOLA • *Faculty of Medicine, Molecular Medicine Section, National Heart and Lung Institute, Imperial College London, London, UK*

ESBEN PEDERSEN • *Molecular Pathology Section, Department of Biomedical Sciences, BRIC, University of Copenhagen, Copenhagen, Denmark*

OLIVIER PERTZ • *Department of Biomedicine, Institute of Biochemistry and Genetics, University of Basel, Basel, Switzerland*

KARINE PEYROLLIER • *Molecular Pathology Section, Department of Biomedical Sciences, BRIC, University of Copenhagen, Copenhagen, Denmark*

JOÃO BETTENCOURT RELVAS • *Instituto de Biologia Molecular e Celular, Universidade do Porto, Porto, Portugal*

NICOLAS REYMOND • *Randall Division of Cell and Molecular Biophysics, King's College London, London, UK*

ANNE J. RIDLEY • *Randall Division of Cell and Molecular Biophysics, King's College London, London, UK*

STEFANIE RIMMELE • *Department of Biochemistry, NMI Natural and Medical Sciences Institute at the University of Tübingen, Reutlingen, Germany*

PHILIPPE RIOU • *Randall Division of Cell and Molecular Biophysics, King's College London, London, UK*

FRANCISCO RIVERO • *Centre for Cardiovascular and Metabolic Research, The Hull York Medical School, University of Hull, Hull, UK*

PATRICK J. ROBERTS • *Lineberger Comprehensive Cancer Center, University of North Carolina, Chapel Hill, NC, USA*

CHRISTOPHER J. SAMPSON • *Institute of Biological and Environmental Sciences, University of Aberdeen, Aberdeen, UK*

GUDULA SCHMIDT • *Institute for Experimental and Clinical Pharmacology and Toxicology, University of Freiburg, Freiburg, Germany*

MICHAEL SCHMOHL • *Department of Biochemistry, NMI Natural and Medical Sciences Institute at the University of Tübingen, Reutlingen, Germany*

NICOLE SCHNEIDERHAN-MARRA • *Department of Biochemistry, NMI Natural and Medical Sciences Institute at the University of Tübingen, Reutlingen, Germany*

LARRY A. SKLAR • *New Mexico Molecular Libraries Screening Center, Albuquerque, NM, USA; Department of Pathology, Cancer Research and Treatment Center, University of New Mexico School of Medicine, Albuquerque, NM, USA*

DÉSIRÉE SPIERING • *Department of Anatomy and Structural Biology and Gruss-Lipper Biophotonics Center, Albert Einstein College of Medicine of Yeshiva University, Bronx, NY, USA*

HIDEKI SUMIMOTO • *Department of Biochemistry, Kyushu University Graduate School of Medical Sciences, Fukuoka, Japan*

ZURAB SURVILADZE • *New Mexico Molecular Libraries Screening Center, Albuquerque, NM, USA; Cancer Research and Treatment Center, University of New Mexico School of Medicine, Albuquerque, NM, USA*

MICHAEL J. WILLIAMS • *Institute of Biological and Environmental Sciences, University of Aberdeen, Aberdeen, UK*

BERND WOLLSCHEID • *Institute of Molecular Systems Biology, NCCR Neuro Center for Proteomics, ETH Zürich, Zürich, Switzerland*

HUAJIANG XIONG • *Department of Zoophysiology, Zoological Institute, University of Kiel, Kiel, Germany*

AN YAN • *Department of Botany and Plant Sciences, Center for Plant Cell Biology, University of California, Riverside, CA, USA*

ZHENBIAO YANG • *Department of Botany and Plant Sciences, Center for Plant Cell Biology, University of California, Riverside, CA, USA*

SUSAN M. YOUNG • *New Mexico Molecular Libraries Screening Center, Albuquerque, NM, USA; Cancer Research and Treatment Center, University of New Mexico School of Medicine, Albuquerque, NM, USA*

SHIZHEN ZHU • *Department of Pediatric Oncology, Dana-Farber Cancer Institute, Harvard Medical School, Boston, MA, USA*

Part I

History and Phylogenesis

Chapter 1

Historical Overview of Rho GTPases

Anne J. Ridley

Abstract

In 1985, the first members of the Rho GTPase family were identified. Over the next 10 years, rapid progress was made in understanding Rho GTPase signalling. Multiple Rho GTPases were discovered in a wide range of eukaryotes, and shown to regulate a diverse range of cellular processes, including cytoskeletal dynamics, NADPH oxidase activation, cell migration, cell polarity, membrane trafficking, and transcription. The Rho regulators, guanine nucleotide exchange factors (GEFs), GTPase-activating proteins (GAPs), and guanine nucleotide dissociation inhibitors (GDIs), were found through a combination of biochemistry, genetics, and detective work. Downstream targets for Rho GTPases were also rapidly identified, and linked to Rho-regulated cellular responses. In parallel, a wide range of bacterial proteins were found to modify Rho proteins or alter their activity in cells, many of which turned out to be useful tools to study Rho functions. More recent work has delineated where Rho GTPases act in cells, the molecular pathways linking some of them to specific cellular responses, and their functions in the development of multicellular organisms.

Key words: Cytoskeleton, Signal transduction, Adhesion, Transcription, GEF, GAP, GDI, Trafficking

1. Introduction

Rho GTPases transduce signals from cell surface receptors to intracellular signalling pathways. Most Rho GTPases switch between an active GTP-bound conformation and an inactive GDP-bound conformation. They are activated by guanine nucleotide exchange factors (GEFs), which stimulate the release of GDP, allowing GTP to bind. They are inactivated by GTPase-activating proteins (GAPs), which catalyse GTP hydrolysis and hence conversion to the GDP-bound conformation. When bound to GTP, they interact with their downstream partners and induce cellular responses. The atypical Rho family members, RhoH, Rnd1, Rnd2,

Francisco Rivero (ed.), *Rho GTPases: Methods and Protocols*, Methods in Molecular Biology, vol. 827,
DOI 10.1007/978-1-61779-442-1_1, © Springer Science+Business Media, LLC 2012

and Rnd3, do not hydrolyse GTP and are therefore constitutively GTP-bound. RhoBTB1 and RhoBTB2 are also atypical: they are much larger than other Rho family members, and are also predicted not to hydrolyse GTP (1, 2).

Rho, Rac, and Cdc42 have been shown to have a plethora of functions, from regulation of the cytoskeleton to cell polarity, vesicle trafficking, and transcription. In some cases, these functions are separate from their effects on the cytoskeleton whereas in other cases they are clearly linked. A particularly fascinating part of the Rho GTPase story is the identification of multiple bacterial toxins and enzymes that modify or interact with one or more Rho protein and thereby alter cellular functions to promote bacterial survival and proliferation in the host.

In this overview, I describe how the late 1980s and early 1990s were a hotbed of activity when Rho GTPases and their interacting partners were discovered and characterized. More recent studies that have shed new light on the function of these proteins are also described.

2. Identification of Rho GTPases

The first *Rho* GTPase gene to be isolated was the marine snail Aplysia californica *Rho*, which was then used to identify the human *RhoA, RhoB,* and *RhoC* genes (3). Subsequently, two *Rho* genes, *Rho1* and *Rho2*, were found by homology to the human Rho isoforms in *Saccharomyces cerevisiae* (4). An exciting discovery came with the cloning of *S. cerevisiae Cdc42*, a gene required for budding and cell polarity, which was found to be homologous to Rho proteins (5). Subsequent studies of Rho proteins *in S. cerevisiae* have been central to our understanding of how they are activated and signal during polarity development. A flurry of new family members followed, found through homology cloning in multiple organisms, from fungi to plants (6–8). Later, other Rho family members were identified from genome sequence data in various organisms (e.g. RhoBTBs; (9)). Miro GTPase were initially assigned to the Rho family (10), but are now considered a unique branch of the Ras superfamily, equally homologous to Rab and Rho (11). Similarly, the GTP-binding domain of RhoBTB3 is very divergent and does not cluster in the Rho family (11, 12). Indeed, RhoBTB3 binds and hydrolyses ATP rather than GTP (13).

Based on analysis of genome sequences, there are 20 Rho GTPases in most mammals, some with splice variants (11). RhoS is an additional member in rodents that is expressed in the testis (14). Evolutionary analysis of Rho GTPases has identified which Rho family members are most conserved versus those that have evolved more recently (11, 12). Rho, Rac, and Cdc42 are found in nearly all animal kingdoms while plants have a distinct set of Rops (Rho in plants) that are most closely related to Racs (11, 15). This suggests that Rac was the original Rho GTPase, generating Rho and Cdc42 by duplication and then subsequently all the other Rho family members.

3. Rho GTPases as Regulators of the Actin Cytoskeleton

The ability of Rho GTPases to alter the arrangement of actin filaments in cells was first suggested by the observation that the enzyme C3 transferase from *Clostridium botulinum* modified Rho isoforms (RhoA, RhoB, and RhoC) by ADP-ribosylation in vitro and induced loss of actin stress fibres in Vero cells (16, 17). ADP-ribosylation of Rho proteins was then found to inhibit their function, whereas microinjection of RhoA protein into cells induced stress fibres (18).

The extensive characterization of Ras proteins and genes helped to inform the early studies of Rho GTPases. For example, it was known that Ras genes were mutated in human cancers, and that they encoded constitutively active GTPase-deficient proteins (19). In addition, it had been shown that mutating a key amino acid in the GTP-binding site allowed Ras to act as a dominant-negative protein (20). By putting the same mutations into Rho GTPases, it was possible to analyse how they acted in cells. The first to be studied was RhoA, for which the constitutively active form induces stress fibres and cell contraction (18). Subsequent studies showed that constitutively active Rac1 induced membrane ruffling and lamellipodia (21) Endogenous Rho and Rac1 were shown to mediate actin cytoskeletal changes in response to extracellular stimuli by using C3 transferase to inhibit Rho and dominant negative Rac1 (21, 22). Soon after, similar methods were used to show that Cdc42 was linked to filopodium extension (23, 24). Subsequently, a screen of all known Rho GTPases showed that most of them affected the actin cytoskeleton in some way in endothelial cells (25).

The actin cytoskeleton is central to cell migration (26), and based on studies using C3 transferase and constitutively active and dominant negative Rho GTPase mutants, a model was developed to explain how the Rho GTPases co-ordinately regulated cell migration: Rac1 drove lamellipodium extension at the front of migrating cells, Rho isoforms stimulated actomyosin contractility at the back, and Cdc42 was required for determining the direction of migration (chemotaxis).

Over the past 10 years, it has become clear that the roles of Rho GTPases in migration are much more complex. In particular, the development of biosensors for Rho GTPases has provided new information about the functions of these proteins in actin cytoskeletal regulation. The first biosensor reported Rac1 activity (27), and subsequently biosensors for RhoA, Rac1, and Cdc42 were designed. Unexpectedly, RhoA was found to be active predominantly in lamellipodia and membrane ruffles at the front of fibroblasts as well as in the tail at the back (28, 29). This work has been central to our current understanding of how each Rho GTPase acts in multiple places in the cell, within different protein complexes, to coordinate cell migration (30).

4. Regulation of Rho GTPases

The three main classes of Rho GTPase regulators (GAPs, GEFs, and GDIs) were all discovered in the early 1990s, during the same time period as the functions of Rho proteins were being unravelled. A RhoGDI was the first Rho regulator to be discovered, with the unique property of inhibiting GDP dissociation from RhoA and RhoB (31). RhoGDIs were subsequently shown to induce dissociation of Rho proteins from membranes by binding to their C-terminal prenyl groups (32, 33).

The first GEF for Rho GTPases to be identified was Dbl (34). This involved some clever detective work: in 1985, Dbl was cloned from a diffuse B-cell lymphoma as an "oncogene" that transformed fibroblasts. Dbl was found to share a region of sequence homology with Cdc24 from *S. cerevisiae*, which was known to act on the same pathway as Cdc42 (34). Dbl and Cdc24 were subsequently shown to act as GEFs for Cdc42. A surprising discovery in 2002 was a completely different family of RhoGEFs, the DOCK proteins (35). DOCK1/DOCK180 homologues were already known to act upstream of Rac through genetic analysis in *Caenorhabditis elegans* and *Drosophila* (36).

A RhoGAP for RhoA was initially purified biochemically (37) and then Bcr was found to have a similar domain that acted as a GAP for Rac but not Rho (38). Multiple new RhoGAPs and RhoGEFs have been found in mammals through sequence homology (39, 40). Structural analysis of Rho GTPases with GEFs, GAPs, and GDIs in the late 1990s was also essential for understanding how they regulate Rho function (41–43).

We now know that many protein complexes contain multiple Rho GTPase interactors, such as a RhoGAP and/or RhoGEF as well as downstream target(s) for Rho GTPases. For example, the Cdc42/Rac1-activated PAK kinases themselves bind to the Cdc42/Rac1 GEF PIX (44). In addition, the Cdc42-interacting protein N-WASP binds to and activates the Cdc42 GEF intersectin (45). These complexes create positive feedback loops between the components that act together to mediate a specific cellular function.

5. Rho GTPase Targets

The first downstream target of a Rho GTPase to be identified was the tyrosine kinase ACK1 for Cdc42, using an elegant overlay assay with cell extracts (46), followed closely by PAK (47). Multiple targets for RhoA were then identified, using either biochemical purification or yeast two-hybrid screening. This work identified ROCKs and mDia1, which have been widely studied for their functions downstream of RhoA, but also citron kinase, PKNs, rhophilin and rhotekin, for which we still know comparatively little about their cellular functions (48).

Many targets for Rho, Rac, and Cdc42 have now been identified (49). The discovery in 2000 of Par6 in the Par polarity complex as a target for Cdc42 was a particularly exciting development (50): Par complex proteins had been studied since 1985 as regulators of asymmetric division in *C. elegans* embryos, and Cdc42 was known to regulate polarity in other systems (51). This initiated a whole new field of research on Cdc42, the Par complex and polarity in multiple organisms and cell types (see Subheading 7). Interestingly, some Rho targets in turn modify Rho members posttranslationally, such as the ROCK phosphorylation of Rnd3 (52) or PIAS1-mediated sumoylation of Rac1 (53).

In addition to Rho, Rac, and Cdc42, targets for several but not all other Rho family members have been identified (1, 2). Newly identified family members have often been tested for their interaction with known Rho/Rac/Cdc42 targets. For example, RhoU (originally called Wrch-1) was shown to bind to PAK1 (54). In addition, yeast two-hybrid screening has been used for Rnd proteins and RhoS to find potential targets (2, 14). However, our knowledge of how these non-canonical Rho proteins act at a molecular level to regulate responses in cells is for the most part still far from clear.

6. Rho GTPases and Transcription

In 1995, new insight into Rho function was revealed when RhoA, Rac1, and/or Cdc42 were reported to stimulate the activity of the serum response factor (SRF) and Jun transcription factors (55). The best characterized pathway between Rho GTPases and transcription is that for SRF. Actin monomers bind to and inhibit the nuclear import of the SRF cofactor MAL (56). Rho GTPases reduce levels of actin monomers leading to nuclear translocation of MAL and increased SRF-mediated transcription.

Rho GTPases were later reported to regulate several other transcription factors, including NFκB and Stat3, although the signalling mechanisms underlying these effects remain controversial (49, 57). Racs can act via PAKs to stimulate the activity of several transcription factors (58). Rac generation of reactive oxygen species through NADPH oxidases also affects transcription (59).

The activation of transcription by Rho GTPases contributes to cell migration. For example, Rac1 can stimulate expression of matrix-degrading metalloproteases (MMPs) which then contribute to cell migration through tissues (60). In addition, MAL and SRF induce the upregulation of genes that contribute to cancer cell invasion and metastasis in animal models (61). Transcriptional changes induced by Rho GTPases can therefore be considered as long-term responses designed to support their short-term roles in initiating and coordinating cell migration.

7. Multiple Other Functions of Rho GTPases

As soon as reagents for studying Rho GTPases became available, they were studied by many groups in many different cell assay systems, resulting in an ever-growing list of functions for these proteins. It is now clear that they affect not only the actin cytoskeleton but microtubules, intermediate filaments and septins (62–64). The cytoskeleton is intricately linked to cell–cell and cell–extracellular matrix adhesions. Some Rho GTPase targets have direct links to adhesion components as well as modulating adhesion-cytoskeleton dynamics. Regulation of cytoskeletal dynamics by Rho GTPase explains at least some of their effects on membrane trafficking (65). Cdc42 is also critical for cell polarity, which in part reflects its ability to polarize the cytoskeleton as well as regulate vesicle trafficking (66).

8. Rho GTPases as Targets of Bacterial Toxins and Viral Proteins

The close links between bacterial infection and changes to Rho GTPases is a fascinating and continually evolving story. After the identification of RhoA/B/C isoforms as targets for C3 transferase, multiple bacterial enzymes have been identified that post-translationally modify Rho GTPases and either activate or inhibit them (17, 67). In addition, the identification in 1998 of a *Salmonella* protein, SopE, that acted as a GEF for Rac1 and Cdc42 (68) was the starting point of studies identifying multiple bacterial and viral proteins that alter Rho GTPase activity in mammalian cells by a plethora of different mechanisms (69). These proteins and toxins from micro-organisms have proven to be useful tools for manipulating and Rho GTPase activity and studying Rho GTPase function in cells.

9. Rho GTPases in Development

After the biochemical and cell biological studies of Rho GTPase functions and regulation, the late 1990s saw a big push to study their roles in multicellular organisms in vivo. For example, mutants and/or expression of dominant negative proteins in *Drosophila* and *C. elegans* have been used to characterize the functions of Rho family members (70–73). So far nine Rho GTPases have been knocked out in mice, as well as a variety of GEFs, GAPs and downstream targets (74). The first mouse knockout was Rac2 in 1999 (75), which has primarily been studied for defects in the haematopoietic system. Global deletion of Rac1 or Cdc42 is lethal at a very early stage of development (76, 77), and so far global RhoA deletion has not been reported, but other Rho GTPase deletions are viable (74). Conditional tissue-specific deletions of Cdc42 and Rac1 in mice have yielded a plethora of information about their functions in vivo (74). Surprisingly, RhoA depletion in skin keratinocytes had no phenotype in vivo (78). It will be fascinating to know the phenotypes of mice lacking other Rho GTPase family members, and whether RhoA depletion has a phenotype in other tissues.

10. Perspectives

Our understanding of Rho GTPases has come a long way since RhoA was first shown to induce stress fibres in fibroblasts, but there are still some important unanswered questions. So far, there is little information on where Rho GTPases interact with their activators or targets either in single cells or in tissues. There are also many unstudied or little studied RhoGEFs and RhoGAPs. How they link up to any of the Rho GTPases, and what protein complexes they are components of, remains to be established. In addition, we still know very little about the physiological functions of most Rho GTPases apart from Cdc42 and the Rho and Rac isoforms. This requires detailed studies of their functions using genetically modified organisms. These areas should keep the Rho GTPase field active for many years to come.

References

1. Aspenstrom, P., Ruusala, A., and Pacholsky, D. (2007) Taking Rho GTPases to the next level: the cellular functions of atypical Rho GTPases. *Exp Cell Res* 313, 3673–9.

2. Riou, P., Villalonga, P., and Ridley, A. J. (2010) Rnd proteins: multifunctional regulators of the cytoskeleton and cell cycle progression. *Bioessays* 32, 986–92.

3. Madaule, P., and Axel, R. (1985) A novel ras-related gene family. *Cell* 41, 31–40.

4. Madaule, P., Axel, R., and Myers, A. M. (1987) Characterization of two members of the rho gene family from the yeast *Saccharomyces cerevisiae*. *Proc Natl Acad Sci USA* 84, 779–83.

5. Johnson, D. I., and Pringle, J. R. (1990) Molecular characterization of CDC42, a

Saccharomyces cerevisiae gene involved in the development of cell polarity. *J Cell Biol* 111, 143–52.

6. Yang, Z., and Watson, J. C. (1993) Molecular cloning and characterization of rho, a ras-related small GTP-binding protein from the garden pea. *Proc Natl Acad Sci USA* 90, 8732–6.

7. Nakano, K., and Mabuchi, I. (1995) Isolation and sequencing of two cDNA clones encoding Rho proteins from the fission yeast *Schizosaccharomyces pombe*. *Gene* 155, 119–22.

8. Bush, J., Franek, K., and Cardelli, J. (1993) Cloning and characterization of seven novel *Dictyostelium discoideum* rac-related genes belonging to the rho family of GTPases. *Gene* 136, 61–8.

9. Rivero, F., Dislich, H., Glockner, G., and Noegel, A. A. (2001) The *Dictyostelium discoideum* family of Rho-related proteins. *Nucleic Acids Res* 29, 1068–79.

10. Fransson, A., Ruusala, A., and Aspenstrom, P. (2003) Atypical Rho GTPases have roles in mitochondrial homeostasis and apoptosis. *J Biol Chem* 278, 6495–502.

11. Boureux, A., Vignal, E., Faure, S., and Fort, P. (2007) Evolution of the Rho family of ras-like GTPases in eukaryotes. *Mol Biol Evol* 24, 203–16.

12. Wherlock, M., and Mellor, H. (2002) The Rho GTPase family: a Racs to Wrchs story. *J Cell Sci* 115, 239–40.

13. Espinosa, E. J., Calero, M., Sridevi, K., and Pfeffer, S. R. (2009) RhoBTB3: a Rho GTPase-family ATPase required for endosome to Golgi transport. *Cell* 137, 938–48.

14. Zhang, N., Liang, J., Tian, Y., Yuan, L., Wu, L., Miao, S., Zong, S., and Wang, L. (2010) A novel testis-specific GTPase serves as a link to proteasome biogenesis: functional characterization of RhoS/RSA-14-44 in spermatogenesis. *Mol Biol Cell* 21, 4312–4324.

15. Valster, A. H., Hepler, P. K., and Chernoff, J. (2000) Plant GTPases: the Rhos in bloom. *Trends Cell Biol* 10, 141–6.

16. Chardin, P., Boquet, P., Madaule, P., Popoff, M. R., Rubin, E. J., and Gill, D. M. (1989) The mammalian G protein rhoC is ADP-ribosylated by *Clostridium botulinum* exoenzyme C3 and affects actin microfilaments in Vero cells. *EMBO J* 8, 1087–92.

17. Aktories, K., and Barbieri, J. T. (2005) Bacterial cytotoxins: targeting eukaryotic switches. *Nat Rev Microbiol* 3, 397–410.

18. Paterson, H. F., Self, A. J., Garrett, M. D., Just, I., Aktories, K., and Hall, A. (1990) Microinjection of recombinant p21rho induces

rapid changes in cell morphology. *J Cell Biol* 111, 1001–7.

19. Bos, J. L. (1989) ras oncogenes in human cancer: a review. *Cancer Res* 49, 4682–9.

20. Feig, L. A., and Cooper, G. M. (1988) Inhibition of NIH 3 T3 cell proliferation by a mutant ras protein with preferential affinity for GDP. *Mol Cell Biol* 8, 3235–43.

21. Ridley, A. J., Paterson, H. F., Johnston, C. L., Diekmann, D., and Hall, A. (1992) The small GTP-binding protein rac regulates growth factor-induced membrane ruffling. *Cell* 70, 401–10.

22. Ridley, A. J., and Hall, A. (1992) The small GTP-binding protein rho regulates the assembly of focal adhesions and actin stress fibers in response to growth factors. *Cell* 70, 389–99.

23. Kozma, R., Ahmed, S., Best, A., and Lim, L. (1995) The Ras-related protein Cdc42Hs and bradykinin promote formation of peripheral actin microspikes and filopodia in Swiss 3 T3 fibroblasts. *Mol Cell Biol* 15, 1942–52.

24. Nobes, C. D., and Hall, A. (1995) Rho, rac, and cdc42 GTPases regulate the assembly of multimolecular focal complexes associated with actin stress fibers, lamellipodia, and filopodia. *Cell* 81, 53–62.

25. Aspenstrom, P., Fransson, A., and Saras, J. (2004) Rho GTPases have diverse effects on the organisation of the actin filament system. *Biochem J* 377, 327–37.

26. Ridley, A. J., Schwartz, M. A., Burridge, K., Firtel, R. A., Ginsberg, M. H., Borisy, G., Parsons, J. T., and Horwitz, A. R. (2003) Cell migration: integrating signals from front to back. *Science* 302, 1704–9.

27. Kraynov, V. S., Chamberlain, C., Bokoch, G. M., Schwartz, M. A., Slabaugh, S., and Hahn, K. M. (2000) Localized Rac activation dynamics visualized in living cells. *Science* 290, 333–7.

28. Pertz, O., Hodgson, L., Klemke, R. L., and Hahn, K. M. (2006) Spatiotemporal dynamics of RhoA activity in migrating cells. *Nature* 440, 1069–72.

29. Kurokawa, K., and Matsuda, M. (2005) Localized RhoA activation as a requirement for the induction of membrane ruffling. *Mol Biol Cell* 16, 4294–303.

30. Heasman, S. J., and Ridley, A. J. (2010) Multiples roles of RhoA during T cell transendothelial migration. *Small GTPases* 1, 174–9.

31. Ueda, T., Kikuchi, A., Ohga, N., Yamamoto, J., and Takai, Y. (1990) Purification and characterization from bovine brain cytosol of a novel regulatory protein inhibiting the dissociation of GDP from and the subsequent binding

of GTP to rhoB p20, a ras p21-like GTP-binding protein. *J Biol Chem* 265, 9373–80.

32. Isomura, M., Kikuchi, A., Ohga, N., and Takai, Y. (1991) Regulation of binding of rhoB p20 to membranes by its specific regulatory protein, GDP dissociation inhibitor. *Oncogene* 6, 119–24.

33. DerMardirossian, C., and Bokoch, G. M. (2005) GDIs: central regulatory molecules in Rho GTPase activation. *Trends Cell Biol* 15, 356–63.

34. Hart, M. J., Eva, A., Evans, T., Aaronson, S. A., and Cerione, R. A. (1991) Catalysis of guanine nucleotide exchange on the CDC42Hs protein by the dbl oncogene product. *Nature* 354, 311–4.

35. Brugnera, E., Haney, L., Grimsley, C., Lu, M., Walk, S. F., Tosello-Trampont, A. C., Macara, I. G., Madhani, H., Fink, G. R., and Ravichandran, K. S. (2002) Unconventional Rac-GEF activity is mediated through the Dock180-ELMO complex. *Nat Cell Biol* 4, 574–82.

36. Cote, J. F., and Vuori, K. (2007) GEF what? Dock180 and related proteins help Rac to polarize cells in new ways. *Trends Cell Biol* 17, 383–93.

37. Garrett, M. D., Self, A. J., van Oers, C., and Hall, A. (1989) Identification of distinct cytoplasmic targets for ras/R-ras and rho regulatory proteins. *J Biol Chem* 264, 10–3.

38. Diekmann, D., Brill, S., Garrett, M. D., Totty, N., Hsuan, J., Monfries, C., Hall, C., Lim, L., and Hall, A. (1991) Bcr encodes a GTPase-activating protein for p21rac. *Nature* 351, 400–2.

39. Tcherkezian, J., and Lamarche-Vane, N. (2007) Current knowledge of the large RhoGAP family of proteins. *Biol Cell* 99, 67–86.

40. Rossman, K. L., Der, C. J., and Sondek, J. (2005) GEF means go: turning on RHO GTPases with guanine nucleotide-exchange factors. *Nat Rev Mol Cell Biol* 6, 167–80.

41. Rittinger, K., Walker, P. A., Eccleston, J. F., Nurmahomed, K., Owen, D., Laue, E., Gamblin, S. J., and Smerdon, S. J. (1997) Crystal structure of a small G protein in complex with the GTPase-activating protein rhoGAP. *Nature* 388, 693–7.

42. Worthylake, D. K., Rossman, K. L., and Sondek, J. (2000) Crystal structure of Rac1 in complex with the guanine nucleotide exchange region of Tiam1. *Nature* 408, 682–8.

43. Scheffzek, K., Stephan, I., Jensen, O. N., Illenberger, D., and Gierschik, P. (2000) The Rac-RhoGDI complex and the structural basis for the regulation of Rho proteins by RhoGDI. *Nat Struct Biol* 7, 122–6.

44. Manser, E., Loo, T. H., Koh, C. G., Zhao, Z. S., Chen, X. Q., Tan, L., Tan, I., Leung, T., and Lim, L. (1998) PAK kinases are directly coupled to the PIX family of nucleotide exchange factors. *Mol Cell* 1, 183–92.

45. Hussain, N. K., Jenna, S., Glogauer, M., Quinn, C. C., Wasiak, S., Guipponi, M., Antonarakis, S. E., Kay, B. K., Stossel, T. P., Lamarche-Vane, N., and McPherson, P. S. (2001) Endocytic protein intersectin-l regulates actin assembly via Cdc42 and N-WASP. *Nat Cell Biol* 3, 927–32.

46. Manser, E., Leung, T., Salihuddin, H., Tan, L., and Lim, L. (1993) A non-receptor tyrosine kinase that inhibits the GTPase activity of p21cdc42. *Nature* 363, 364–7.

47. Manser, E., Leung, T., Salihuddin, H., Zhao, Z. S., and Lim, L. (1994) A brain serine/threonine protein kinase activated by Cdc42 and Rac1. *Nature* 367, 40–6.

48. Van Aelst, L., and D'Souza-Schorey, C. (1997) Rho GTPases and signaling networks. *Genes Dev* 11, 2295–322.

49. Jaffe, A. B., and Hall, A. (2005) Rho GTPases: biochemistry and biology. *Annu Rev Cell Dev Biol* 21, 247–69.

50. Joberty, G., Petersen, C., Gao, L., and Macara, I. G. (2000) The cell-polarity protein Par6 links Par3 and atypical protein kinase C to Cdc42. *Nat Cell Biol* 2, 531–9.

51. Etienne-Manneville, S., and Hall, A. (2003) Cell polarity: Par6, aPKC and cytoskeletal crosstalk. *Curr Opin Cell Biol* 15, 67–72.

52. Riento, K., Totty, N., Villalonga, P., Garg, R., Guasch, R., and Ridley, A. J. (2005) RhoE function is regulated by ROCK I-mediated phosphorylation. *EMBO J* 24, 1170–80.

53. Castillo-Lluva, S., Tatham, M. H., Jones, R. C., Jaffray, E. G., Edmondson, R. D., Hay, R. T., and Malliri, A. (2010) SUMOylation of the GTPase Rac1 is required for optimal cell migration. *Nat Cell Biol* 12, 1078–85.

54. Tao, W., Pennica, D., Xu, L., Kalejta, R. F., and Levine, A. J. (2001) Wrch-1, a novel member of the Rho gene family that is regulated by Wnt-1. Genes Dev 15, 1796–807.

55. Ridley, A. J. (1996) Rho: theme and variations. *Curr Biol* 6, 1256–64.

56. Miralles, F., Posern, G., Zaromytidou, A. I., and Treisman, R. (2003) Actin dynamics control SRF activity by regulation of its coactivator MAL. *Cell* 113, 329–42.

57. Benitah, S. A., Valeron, P. F., van Aelst, L., Marshall, C. J., and Lacal, J. C. (2004) Rho GTPases in human cancer: an unresolved link to upstream and downstream transcriptional regulation. *Biochim Biophys Acta* 1705, 121–32.

58. Bokoch, G. M. (2003) Biology of the p21-activated kinases. *Annu Rev Biochem* 72, 743–81.

59. Wu, W. S. (2006) The signaling mechanism of ROS in tumor progression. *Cancer Metastasis Rev* 25, 695–705.

60. Kheradmand, F., Werner, E., Tremble, P., Symons, M., and Werb, Z. (1998) Role of Rac1 and oxygen radicals in collagenase-1 expression induced by cell shape change. *Science* 280, 898–902.

61. Medjkane, S., Perez-Sanchez, C., Gaggioli, C., Sahai, E., and Treisman, R. (2009) Myocardin-related transcription factors and SRF are required for cytoskeletal dynamics and experimental metastasis. *Nat Cell Biol* 11, 257–68.

62. Braga, V. M. (2002) Cell-cell adhesion and signalling. *Curr Opin Cell Biol* 14, 546–56.

63. Ito, H., Iwamoto, I., Morishita, R., Nozawa, Y., Narumiya, S., Asano, T., and Nagata, K. (2005) Possible role of Rho/Rhotekin signaling in mammalian septin organisation. *Oncogene* 24, 7064–72.

64. Hall, A. (2009) The cytoskeleton and cancer. *Cancer Metastasis Rev* 28, 5–14.

65. Ridley, A. J. (2006) Rho GTPases and actin dynamics in membrane protrusions and vesicle trafficking. *Trends Cell Biol* 16, 522–9.

66. Harris, K. P., and Tepass, U. (2010) Cdc42 and vesicle trafficking in polarized cells. *Traffic* 11, 1272–9.

67. Visvikis, O., Maddugoda, M. P., and Lemichez, E. (2010) Direct modifications of Rho proteins: deconstructing GTPase regulation. *Biol Cell* 102, 377–89.

68. Hardt, W. D., Chen, L. M., Schuebel, K. E., Bustelo, X. R., and Galan, J. E. (1998) *S. typhimurium* encodes an activator of Rho GTPases that induces membrane ruffling and nuclear responses in host cells. *Cell* 93, 815–26.

69. Finlay, B. B. (2005) Bacterial virulence strategies that utilize Rho GTPases. *Curr Top Microbiol Immunol* 291, 1–10.

70. Luo, L., Liao, Y. J., Jan, L. Y., and Jan, Y. N. (1994) Distinct morphogenetic functions of similar small GTPases: *Drosophila* Drac1 is involved in axonal outgrowth and myoblast fusion. *Genes Dev* 8, 1787–802.

71. Eaton, S., Auvinen, P., Luo, L., Jan, Y. N., and Simons, K. (1995) CDC42 and Rac1 control different actin-dependent processes in the *Drosophila* wing disc epithelium. *J Cell Biol* 131, 151–64.

72. Strutt, D. I., Weber, U., and Mlodzik, M. (1997) The role of RhoA in tissue polarity and Frizzled signalling. *Nature* 387, 292–5.

73. Reddien, P. W., and Horvitz, H. R. (2000) CED-2/CrkII and CED-10/Rac control phagocytosis and cell migration in *Caenorhabditis elegans*. *Nat Cell Biol* 2, 131–6.

74. Heasman, S. J., and Ridley, A. J. (2008) Mammalian Rho GTPases: new insights into their functions from in vivo studies. *Nat Rev Mol Cell Biol* 9, 690–701.

75. Roberts, A. W., Kim, C., Zhen, L., Lowe, J. B., Kapur, R., Petryniak, B., Spaetti, A., Pollock, J. D., Borneo, J. B., Bradford, G. B., Atkinson, S. J., Dinauer, M. C., and Williams, D. A. (1999) Deficiency of the hematopoietic cell-specific Rho family GTPase Rac2 is characterized by abnormalities in neutrophil function and host defense. *Immunity* 10, 183–96.

76. Chen, F., Ma, L., Parrini, M. C., Mao, X., Lopez, M., Wu, C., Marks, P. W., Davidson, L., Kwiatkowski, D. J., Kirchhausen, T., Orkin, S. H., Rosen, F. S., Mayer, B. J., Kirschner, M. W., and Alt, F. W. (2000) Cdc42 is required for PIP(2)-induced actin polymerization and early development but not for cell viability. *Curr Biol* 10, 758–65.

77. Sugihara, K., Nakatsuji, N., Nakamura, K., Nakao, K., Hashimoto, R., Otani, H., Sakagami, H., Kondo, H., Nozawa, S., Aiba, A., and Katsuki, M. (1998) Rac1 is required for the formation of three germ layers during gastrulation. *Oncogene* 17, 3427–33.

78. Jackson, B., Peyrollier, K., Pedersen, E., Basse, A., Karlsson, R., Wang, Z., Lefever, T., Ochsenbein, A. M., Schmidt, G., Aktories, K., Stanley, A., Quondamatteo, F., Ladwein, M., Rottner, K., van Hengel, J., and Brakebusch, C. (2011) RhoA is dispensable for skin development, but crucial for contraction and directed migration of keratinocytes. *Mol Biol Cell* 22, 593–605.

Chapter 2

Rho GTPases: Deciphering the Evolutionary History of a Complex Protein Family

Marek Eliáš and Vladimír Klimeš

Abstract

Rho GTPases constitute a significant subgroup of the eukaryotic Ras superfamily of small GTPases implicated in the regulation of diverse cellular processes, such as the dynamics of the actin cytoskeleton, establishment, and maintenance of cell polarity and membrane trafficking. Whereas a few eukaryotes lack Rho genes, a majority of species typically bear multiple Rho paralogs, raising a question about the origin of the family and the paths of its diversification in individual eukaryotic lineages. In this chapter, we ruminate on several aspects of the evolutionary history of the Rho family and methodological challenges of its reconstruction. First, we provide an updated survey of Rho GTPases in diverse eukaryotic branches, demonstrating almost ubiquitous occurrence of Rho genes across the eukaryotic phylogeny most consistent with the presence of at least one Rho gene already in the last eukaryotic common ancestor. Second, we discuss the obstacles in reconstructing the history of gene duplications giving rise to the extant diversity of Rho paralogs in different species, and point to numerous limitations posed by the current phylogenetic methodology. Third, as a case study demonstrating various issues of data collection, phylogenetic analyses and interpretations of trees, we present an analysis of the Rho family in the fungal kingdom, revealing the existence of at least four separate paralogs (Cdc42, Rac, Rho1, and Rho4) in early fungi and subsequent potentially independent expansions of the family in different fungal subgroups. We conclude with the warning that the currently dominating perception of the Rho phylogeny is biased by the metazoan (and especially vertebrate) perspective, and a new, more global view is to be worked out when a better genome sampling and more adequate methods of phylogenetic inference are employed.

Key words: Rho GTPases, Phylogeny, Evolution, Eukaryotic cell, Fungi

1. Introduction: The Rho Family and Its Origin

Rho GTPases are said to form a protein family (1–4), which by definition means that all of them are derived from a single ancestral gene in an ancient predecessor of current organisms. Reconstructing the series of evolutionary events that multiplied the offspring of this first Rho gene to its present multitude is a desirable and intellectually

Francisco Rivero (ed.), *Rho GTPases: Methods and Protocols*, Methods in Molecular Biology, vol. 827,
DOI 10.1007/978-1-61779-442-1_2, © Springer Science+Business Media, LLC 2012

appealing task of its own, but it also has a potential to provide fundamental insights into the functional aspects of Rho proteins in diverse species. The purpose of this chapter is to offer a broader perspective on the evolution of the Rho family, with the special aim of pointing toward various methodological issues associated with evolutionary analyses of such a complex group of proteins.

We may start by asking about the very first Rho GTPase – where did it come from and what was the organism it resided in? The expanding space of sequenced prokaryotic genomes has already brought to light a number of bacterial and archaeal GTPases that are surprisingly similar to eukaryotic "small" GTPases of the Ras superfamily, but neither seem specifically related to the Rho family to the exclusion of other GTPase families (5, 6). In other words, it seems that the Rho family probably shares a specifically eukaryotic ancestry with at least some other traditionally recognized small GTPase families, such as Rab, Ran, and Ras. The restricted size of the GTPase domain conserved among the different eukaryotic GTPase families (in terms of the number of phylogenetically informative amino acid positions) makes reconstruction of their actual phylogenetic relationships inherently difficult, yielding poorly resolved trees regardless of the method of phylogenetic inference employed (unpublished observations). Nevertheless, it seems that the actual sister group to Rho is represented by Miro proteins, or at least by their less divergent N-terminal GTPase domain (the origin of the even more divergent C-terminal GTPase domain is much less obvious). Miro, originally described as "mitochondrial Rho" (7) and sometimes treated as a member of the Rho family (1), is probably better considered as a separate, structurally unique and functionally specialized family of eukaryotic GTPases that most likely existed already in the last eukaryotic common ancestor (LECA) (8).

Can the same antiquity be ascribed to the Rho family proper? To aid answering this question, we have developed a simple program, *rhosearch*, which automatically extracts Rho sequences from protein or EST databases on the basis of searches with *blast* (9). The program works in two steps: (1) it starts by searching for candidate Rho homologs in a user-defined database with a set of Rho "probes" (representative Rho sequences) as queries; (2) it then filters the candidates for true members of the Rho family by using them as queries in searches against a predefined database of GTPases sequences and looking for those giving better hits to known Rho sequences than to sequences of other GTPase groups. (The program along with detailed documentation is available at http://www.1.osu.cz/~elias/GTPases/rhosearch/). Using this tool complemented with further manual checking (e.g., for potential unannotated genes missing in protein databases) enabled us to easily extend previous surveys of the Rho family (3, 4) to more recently sequenced genomes filling important gaps in the phylogenetic

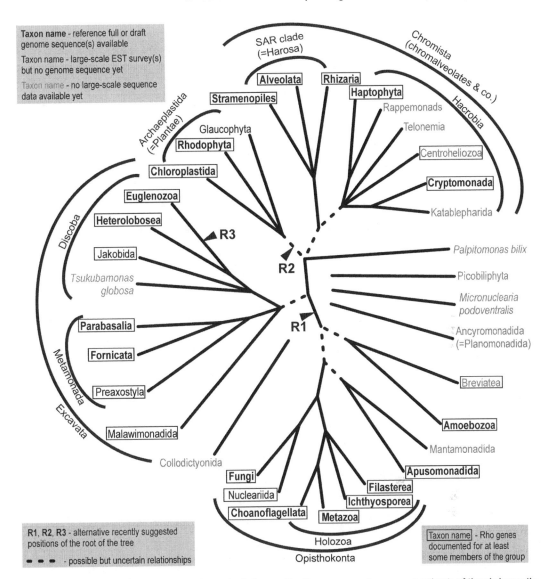

Fig. 1. Rho GTPases in the context of the eukaryotic phylogeny. The figure represents a current estimate of the phylogenetic relationships among major eukaryotic lineages based on recent phylogenetic and phylogenomic analyses (46–52). Three possible positions of the root of the tree that were suggested during the last few years are shown: R1 – "unikonts-bikonts" rooting (53), R2 – Archaeplastida rooting (54), and R3 – Euglenozoa rooting (55). The phylogenetic position of some minor eukaryotic groups with limited sequence data remains uncertain, so their branches do not connect to any particular position within the tree. Taxa in which the presence of Rho GTPase genes is documented in at least some members are boxed.

sampling (see Fig. 1, Table 1). The results attest an almost ubiquitous occurrence of Rho genes in eukaryotic genomes with just a handful of exceptions, namely diatoms, a subgroup of green algae ("core" chlorophytes), such as *Chlamydomonas reinhardtii* and *Chlorella variabilis* (but not the prasinophyte green algae *Ostreococcus* spp. or *Micromonas* spp.), Myzozoa (apicomplexans, *Perkinsus marinus*), and some trypanosomatids except *Trypanosoma cruzi*. Furthermore, by analyzing expressed sequence tag (EST)

Table 1
The Rho family in various eukaryotic species

Eukaryotic group	Species	Number of Rho genes
Metazoa	*Homo sapiens*	21
	Drosophila melanogaster	8
	Nemastostella vectensis	15
Choanoflagellata	*Monosiga brevicollis*	5
Ichthyosporea	*Sphaeroforma arctica*	4
Filasterea	*Capsaspora owczarzaki*	6
Fungi	Microsporidia (*Encephalitozoon cuniculi, Nosema ceranae, Nematocida parisii*) Other fungi see Fig. 3	2–3
Nucleariida	*Nuclearia simplex* (strain 2)	≥1
Apusomonadida	*Thecamonas trahens*	10
Amoebozoa	*Dictyostelium discoideum*	21
	Entamoeba histolytica	29
Breviatea	*Breviata anathema*	≥1
Malawimonadida	*Malawimonas jakobiformis*	≥4
Preaxostyla	*Trimastix pyriformis*	≥2
Fornicata	*Giardia intestinalis*	1
Parabasalia	*Trichomonas vaginalis*	≥12
Jakobida	*Reclinomonas americana*	≥6
Heterolobosea	*Naegleria gruberi*	≥60
Euglenozoa	*Trypanosoma cruzi*	1
	Trypanosoma brucei	–
	Leishmania major	–
Chloroplastida	*Arabidopsis thaliana*	11
	Selaginella moellendorffii	2
	"core" chlorophytes (*Chlamydomonas reinhardtii, Volvox carteri, Chlorella variabilis, Coccomyxa* sp. C-169)	–
	Mamiellales (*Ostreococcus* spp., *Micromonas* spp.)	1
Rhodophyta	*Cyanidioschyzon merolae*	2

(continued)

Table 1
(continued)

Eukaryotic group	Species	Number of Rho genes
Stramenopiles	*Phytophthora sojae*	1
	Ectocarpus siliculosus	1
	Diatoms (*Thalassiosira pseudonana*, *Phaeodactylum tricornutum*)	–
Alveolata	*Tetrahymena thermophila*	2
	Paramecium tetraurelia	13
	Myzozoa (*Plasmodium falciparum*, *Theileria parva*, *Toxoplasma gondii*, *Cryptosporidium parvum*, *Perkinsus marinus*)	–
Rhizaria	*Bigelowiella natans*	≥12
Haptophyta	*Emiliania huxleyi*	1
Centroheliozoa	*Raphidiophrys contractilis*	≥1
Cryptomonada	*Guillardia theta*	1

The numbers for some genome-sequenced species (*Trichomonas vaginalis*, *Naegleria gruberi*, *Bigelowiella natans*) are uncertain, since some Rho-like loci may actually represent pseudogenes. Only data from EST survey are available for Nucleariida, Malawimonada, Preaxostyla, and Jakobida, hence the numbers of Rho genes represent minimal estimates. For Breviatea and Centroheliozoa, no large-scale sequence resources are available, but individual Rho sequences could nevertheless be found in GenBank. Accession numbers of all sequences reported in the table are available upon request

sequences available for some important major eukaryotic lineages still lacking genome-sequenced representatives, we have documented the presence of Rho genes in groups, such as Nucleariida, Jakobida, Preaxostyla, and Malawimonadida, leaving the Glaucophyta as the sole major eukaryotic group with a fair sample of ESTs not recording a Rho gene (see Fig. 1, Table 1). Regardless of the uncertainties surrounding the actual position of the root of the eukaryotic tree (see Fig. 1; refs. 10, 11), neither of the Rho-lacking groups can realistically be considered as sister or even ancestral to the remaining eukaryotes, implying that the Rho family is an ancestral feature of extant eukaryotes and that the rare Rho absence always means its secondary loss. It can therefore be safely inferred that the first Rho gene emerged before LECA in the eukaryotic stem lineage, most likely by duplication of an "undifferentiated" eukaryotic small GTPase gene giving rise to the ancestor of the Miro family as its second product. However, we can presently only speculate about the genomic and cellular context into which the first Rho was born.

2. Challenges of Reconstructing the History of Gene Duplications and Losses Within the Rho Family

With very few exceptions characterized by possessing only one Rho gene (e.g., the diplomonad *Giardia intestinalis*, the kinetoplastid *Trypanosoma cruzi*, prasinophyte green algae, the haptophyte *Emiliania huxleyi*, the oomycete *Phytophthora sojae*, or the brown alga *Ectocarpus siliculosus*), most eukaryotic species harbor more or less expanded families of Rho genes, with the current record holder being the heterolobosean amoeboflagellate *Naegleria gruberi* harboring a complement of at least 60 Rho and Rho-like genes (see Table 1). Whereas gene families in prokaryotes seem to expand predominantly via acquiring xenologous genes by horizontal gene transfer (HGT) (12), gene duplication is obviously the major driver of the origin of new genes in eukaryotic genomes (13–15). Although it was suggested that the vertebrate-specific RhoH was gained by a vertebrate ancestor from a parasite or a retrovirus (4), the evidence for this hypothesis seems inconclusive at best. Regardless, it is conceivable that the varying Rho gene inventories in different species result primarily from gene duplications and losses. Phylogenetic analyses of Rho sequences and reconciliation of the resulting trees with the organismal phylogeny should enable to reconstruct individual gene duplication, loss, and transfer events in principle, but in practice this remains a challenging task for the following reasons:

1. Obstacles may come from uncertainties in the actual organismal phylogeny. For example, our survey of Rho genes in eukaryotic genomes revealed that the RhoBTB subfamily, so far known only from metazoans and the slime mold *Dictyostelium* (16), occurs also in the heterolobosean *Naegleria gruberi*, but not in other excavates, archaeplastids or chromists (unpublished observation). If we assume the "unikonts-bikonts" rooting of the eukaryotic phylogeny and provided that the phylogenetic distribution of RhoBTB genes is not affected by HGT events, we would infer that a RhoBTB gene existed already in the LECA and was lost several times independently during subsequent eukaryotic evolution (see Fig. 1). However, assuming other possible positions of the eukaryotic root, such as the other two indicated in Fig. 1, would be more compatible with a post-LECA origin of RhoBTB, necessitating a lower number of RhoBTB losses.

2. Inferences on the evolutionary history of any gene family may be misled by an incomplete taxon sampling. As an example, consider the Rnd subgroup of the metazoan Rho family. Based on the

most comprehensive phylogenetic analysis of the Rho family published to date (4), Boureux et al. concluded that Rnd emerged in chordates, but our survey of more recently released genome sequences revealed the presence of probable Rnd orthologs in other phyla, such as molluscs or cnidarians (unpublished observations), which would mean a much earlier origin of Rnd than conceived previously and independent losses by some metazoan lineages (echinoderms, ecdysozoans, platyhelminths).

3. Inferring the phylogenetic relationships among the Rho genes themselves is inherently difficult, in part owing to the nature of their sequences and the evolutionary process operating on them: (a) a relatively small length of the conserved GTPase domain, limiting the number of phylogenetically informative characters and hence increasing the risk of statistical errors in the tree inference; (b) a highly uneven rate and mode of evolution of different positions, with some of them strongly constrained and hence essentially invariant and phylogenetically noninformative, while with others rapidly evolving and potentially substitutionally saturated (with the record of the repeated substitutions effectively erased); (c) a frequent gene duplication, with a potentially highly asymmetric rate and a different mode of evolution between the paralogs in the case of neofunctionalization (emergence of "novel" Rho types associated with fast and perhaps extensive modifications of the original sequence).

A combination of these effects and other confounding issues causes that trees inferred for a set of Rho sequences, and/or interpretations thereof, are often, if not by rule, inaccurate to at least certain degree. As an example, let us look more closely on a phylogenetic tree of eukaryotic Rho GTPases reported in ref. 4, here adopted with slight modifications as Fig. 2. This tree features a number of striking or even odd aspects. For instance, it is now very well established that stramenopiles (including *Phythophtora*) and alveolates (including *Tetrahymena*) are among the most closely related eukaryotic groups (see Fig. 1), but the *Phytophthora* Rac sequence and the *Tetrahymena* Rac sequence are placed in the Fig. 2 tree at very distant positions rather than in a common clade. Similarly, the number of Rho sequences from Amoebozoa (*Dictyostelium discoideum* and *Entamoeba histolytica*) included in the tree do not form a single clade or perhaps a few clades that would exhibit obvious orthologous relationships to Rho clades from opisthokonts (metazoans and fungi) and/or other eukaryotes, but instead branch off in an extensively paraphyletic manner across the whole tree. This topology is simply very different from a presumed species tree (organismal relationships) and if interpreted at face value, it would indicate that there was a large number of Rho paralogs in the common ancestor of diverse eukaryotic groups and that most of these paralogs were massively lost selectively by different eukaryotic lineages. However, this scenario is rather implausible.

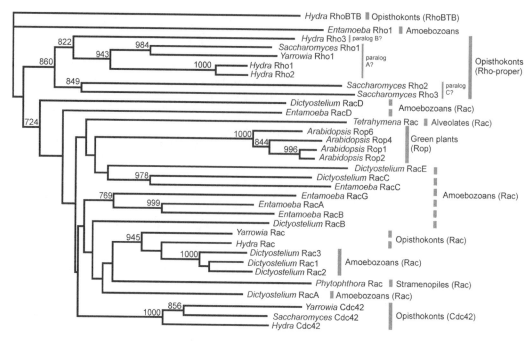

Fig. 2. How reliable are phylogenetic trees of Rho GTPases? The scheme shows a neighbor joining phylogenetic tree of a selection of eukaryotic Rho GTPase sequences adopted (with slight modifications concerning the labels of the sequences and of the different groups on the right) from ref. 4. The numbers at branches are bootstrap support values based on 1,000 replicates. Taxon-specific clades are delimited with solid vertical lines, paralogous segments of the tree comprising sequences from the same taxonomic group are labeled by a dashed vertical line. The hypothetical paralogs A, B, and C labeled within the opisthokont Rho-proper group are discussed in the text. The assignment of the sequences into the five categories Rac, Cdc42, Rho(-proper), RhoBTB, and Rop follows the original figure by Boureux et al. (4).

Instead, the limitations of phylogenetic analyses described above and the absence of statistical support for most branches in the tree suggest that the tree topology may be inaccurate and has failed to reconstruct correctly the actual existence of a stramenopile/alveolate Rho clade and of just a few (or possibly even only one) purely amoebozoan Rho clades.

However, it is very well known that at some circumstances even strongly supported topologies may be wrong (17, 18) and this may be the case of the topology within the fungal/metazoan Rho-proper group in the Fig. 2 tree. For this topology to be true, it would need the existence of three separate Rho-proper paralogs in the common ancestor of fungi and metazoans, with one of the paralogs (A) represented by the yeast Rho1 and *Hydra* Rho1/2, the other (B) by the *Hydra* Rho3, and the last (C) by the yeast Rho2/3 genes. It would then follow that the paralog B was lost from the yeast lineage and the paralog C from the *Hydra* lineage, or at least that the yeast and *Hydra* genes representing the missing paralogs were not included in this analysis. Alternatively, the topology within the Rho-proper clade in Fig. 2 tree may be artificial despite the relatively high bootstrap support values, allowing for a simpler scenario with just a single Rho-proper gene in the ancestor of fungi and metazoans.

A fair discussion of the Fig. 2 tree must consider both possibilities, but note that the "three paralogs" scenario was completely omitted by the authors of the Fig. 2 tree (4). Such an inconsequent approach to interpreting phylogenetic trees seems to be a practice unfortunately too frequent in the current literature.

All these challenges notwithstanding, we are starting to see at least the gross outlines of the Rho phylogeny. Most attention has been paid to inferring the phylogeny of metazoan Rhos, leading to a view of a starting complement of at least three paralogs – Rac, Cdc42, and Rho(-proper) – in the opisthokont ancestor progressively elaborated by adding new and new paralogs during the metazoan phylogeny, particularly in the lineage leading to vertebrates (4). It is possible that a similar evolutionary path has been followed by Rho families in other eukaryotic groups, at least in those characterized by extensive sets of Rho genes (e.g., amoebozoans, heteroloboseans, rhizarians). It is beyond the scope of this chapter to deliver a fully worked-out picture of the Rho phylogeny, but to further document various methodological issues associated with reconstructing the evolutionary history of the Rho family, we have conducted an analysis of Rho GTPases in one particular group of eukaryotes, the kingdom Fungi.

3. A Case Study: Reconstructing the Phylogeny of Rho GTPases in Fungi

Rho genes have been studied in many fungal species (19, 20) and the coverage of the fungal phylogenetic diversity by complete or draft genome sequences has improved considerably during last few years, but to our knowledge no comprehensive phylogenetic analysis of fungal Rhos have been reported. To get an initial insight into the phylogeny of the fungal Rho family, we followed the relatively well-developed fungal phylogenetic classification (21–23) and selected a target set of 20 species representing as many major fungal lineages as possible (see Fig. 3), although we decided to omit Rho GTPases from microsporidia, a probable fungal group characterized by generally extremely divergent gene sequences (24).

3.1. Assembling a Sequence Dataset

For 18 species, annotated complete or high-quality draft genome sequences were available, so we could start building the respective Rho inventories using *rhosearch* (see Subheading 1). Using this procedure, we found between 4 and 11 Rho proteins in each genome. However, it is important to keep in mind that the quality and completeness of available genome annotations vary, and it is always useful to check any genome by searching directly against the genomic sequence using *tblastn*. Indeed, in some genomes we found additional Rho genes that for some reasons had escaped annotation and were thus not represented by a corresponding product in protein databases (see Fig. 4). In addition, it is not

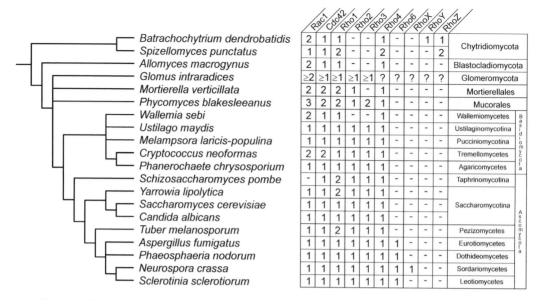

Fig. 3. The fungal Rho family. The 20 fungal species analyzed in this study are shown with the likely species tree as reconstructed in recent phylogenetic analyses of the fungal kingdom (21–23). The classification of the species into higher-order taxa is shown on the right. For each major fungal Rho paralog (indicated at the *top*), the number of copies (*in-paralogs*) residing in the genomes of individual species is specified; a dash (–) means an apparent absence (primary or secondary) of a gene representing the paralog. For *G. intraradices* analyzed on the basis of an EST survey only, the set of Rho genes retrieved is not necessarily complete, so the numbers are only minimal estimates, whereas the absence of paralog representatives (*question mark*) cannot be viewed as definitive. Note that the number of Rho loci identified in the *A. macrogynus* genome is higher than those indicated in the table, but these extra copies seem to represent allelic variants or very recently emerged duplicates (possibly reflecting polyploidy or a recent whole genome duplication event). Accession numbers for the respective gene or protein sequences are provided in Fig. 4.

Fig. 4. A maximum likelihood phylogenetic tree of the fungal Rho family. The tree was obtained using PhyML 3.0 and employing the LG + Γ4 + I substitution model (with empirical amino acid frequencies) and the NNI + SPR heuristics (see Subheading 3.3 for details on the tree search procedure). Some fungal Rho proteins with incomplete or divergent sequences were omitted from the main analysis presented, but were included in a second analysis based on a shorter alignment; in the latter tree, they branched off as three separate lineages, here shown at the bottom of the tree, at positions approximately indicated in the main tree by dashed ovals (the exact positions could not be specified, since the tree inferred from the shorter alignment is topologically different in some regions). The length of the basal branch of the Rho6 group and of the terminal branch of NcrRhoX was reduced to their halves to fit them into the figure. Numbers above and below branches denote, respectively, SH-like aLRT support values calculated by PhyML and bootstrap support values (from 100 replicates) calculated using the rapid bootstrap algorithm implemented in RAxML 7.0.3. Only values above 0.900 and 50, respectively, are shown. For the sake of clarity, branch support within the major fungal Rho clades is indicated only by black dots (shown only for those branches receiving support values >0.900/>50). The tree was arbitrarily rooted with the unique RhoY sequence from *Batrachochytrium dendrobatidis*, but the tree is to be viewed as unrooted. Species abbreviations: Afu *Aspergillus fumigatus* (strain Af293), Ama *Allomyces macrogynus*, Bde *Batrachochytrium dendrobatidis* (strain JEL423), Cal *Candida albicans* (strain SC5314), Cne *Cryptococcus neoformans* (var. *neoformans*, strain JEC21), Gir *Glomus intraradices*, Mla *Melampsora laricis-populina*, Mve *Mortierella verticillata*, Ncr *Neurospora crassa* (strain OR74A), Pbl *Phycomyces blakesleeanus* (strain NRRL 1555), Pch *Phanerochaete chrysosporium*, Pno *Phaeosphaeria nodorum* (strain SN15), Sce *Saccharomyces cerevisiae* (strain S288c), Spo *Schizosaccharomyces pombe* (strain 972 h-), Spu *Spizellomyces punctatus*, Ssc *Sclerotinia sclerotiorum* (strain 1980), Tme *Tuber melanosporum* (strain Mel28), Uma *Ustilago maydis* (strain 521), Wse *Wallemia sebi*, Yli *Yarrowia lipolytica* (strain CLIB99). Sequence accession numbers are valid for the following databases: species-specific genome databases at the Joint Genome Institute (JGI) – *M. laricis-populina*, *P. blakesleeanus*, *P. chrysosporium*, *W. sebi*; species-specific genome databases at the Broad Institute – *A. macrogynus*, *B. dendrobatidis*, *M. verticillata*, *N. crassa*, *S. punctatus*; other entries come from NCBI databases (GenPept, RefSeq, GenBank). * – incorrectly predicted protein sequences, # – unannotated genes, & – EST assemblies (only one representative EST sequence per contig is indicated by the accession number); the *C. albicans* Rho2 and the *P. nodorum* Rho1 gene sequences are affected by sequencing errors. Corrected and newly predicted protein sequences of fungal Rhos are available upon request.

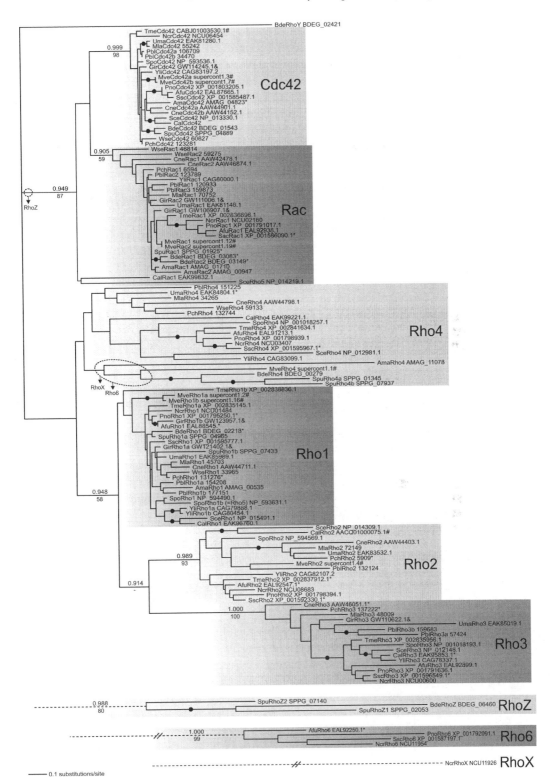

always properly acknowledged that protein sequences predicted from genome assemblies and available in public databases are often incorrect, affected by a variety of errors (incorrect exon/intron boundaries, missing or artificial supernumerary exons, artificial fusions with adjacent independent genes, etc.). Indeed, a closer look on the predicted protein sequences of fungal Rhos by inspecting pair-wise or multiple alignments with confirmed sequences revealed obvious errors in at least 19 of the sequences analyzed (see Fig. 4); the respective gene models could be manually corrected with high confidence in most cases and the revised protein sequences were used in subsequent analyses. In two genes (see Fig. 4), errors were even found in the gene sequence itself (the genomic assembly) and corrections introduced by inspecting the original unassembled sequencing reads (available from Trace Archive; http://www.ncbi.nlm.nih.gov/Traces) restored the highly conserved coding sequence.

For Mortierellales, an isolated group of the traditional paraphyletic "zygomycetes" (23), a draft genome assembly had been released for *Mortierella verticillata* just before our analyses, but without the accompanying annotation. We therefore searched the genome sequence with *tblastn*, identified all Rho loci (eight in total) and predicted manually the corresponding gene models. Finally, no genome sequence was available for the whole fungal phylum Glomeromycota (also a former member of "zygomycetes"), but to include it into our analysis we exploited a relatively extensive set of EST sequences (almost 47,000) from *Glomus intraradices*. Using *rhosearch*, which implements a clustering step after searching an EST database to get a nonredundant set of sequences, we obtained assembled contigs and one singleton EST representing cDNAs of seven Rho genes, all but one with obviously complete coding sequences.

3.2. Building a Multiple Alignment

The next step was to build a multiple alignment of the polished protein sequences. A number of different tools are available for this task (e.g., see refs. 25–29). We tested three different programs (all with default settings), Clustal X 2.0 (25), MUSCLE 3.7 (26), and MAFFT_TG 6.822 (29), the latter two run using the service of the CIPRES Science Gateway (http://www.phylo.org/sub_sections/portal/). The alignments constructed by MUSCLE and MAFFT_TG seemed better than that produced by Clustal, which showed lesser accuracy in some less conserved regions. For subsequent analyses, we selected the MAFFT alignment. There was minimal need for its manual curation, but some regions of some more divergent Rho proteins could not be aligned reliably by any method. In addition, the region of the so-called Rho insert is so poorly conserved even among representatives of different "standard" Rho subfamilies that no reliable alignment across the whole family can be obtained. For phylogenetic analyses, this region as well as the

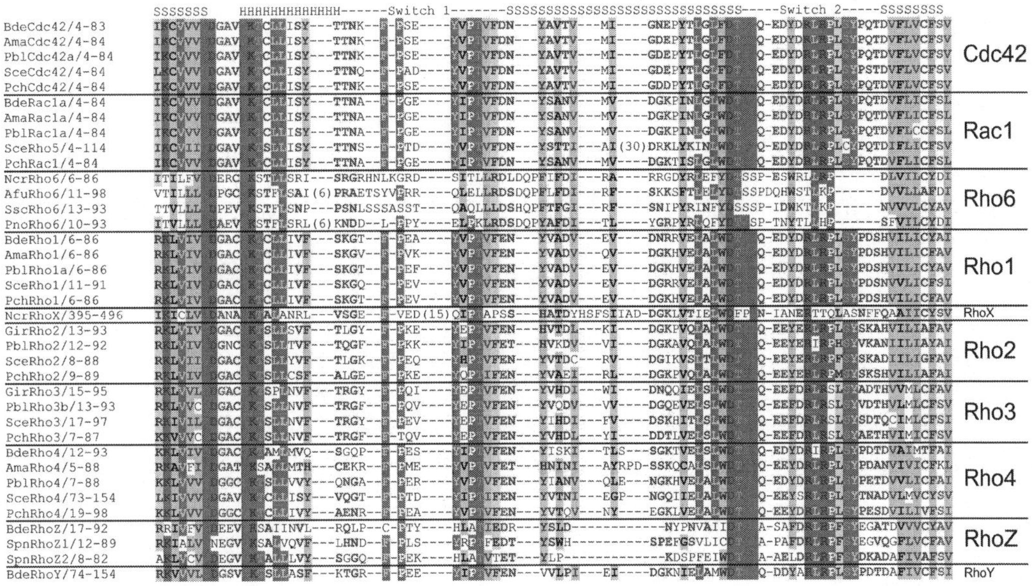

Fig. 5. Variations in the GTPase domain of fungal Rho proteins. Due to space limitations only the N-terminal half of the domain including the most significant differences is presented here. The accession numbers of the sequences can be found in Fig 4. The region of each sequence included in the alignment is delimited after a slash. The extent of elements of the secondary structure of typical Rho GTPases and the switch 1 and 2 regions are shown at the top (H – α-helix, S – β-strand). The alignment was processed using CHROMA (56) to highlight the degree of conservation of individual positions.

highly variable regions at the very N-terminus and at the C-terminal tail downstream of the GTPase domain must be omitted, since they do not confer any meaningful phylogenetic signal. Programs designed to automatically detect and mask unreliable or divergent regions of multiple alignments are available (30, 31), but in the case of the Rho alignment this can be satisfactorily achieved by eye. Furthermore, the accuracy of a phylogenetic tree inferred from an alignment may be negatively affected by the inclusion of extremely divergent sequences, hence for our initial phylogenetic analysis we removed from the alignment a series of sequences deviating most from the conserved sequence pattern (see Fig. 5): a unique sequence from *Neurospora crassa* hereafter referred to as RhoX, a group of very unusual Rho-like sequences from some ascomycetes we here denote Rho6, and a group of rather divergent sequences from chytrids denoted RhoZ. Finally, we excluded from the analysis the Rho2 sequence from *G. intraradices* because it was incomplete (derived from a single EST read GW090856.1). All manipulations with the alignment were done with the aid of GeneDoc (32).

3.3. Inferring a Tree

The resulting alignment comprising 129 sequences and 157 reliably aligned positions was analyzed using the maximum likelihood (ML) method of phylogenetic inference. Previous phylogenetic analyses of the Rho family (e.g., see refs. 1, 3, 4, 33) generally utilized a relatively simplistic approach employing simple substitution models and the neighbor joining (NJ) clustering algorithm, but it is now very well established that ML and other more advanced methods (e.g., Bayesian inference) combined with a more complex substitution models more efficiently extract the actual phylogenetic signal from the sequences and are less prone (although not immune) to various phylogenetic artifacts (34). To carry out the analysis, we employed the program PhyML 3.0 (35), which enables analyzing datasets of our size in a reasonable time and allows for using the latest general substitution matrix for protein sequences, LG (36). The tree search procedure further utilized a correction for rate heterogeneity among sites (uneven frequency of substitutions at different positions along the protein sequence) modeled as an approximation of a Γ distribution with four rate categories ($\Gamma4$) plus a category of invariant sites (I). This correction is now viewed as a mandatory part of evolutionary models for phylogenetic analyses and ignoring it may lead to highly inaccurate results (37). Finally, we employed a distance (BioNJ) tree as a starting point for a heuristic search through the tree space utilizing both the nearest neighbor interchange (NNI) and subtree pruning and regrafting (SPR) searches to minimize the chance of missing the true tree due to the search becoming trapped in a local likelihood maximum. To evaluate the tree topology, we let PhyML run the SH-like approximate likelihood ratio test (aLRT) (38), and to get an independent measure of the tree robustness we also carried out a bootstrap analysis using the recently developed rapid bootstrap algorithm implemented in RAxML 7.0.3 (39).

In order to investigate the phylogenetic position of the Rho6, RhoX, and RhoZ sequence omitted from the analysis above, we prepared a second multiple alignment retaining these sequences but excluding the region, where these sequences tend to be particularly divergent and where their alignment is rather problematic (i.e., the switch 1 and the region directly downstream, see Fig. 5). The alignment of the remaining 129 position was analyzed using the same approach as the first dataset.

To test for the differences in performance between the more complex ML approach and a simpler phylogenetic method, we used the Phylip 3.63 package (40) to calculate a distance matrix from the same alignment as used for inferring the ML tree (with selecting the JTT substitution matrix and without correcting for among-site rate heterogeneity) and to built an NJ tree.

3.4. From a Phylogenetic Tree to an Evolutionary Scenario

We dissected the ML tree obtained (see Fig. 4) to reconstruct the basic outlines of the actual phylogeny of the fungal Rho family. Our logic was based on considering the tree topology from the perspective of the presumed species tree (see Fig. 3) and balancing between the complexity of the evolutionary scenario implied by the tree as such and between introducing assumptions about possible artifacts or inaccuracies of the topology inferred. Note that the nomenclature of the Rho genes in different fungal species as used previously in the literature may differ or even be confusing by applying the same name to non-orthologous genes. For the sake of simplicity we label here most genes following orthology assignment to *Saccharomyces cerevisiae* genes, except the Rac group (see Subheading 3.4.2) and the novel Rho type without obvious counterparts in *S. cerevisiae* (see Subheading 3.4.6).

3.4.1. Cdc42

One of the strongly supported groups in the tree can be readily recognized as the ancient opisthokont paralog Cdc42. Genes representing this paralog can be found in every fungal species examined (see Fig. 3), attesting its probable functional essentiality throughout the whole fungal phylogeny. Some species possess two Cdc42 in-paralogs that appear to reflect lineage-specific duplications, since the two copies from the same species always cluster together. However, the internal topology within the Cdc42 group at many points deviates from relationships among the species (depicted in Fig. 3). Rather than indication of HGT or hidden paralogy, this is perhaps primarily due to stochastic errors caused by weak phylogenetic signal in variable regions of Cdc42 genes. The same interpretation of the discordance between the species tree and the gene tree can perhaps be considered for the other paralogous groups of Rho genes discussed below (note the lack of significant statistical support for most branches within the subtrees representing individual major paralogs).

3.4.2. Rac

A clade next to Cdc42 but much more weakly supported includes known Rac (Rac1) genes and can be therefore considered equal to the opisthokont Rac paralog. However, we posit that the tree actually fails to show the whole extent of this paralog, as it should also embrace the *S. cerevisiae* Rho5 and the *Candida albicans* Rac1 clustering together (without support) at a basal position within a broader very well supported Rac/Cdc42 clade. Our argument is based on applying parsimony reasoning on the taxonomic composition of individual branches of the tree. Would the yeast Rho5/Rac1 be phylogenetically really separate from the Rac and Cdc42 groups, they would have to be as ancient, i.e., established as a separate paralog before the divergence of the major fungal phyla. This would necessarily mean an unreasonably high number of independent losses in the different lineages beside the yeasts, so it is simpler to assume that the tree does not correctly show their actual affinity. Indeed, the "core" Rac clade misses genes exactly from *S. cerevisiae* and *C. albicans* (plus *Schizosaccharomyces pombe*, which appears genuinely devoid of even a divergent version of the Rac paralog). This complementarity in taxonomic distribution, not seen for other theoretical "homes" (Cdc42, Rho1, 2, etc.) for the two orphan yeast genes, strongly suggests that they cannot be anything else than the "missing" Rac orthologs misplaced in the tree due to being too divergent (note that the *C. albicans* gene has been identified as a Rac ortholog previously (41)). It follows that the previous claims in the literature about the nonexistence of Rac orthologs (let alone homologs!) in the budding yeast (41, 42) needs to be revisited.

3.4.3. Rho2 and Rho3

Two more major clades in our ML tree are strongly supported and may be interpreted as fungal-specific paralogs Rho2 and Rho3. The phylogenetic distribution of Rho2 and Rho3 is most compatible with the notion that they evolved after the divergence of the chytrid lineages (Chytridiomycota and Blastocladiomycota); Rho3 appears to be missing from the *Mortierella verticillata* genome, but whether this absence is primary or secondary cannot be resolved at present owing to the uncertain phylogenetic position of Mortierellales with respect to Mucorales, Glomeromycota, and Dikarya (Ascomycota and Basidiomycota) (23). However, the striking absence of both Rho2 and Rho3 from *Wallemia sebi* (a member of a poorly known deep lineage of Basidiomycota), if not due to incompleteness of the genome sequence available, is undoubtedly a result of secondary loss.

3.4.4. Rho1

Another cluster in our tree is annotated as Rho1, but in this case there is no significant statistical support for monophyly of this group (see Fig. 4). Indeed, we suggest that its monophyly in our tree may actually be an artifact, since the Rho1 group may be phylogenetically ancestral to the Rho2 and Rho3 groups rather than being their sister group as shown in the tree. This hypothesis is based on considering the taxonomic provenance of genes of these groups. The Rho1 group comprises sequences from all fungal taxa, including chytrids (Chytridiomycota and Blastocladiomycota), whereas there are no chytrid genes related specifically to the Rho2 and Rho3 groups. Hence, the topology as reconstructed in the Fig. 4 tree would necessitate secondary losses of Rho2/3 from the two chytrid lineages – either already as separated paralogs or at least as an undifferentiated Rho2/3 paralog if Rho2 and Rho3 separated via a duplication of a proto-Rho2/3 gene only after the divergence of chytrid lineages. It is, however, more parsimonious to assume that Rho2 and Rho3 emerged as "neofunctionalized" paralogs in "post-chytrid" fungi only (i.e., after the chytrid lineages had separated from the "main" fungal lineage). The extensive divergence of Rho2 and Rho3 sequences (compare their long branches) may then be responsible for the failure of the phylogenetic analysis to correctly place them within the Rho1 group in a sister position of Rho1 sequences from post-chytrid fungi. The lack of exclusive clustering of Rho1 sequences from post-chytrid fungi themselves and other departures from species tree within the

Rho1 group may generally result from a stochastic error (see Subheading 3.4.1). However, an unexpectedly strong support for clustering of the GirRho1b sequence from the glomeromycete *Glomus intraradices* with the Rho1 gene from an ascomycete *Aspergillus fumigatus* is suspicious. Indeed, a taxonomically broader analysis of fungal Rho1 sequences (results not shown) support the unexpected position of GirRho1b among aspergilli and related genera, suggesting either an HGT event into the *G. intraradices* lineage, or (perhaps more likely) a contamination of the *G. intraradices* cDNA library used for generating the EST dataset.

3.4.5. Rho4

The clade in our tree labeled as Rho4 is best considered as a true clade of orthologous genes despite the lack of statistical support, since this is the most parsimonious interpretation of its actual taxonomic composition, generally just one gene for each fungal species (except the recently duplicated paralogs in *Spizellomyces punctatus* and the lack of a representative sequence from possibly incomplete EST set of *Glomus intraradices*). As such, it would imply the existence of a distinct Rho4 paralog in the common ancestor of the fungal kingdom with subsequent vertical dissemination. Other possibilities, such as several independent origins of the genes included in the Rho4 group by taxon-specific duplication of Rho genes of other paralogs (Rho1, Rho2, or others) followed by extreme divergence obscuring these true relationships and artificially causing these genes to be clustered into a monophyletic group, cannot be excluded, but the dictum of Occam's razor tells us that given the data in hand we should prefer the former interpretation.

3.4.6. Novel Fungal Rho Groups

Whereas the majority of fungal Rho genes can be more or less readily classified into one of the six principal widely conserved fungal Rho paralogs, some fungal species harbor apparently unique and/or divergent Rho genes with a rather enigmatic origin (see Fig. 4). Nevertheless, all of them show higher sequence similarity to "Rho-proper," i.e., members of the opisthokont-specific Rho subfamily, than to Cdc42/Rac1 proteins. Little more can be presently said about the RhoY gene from *Batrachochytrium dendrobatidis* and the RhoZ group more widely conserved in Chytridiomycota. A better genome sampling of basal fungal phyla and extending the analyses to nonfungal taxa may hopefully cast more light on the origin of these paralogs. The sequences of the Rho6 group from some filamentous ascomycetes and of the RhoX gene from *N. crassa*, when included in the phylogenetic analysis, branch off in the region of the Rho4 paralog, but they form extremely long branches and their position has no statistical support, so it cannot be trusted (indeed, an alternative ML analysis using RAxML recovers a different topology, data not shown). To get some clues independently of the protein sequence level, we performed a comparative analysis of intron positions in fungal Rho genes (see http://www1.osu. cz/~elias/GTPases/FungalRho_introns.html/). Mapping of these positions onto a multiple alignment of Rho protein sequences revealed that the Rho6 and RhoX genes share an intron otherwise exclusive for some Rho1 genes from Ascomycota. Duplication of the Rho1 gene followed by an extreme divergence of one of the paralogs is thus the most likely scenario for the emergence of both Rho6 and RhoX. These two groups probably arose independently of each other. Rho6 has severely modified the switch 1-to switch 2 region (see Fig. 5), but otherwise resemble typical Rho proteins by possessing a C-terminal prenylation motif. Rho6 genes are distributed over diverse classes of the Pezizomycotina subphylum, suggesting that Rho6 emerged early in filamentous ascomycetes. On the other hand, *blastp* searches indicate that RhoX is probably restricted only to some members of the class Sordariomycetes. RhoX proteins also depart from the typical Rho GTPase domain, but in a manner different from Rho6 (see Fig. 5). Furthermore, they lack a C-terminal prenylation motif and instead exhibit a very long extension (of almost 400 amino acid residues in case of the *N. crassa* RhoX) N-terminal to the GTPase domain that does not display any readily recognizable homology to other proteins and is poorly conserved even among the different RhoX proteins themselves. Both Rho6 and RhoX proteins may be responsible for executing highly specialized taxon-specific cellular functions and are therefore interesting targets for investigating the frontiers of the Rho biology.

3.4.7. Conclusions

Despite some open issues, our ML-based phylogenetic analysis, combined with carefully considering various aspects of the inferred tree, such as branch support values and taxonomic composition of individual clades, leads to a relatively solid picture of the fungal Rho phylogeny. On the other hand, the tree obtained from the same sequence dataset by the NJ algorithm is more difficult to interpret and probably much less accurate, as it e.g., shows the sequences constituting the Rho4 clade in the ML tree as three separate paraphyletic clades (even separating the obviously orthologous ascomycete and basidiomycete Rho4 genes) or as it fails to cluster together the *S. cerevisiae* Rho5 with the *C. albicans* Rac1 (not shown). Although it is unfair to condemn the simpler phylogenetic methods based on just a sole example, we strongly recommend employing (at least as an alternative) more sophisticated methods, such as ML, now easily executable for at least moderately sized datasets and made even simpler to use with the implementation on computer clusters accessible via the Internet as a service for the community (e.g., http://www.atgc.lirmm.fr/phyml/; http://www.phylo.org/sub_sections/portal/).

4. Outlook

In this chapter, we tried to demonstrate that the phylogeny of the Rho GTPase family (and of any other gene family of a similar complexity) cannot be fully understood without overcoming several important obstacles, including inaccurate gene models and the corresponding protein sequences, gaps in the taxon sampling, and various limitations of the methods of phylogenetic inference. The first problem is perhaps the least severe, as accurate gene models can be typically obtained by manual curation being increasingly simpler with the growth of sequence databases offering a wealth of close homologs for comparison. However, this approach sometimes fails for divergent paralogs restricted to particular recent group of species (in our case, some of the chytrid Rho GTPases), where experimental determination of spliced transcripts may be the only way to verify the actual gene structure.

The limitations stemming from the lack of an appropriate taxon sampling are being mitigated by the swift advances in genome sequencing technologies, which have already allowed for sequencing the genomes from groups that we would not dream about a few years ago (see Fig. 1). Further genome sequences from groups such as amoebozoans, excavates, or rhizarians, will be instrumental in drawing a picture of the actual phylogenetic diversity of their Rho families, which may exceed considerably the complexity of metazoan (including vertebrate) Rho complements. Indeed, we want to underscore that the classification of the Rho family into the Rac, Cdc42, Rho(-proper), and possibly RhoBTB groups, working well for opisthokonts and sometimes applied to the eukaryotic Rho family in general (Fig. 2; see ref. 4), does not by any means reflect accurately the actual diversity of the family. For instance, as correctly pointed out by Vlahou and Rivero (33), Rho genes in *Dictyostelium discoideum* are called Rac (RacA, RacB, etc.) for historical reasons only and the names do not imply that they form a homogeneous group. Quite to the contrary, the sequence divergence between different Rho genes in *D. discoideum* can be the same or even higher than between the various opisthokont Rho subgroups. It is likely that with an improved genome sampling of diverse amoebozoans, it will be possible to delineate novel Rho subfamilies that are widely conserved in, yet restricted to, amoebozoa in a manner parallel to the Cdc42 or Rho(–proper) being ubiquitous in, yet specific for, opisthokonts. Indeed, one such emerging subfamily may be represented by RacC paralogs from both *D. dictyostelium* and *Entamoeba histolytica* (Fig. 2; see ref. 3).

The approach to inferring trees for Rho GTPases applied here on the fungal Rho family is more sophisticated than those used in previous studies (3, 4, 33), but we admittedly have not incorporated some of the recent achievements in the phylogenetic methodology, such as accounting for heterogeneity in the substitution process across sites (43), accounting for site-specific changes in evolutionary rates (heterotachy) (44), or accounting for solvent accessibility and secondary structure in the substitution model employed (45). It is probable that the progress in the phylogenetic methodology will provide much more precise views of the Rho GTPase phylogeny, although the need for careful interpretations of the inferred trees to recognize possible topological artifacts, such as those we saw with the fungal Rho tree, will probably remain.

Acknowledgments

This work was supported by the P305/10/0205 grant from the Czech Science Foundation. We thank Joint Genome Institute, Broad Institute and Baylor College of Medicine HGSC for the possibility of exploring genome sequences of various eukaryotic species prior to publication.

References

1. Wennerberg, K., and Der, C.J. (2004) Rho-family GTPases: it's not only Rac and Rho (and I like it). *J Cell Sci 117*, 1301–1312.

2. Jaffe, A.B., and Hall, A. (2005) Rho GTPases: biochemistry and biology. *Annu Rev Cell Dev Biol 21*, 247–69.

3. Brembu, T., Winge, P., Bones, A.M., and Yang, Z. (2006) A RHOse by any other name: a comparative analysis of animal and plant Rho GTPases. *Cell Res 16*, 435–445.

4. Boureux, A., Vignal, E., Faure, S., and Fort, P. (2007) Evolution of the Rho family of ras-like GTPases in eukaryotes. *Mol Biol Evol 24*, 203–216.

5. Dong, J.H., Wen, J.F., and Tian, H.F. (2007) Homologs of eukaryotic Ras superfamily proteins in prokaryotes and their novel phylogenetic correlation with their eukaryotic analogs. *Gene 396*, 116–124.

6. Yutin, N., Wolf, M.Y., Wolf, Y.I., and Koonin, E.V. (2009) The origins of phagocytosis and eukaryogenesis. *Biol Direct 4*, 9.

7. Fransson, A., Ruusala, A., and Aspenström, P. (2003) Atypical Rho GTPases have roles in mitochondrial homeostasis and apoptosis. *J Biol Chem 278*, 6495–6502.

8. Vlahou, G., Eliáš, M., von Kleist-Retzow, J.C., Wiesner, R.J., and Rivero, F. (2011) The Ras related GTPase Miro is not required for mitochondrial transport in *Dictyostelium discoideum*. *Eur J Cell Biol 90*, 342–355.

9. Altschul, S.F., Madden, T.L., Schäffer, A.A., Zhang, J., Zhang, Z., Miller, W., and Lipman, D.J. (1997) Gapped BLAST and PSI-BLAST: a new generation of protein database search programs. *Nucleic Acids Res 25*, 3389–3402.

10. Roger, A.J., and Simpson, A.G. (2009) Evolution: revisiting the root of the eukaryote tree. *Curr Biol 19*, R165–R167.

11. Koonin, E.V. (2010) The origin and early evolution of eukaryotes in the light of phylogenomics. *Genome Biol 11*, 209.

12. Treangen, T.J., and Rocha, E.P.C. (2011) Horizontal transfer, not duplication, drives the expansion of protein families in prokaryotes. *PloS Genet 7*, e1001284.

13. Ohno, S. (1970) Evolution by gene duplication. Springer, New York.

14. Koonin, E.V. (2009) Darwinian evolution in the light of genomics. *Nucleic Acids Res 37*, 1011–1034.

15. Levasseur, A., and Pontarotti, P. (2011) The role of duplications in the evolution of genomes highlights the need for evolutionary-based approaches in comparative genomics. *Biol Direct 6*, 11.

16. Berthold, J., Schenkova, K., and Rivero, F. (2008) Rho GTPases of the RhoBTB subfamily and tumorigenesis. *Acta Pharmacol Sin 29*, 285–295.

17. Moreira, D., and Philippe, H. (2000) Molecular phylogeny: pitfalls and progress. *Int Microbiol 3*, 9–16.

18. Gribaldo, S., and Philippe, H. (2002) Ancient phylogenetic relationships. *Theor Popul Biol 61*, 391–408.

19. Pérez, P., and Rincón, S.A. (2010) Rho GTPases: regulation of cell polarity and growth in yeasts. *Biochem J 426*, 243–253.

20. Harris, S.D. (2011) Cdc42/Rho GTPases in fungi: variations on a common theme. *Mol Microbiol 79*, 1123–1127.

21. Hibbett, D.S., Binder, M., Bischoff, J.F., Blackwell, M., Cannon, P.F., Eriksson, O.E., Huhndorf, S., James, T., Kirk, P.M., Lücking, R., Thorsten Lumbsch, H., Lutzoni, F., Matheny, P.B., McLaughlin, D.J., Powell, M.J., Redhead, S., Schoch, C.L., Spatafora, J.W., Stalpers, J.A., Vilgalys, R., Aime, M.C., Aptroot, A., Bauer, R., Begerow, D., Benny, G.L., Castlebury, L.A., Crous, P.W., Dai, Y.C., Gams, W., Geiser, D.M., Griffith, G.W., Gueidan, C., Hawksworth, D.L., Hestmark, G., Hosaka, K., Humber, R.A., Hyde, K.D., Ironside, J.E., Kõljalg, U., Kurtzman, C.P., Larsson, K.H., Lichtwardt, R., Longcore, J., Miadlikowska, J., Miller, A., Moncalvo, J.M., Mozley-Standridge, S., Oberwinkler, F., Parmasto, E., Reeb, V., Rogers, J.D., Roux, C., Ryvarden, L., Sampaio, J.P., Schüssler, A., Sugiyama, J., Thorn, R.G., Tibell, L., Untereiner, W.A., Walker, C., Wang, Z., Weir, A., Weiss, M., White, M.M., Winka, K., Yao, Y.J., and Zhang, N. (2007) A higher-level phylogenetic classification of the Fungi. *Mycol Res 111*, 509–547.

22. McLaughlin, D.J., Hibbett, D.S., Lutzoni, F., Spatafora, J.W., and Vilgalys, R. (2009) The search for the fungal tree of life. *Trends Microbiol 17*, 488–497.

23. Liu, Y., Steenkamp, E.T., Brinkmann, H., Forget, L., Philippe, H., and Lang, B.F. (2009) Phylogenomic analyses predict sistergroup relationship of nucleariids and fungi and paraphyly of zygomycetes with significant support. *BMC Evol Biol 9*, 272.

24. Keeling, P.J., and Fast, N.M. (2002) Microsporidia: biology and evolution of highly reduced intracellular parasites. *Annu Rev Microbiol 56*, 93–116.

25. Thompson, J.D., Gibson, T.J., Plewniak, F., Jeanmougin, F., and Higgins, D.G. (1997) The CLUSTAL_X windows interface: flexible strategies for multiple sequence alignment aided by quality analysis tools. *Nucleic Acids Res 25*, 4876–4882.

26. Edgar, R.C. (2004) MUSCLE: multiple sequence alignment with high accuracy and high throughput. *Nucleic Acids Res 32*, 1792–1797.

27. Pei, J., and Grishin, N.V. (2007) PROMALS: towards accurate multiple sequence alignments of distantly related proteins. *Bioinformatics 23*, 802–808.

28. Liu, Y., Schmidt, B., and Maskell, D.L. (2010) MSAProbs: multiple sequence alignment based on pair hidden Markov models and partition function posterior probabilities. *Bioinformatics 26*, 1958–1964.

29. Katoh, K. and Toh, H. (2010) Parallelization of the MAFFT multiple sequence alignment program. *Bioinformatics 26*, 1899–1900.

30. Talavera, G., and Castresana, J. (2007) Improvement of phylogenies after removing divergent and ambiguously aligned blocks from protein sequence alignments. *Syst Biol 56*, 564–577.

31. Kück, P. Meusemann, K., Dambach, J., Thormann, B. von Reumont, B.M., Wägele, J.W., and Misof, B. (2010) Parametric and non-parametric masking of randomness in sequence alignments can be improved and leads to better resolved trees. *Front Zool 7*, 10.

32. Nicholas, K.B., and Nicholas, H.B. (1997) GeneDoc: a tool for editing and annotating multiple sequence alignments. Distributed by the authors at http://www.nrbsc.org/gfx/genedoc/.

33. Vlahou, G., and Rivero, F. (2006) Rho GTPase signaling in *Dictyostelium discoideum*: insights from the genome. *Eur J Cell Biol 85*, 947–959.

34. Felsenstein, J. (2004) Inferring Phylogenies. Sinauer Associates, Sunderland, Massachusetts.

35. Guindon, S., Dufayard, J.F., Lefort, V., Anisimova, M., Hordijk, W., and Gascuel, O. (2010) New algorithms and methods to estimate maximum-likelihood phylogenies: assessing the performance of PhyML 3.0. *Syst Biol 59*, 307–321.

36. Le, S.Q., and Gascuel, O. (2008) An improved general amino acid replacement matrix. *Mol Biol Evol 25*, 1307–1320.

37. Yang, Z. (1996) Among-site rate variation and its impact on phylogenetic analyses. *Trends Ecol Evol 11*, 367–372.

38. Anisimova, M., and Gascuel, O. (2006) Approximate likelihood-ratio test for branches: a fast, accurate and powerful alternative. *Syst Biol 55*, 539–552.

39. Stamatakis, A., Hoover, P and Rougemont, J. (2008) A rapid bootstrap algorithm for theRAxML Web servers. *Syst Biol 57*, 758–771.

40. Felsenstein, J. (2005) PHYLIP (Phylogeny Inference Package) version 3.6. Distributed by the author. http://evolution.genetics.washington.edu/phylip.html.

41. Bassilana, M., and Arkowitz, R.A. (2006) Rac1 and Cdc42 have different roles in *Candida albicans* development. *Eukaryot Cell 5*, 321–329.

42. Hope, H., Schmauch, C., Arkowitz, R.A., and Bassilana, M. (2010) The *Candida albicans* ELMO homologue functions together with Rac1 and Dck1, upstream of the MAP Kinase Cek1, in invasive filamentous growth. *Mol Microbiol 76*, 1572–1590.

43. Lartillot, N., and Philippe, H. (2004) A Bayesian mixture model for across-site heterogeneities in the amino-acid replacement process. *Mol Biol Evol 21*, 1095–1109.

44. Kolaczkowski, B., and Thornton, J.W. (2008) A mixed branch length model of heterotachy improves phylogenetic accuracy. *Mol Biol Evol 25*, 1054–1066.

45. Le, S.Q., and Gascuel, O. (2010) Accounting for solvent accessibility and secondary structure in protein phylogenetics is clearly beneficial. *Syst Biol 59*, 277–287.

46. Burki, F., Inagaki, Y., Bråte, J., Archibald, J.M., Keeling, P.J., Cavalier-Smith, T., Sakaguchi, M., Hashimoto, T., Horak, A., Kumar, S., Klaveness, D., Jakobsen, K.S., Pawlowski, J., and Shalchian-Tabrizi, K. (2009) Large-scale phylogenomic analyses reveal that two enigmatic protist lineages, telonemia and centroheliozoa, are related to photosynthetic chromalveolates. *Genome Biol Evol 1*, 231–238.

47. Hampl, V., Hug, L., Leigh, J.W., Dacks, J.B., Lang, B.F., Simpson, A.G., and Roger, A.J. (2009) Phylogenomic analyses support the monophyly of Excavata and resolve relationships among eukaryotic "supergroups". *Proc Natl Acad Sci USA 106*, 3859–3864.

48. Parfrey, L.W., Grant, J., Tekle, Y.I., Lasek-Nesselquist, E., Morrison, H.G., Sogin, M.L., Patterson, D.J., and Katz, L.A. (2010) Broadly sampled multigene analyses yield a well-resolved eukaryotic tree of life. *Syst Biol 59*, 518–533.

49. Glücksman, E., Snell, E.A., Berney, C., Chao, E.E., Bass, D., and Cavalier-Smith, T. (2011) The Novel Marine Gliding Zooflagellate Genus *Mantamonas* (Mantamonadida ord. n.: Apusozoa). *Protist 162*, 207–221.

50. Kim, E., Harrison, J.W., Sudek, S., Jones, M.D., Wilcox, H.M., Richards, T.A., Worden, A.Z., and Archibald, J.M. (2011) Newly identified and diverse plastid-bearing branch on the eukaryotic tree of life. *Proc Natl Acad Sci USA 108*, 1496–1500.

51. Walker, G., Dorrell, R.G., Schlacht, A., and Dacks, J.B. (2011) Eukaryotic systematics: a user's guide for cell biologists and parasitologists. *Parasitology 15*, 1–26.

52. Yabuki, A., Nakayama, T., Yubuki, N., Hashimoto, T., Ishida, K., and Inagaki, Y. (2011) *Tsukubamonas globosa* n. g., n. sp., a novel excavate flagellate possibly holding a key for the early evolution in "Discoba." *J Euk Microbiol 58*, 319–331.

53. Stechmann, A., and Cavalier-Smith, T. (2003) The root of the eukaryote tree pinpointed. *Curr Biol 13*, R665–R666.

54. Rogozin, I.B., Basu, M.K., Csürös, M., and Koonin, E.V. (2009) Analysis of rare genomic changes does not support the unikont-bikont phylogeny and suggests cyanobacterial symbiosis as the point of primary radiation of eukaryotes. *Genome Biol Evol 1*, 99–113.

55. Cavalier-Smith, T. (2010) Kingdoms Protozoa and Chromista and the eozoan root of the eukaryotic tree. *Biol Lett 6*, 342–345.

56. Goodstadt, L., and Ponting, C.P. (2001) CHROMA: consensus-based colouring of multiple alignments for publication. *Bioinformatics 17*, 845–846.

Part II

Biochemistry of Rho GTPases

Chapter 3

Biochemical Assays to Characterize Rho GTPases

Mamta Jaiswal, Badri N. Dubey, Katja T. Koessmeier, Lothar Gremer, and Mohammad R. Ahmadian

Abstract

Rho GTPases act as tightly regulated molecular switches governing a large variety of critical cellular functions. Their activity is controlled by two different biochemical reactions, the GDP/GTP exchange and the GTP hydrolysis. These very slow reactions require catalysis in cells by two kinds of regulatory proteins. While the guanine nucleotide exchange factors (GEFs) activate small GTPases by stimulating the exchange of bound GDP for the cellular abundant GTP, GTPase-activating proteins (GAPs) accelerate the intrinsic rate of GTP hydrolysis by several orders of magnitude, leading to their inactivation. There are a number of methods that can be used to characterize the specificity and activity of such regulators to understand the effect of binding on the protein structure and, ultimately, to gain insights into their biological functions. This chapter describes (1) detailed protocols for the expression and purification of Rho GTPases, of effector-binding domains, and catalytic domains of GEFs and GAPs; (2) the preparation of nucleotide-free and fluorescent nucleotide-bound Rho GTPases; and (3) methods for monitoring the intrinsic and GEF-catalyzed nucleotide exchange, the intrinsic and GAP-stimulated GTP hydrolysis, and the effector interaction with active GTPase (three alternative approaches).

Key words: Fluorescence spectroscopy, GAP, GEF, GTPase, Guanine nucleotide, Mant, Protein–protein interactions, Rho, Tamra, Effector

1. Introduction

Rho family GTPases act as tightly regulated molecular switches governing a variety of critical cellular functions (1–5). Their activity is controlled by two biochemical reactions, the GDP/GTP exchange and the GTP hydrolysis, which can be catalyzed by two kinds of regulatory proteins (6). While the guanine nucleotide exchange factors (GEFs) activate Rho GTPases by stimulating the slow exchange of bound GDP for the cellular abundant GTP,

Francisco Rivero (ed.), *Rho GTPases: Methods and Protocols*, Methods in Molecular Biology, vol. 827,
DOI 10.1007/978-1-61779-442-1_3, © Springer Science+Business Media, LLC 2012

GTPase-activating proteins (GAPs) accelerate the slow intrinsic rate of GTP hydrolysis by several orders of magnitude, leading to inactivation. The formation of the active GTP-bound state of the GTPase is accompanied by conformational changes mainly at two regions (called switch I and II) that provide a platform for a selective interaction with a multitude of downstream effectors, which in turn initiate downstream signaling (6–8).

Our understanding of Rho GTPase regulation and signaling is becoming increasingly complex since more than 69 GEFs, 80 GAPs, and 90 effectors are considered to be potential interacting partners of the 22 mammalian members of the Rho family (6, 9–11). Only a sparse number of such intermolecular interactions have been primarily investigated in vitro with solid-phase methods like radioactive ligand overlay, pull-down assays, or yeast two-hybrid studies. These methods are often not sufficient to determine the specificity of regulation and to quantify the activity of recombinant proteins. However, many of the potential interactions defined by these methods require a more detailed analysis of their kinetics by appropriate real-time methods. To obtain a detailed picture of the molecular switch function of Rho GTPases and their interaction with regulators and effectors, fluorescent guanine nucleotides are often ideally suited to fulfill these criteria as it is known that they do not grossly disturb the biochemical properties of the GTPase and that the fluorescence reporter group is sensitive to changes in the local environment to produce a sufficiently large fluorescence change (12–14). Furthermore, the reporter group is often sensitive to the interaction with partner proteins that are able to bind in its neighborhood.

This chapter describes the application of two different fluorescently labeled guanine nucleotides in the biochemical analysis of Rho GTPases (see Fig. 1), which can be used to determine the binding affinities of regulators and effectors as well as to evaluate the activities of GEF-catalyzed nucleotide exchange and GAP-stimulated GTP hydrolysis, respectively.

2. Materials

2.1. Bacterial Strains

Different *Escherichia coli* strains BL21(DE3), BL21(DE3) codon plus RIL, BL21(DE3) pLysS, BL21(DE3) Rosetta (Novagen) are used to recombinantly express eukaryotic genes and gene fragments.

2.2. Chemicals and Reagents

1. Isopropyl-β-D-thiogalactopyranoside (IPTG).
2. 6 M guanidinium hydrochloride.

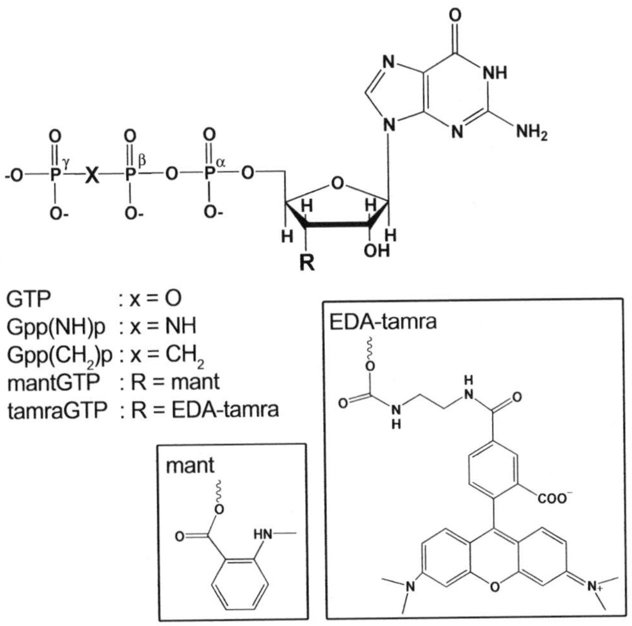

Fig. 1. Chemical structures of the guanosine nucleotide derivatives used in this chapter. Unlabeled fluorescent nucleotides contain an OH group at the position R.

3. Enzymes: Thrombin (Serva), PreScission protease (GE Healthcare), TEV protease (Invitrogen), agarose bead-coupled alkaline phosphatase (Sigma–Aldrich), soluble alkaline phosphatase (Roche Diagnostics), snake venom phosphodiesterase (Sigma–Aldrich).

4. Ponceau Red: 0.1% Ponceau Red S (w/v), 5% acetic acid (v/v) in double-distilled water.

5. 500 mM ethylendiamine-N, N, N', N'-tetraacetic acid (EDTA), pH 8.0, adjusted with 1 N NaOH.

6. Nitrocellulose membrane.

7. Amicon Ultra centrifugal filter units (Millipore) with molecular mass cutoff of 5–100 kDa for concentrating proteins.

8. Bottle-top filter units with 0.2-μm cutoff, 500 mL, for filtering buffer and solutions.

2.3. Nucleotides

The nucleotides used in biochemical assays are adenosine 5′-triphosphahte (ATP), guanosine 5′-diphosphate (GDP), and guanosine 5′-triphosphate (GTP) (Pharma Waldhof; 10 mM in deionized water, pH 7.5), the nonhydrolyzable GTP analog β, γ-methyleneguanosine 5′-triphosphate [Gpp(CH$_2$)p] and guanosine 5′-[β,γ-imido]triphosphate (GppNHp) (Sigma–Aldrich and Jena Biosciences).

Two different fluorescence reporter groups, N-methylanthraniloyl (mant) (12) and tetramethylrhodamine (tamra) (14, 15), attached to the 2′(3′)-hydroxyl group of the ribose moiety of the guanine nucleotides (GDP, GTP, and GppNHp) (see Fig. 1), are used both to monitor protein–ligand as well as protein–protein interactions and to measure catalytic activities of regulatory proteins. The fluorescent nucleotides mantGDP, mantGTP, mantGppNHp (10 mM solution in deionized water, pH 7.5), and tamraGTP (2 mM solution in deionized water, pH 7.5) are synthesized as described in ref. 13 and can be purchased from Jena Biosciences.

2.4. Buffers and Media

Prepare all solutions in double-distilled water at room temperature (25°C), unless indicated otherwise. All buffers should be filtered and degassed.

1. Terrific broth (TB) medium: 12 g/L bacto-tryptone, 24 gL yeast extract, 0.4% (v/v) glycerol, 2.31 g/L KH_2PO_4, 12.54 g/L K_2HPO_4.

2. Lysis buffer: 30 mM Tris–HCl, pH 7.5, 50 mM NaCl, 5 mM $MgCl_2$, 3 mM DTT, 100 mM NaCl, lysozyme (0.5 mg/mL), DNAse I (10 μg/mL), complete EDTA-free protease inhibitor cocktail (Roche) (1 tablet per 200 mL).

3. Wash buffer: 30 mM Tris–HCl, pH 7.5, 50 mM Nacl, 5 mM $MgCl_2$, 3 mM DTT.

4. Standard buffer: 30 mM Tris–HCl, pH 7.5, 5 mM $MgCl_2$, 3 mM DTT, 100 mM NaCl.

5. High-salt ATP buffer: Standard buffer containing 400 mM KCl and 1 mM ATP.

6. Glutathione (GSH) elution buffer: Standard buffer containing 20 mM reduced glutathione (Merck), pH 7.5 (adjusted with 1 N NaOH).

7. Exchange mix (10×): 2 M $(NH_4)_2SO_4$, 10 mM $ZnCl_2$.

8. HPLC buffer: 100 mM K_2HPO_4/KH_2PO_4, pH 6.5, 10 mM tetrabutylammonium bromide, 7.5–25% (v/v) acetonitrile.

9. GEF buffer: 30 mM Tris–HCl, pH 7.5, 5 mM $MgCl_2$, 3 mM DTT, 10 mM K_2HPO_4/KH_2PO_4, pH 7.4.

10. GAP buffer: 30 mM Tris–HCl, pH 7.5, 10 mM $MgCl_2$, 3 mM DTT, 10 mM K_2HPO_4/KH_2PO_4, pH 7.4.

11. Effector buffer: 30 mM Tris–Cl, pH 7.5, 100 mM NaCl, 5 mM $MgCl_2$, 3 mM DTT.

2.5. Chromatography Columns

Following columns are used for protein purification and analysis: Reversed-phase C-18 HPLC column Ultrasphere ODS, 5 μm; 250×4, 6 mm (Beckman Coulter), guard column Nucleosil 100-5-C18, 5 μm (Bischoff Chromatography); glutathione sepharose

4B FF column (GE Healthcare); Superdex 75 or 200, 16/60, or 26/60 columns (GE Healthcare) (column dimensions are given as 16- or 26-mm diameter and 60-cm length); NAP 5 column (GE Healthcare).

2.6. Instruments

1. Shaker incubator (Infors HT).

2. Sonicator (Bandelin electronics).

3. M-110S laboratory microfluidizer processor (Microfluidics).

4. Centrifuge Avanti J-20 XP with 6-L rotor JLA-8.1000 and rotor JA 25.50 (Beckman Coulter) or equivalent.

5. ÄKTA prime and purifier (GE Healthcare).

6. UV–Vis spectrometer (Biophotometer, Eppendorf).

7. HPLC instrument (Beckman Gold, Beckman Coulter).

8. Fluorescence spectrometer (Perkin-Elmer, LS50B; FluoroMax-4, Horiba).

9. Stopped-flow instrument (Applied Photophysics SX18MV or Hi-Tech SF-61 DX2).

10. Quartz cuvettes (Suprasil 108.002F-QS, Hellma).

2.7. Software

Program packages of Grafit (Erithacus Software), Origin (OriginLab), and Sigmaplot (Systat Software Inc) are used for the evaluation of the data.

3. Methods

3.1. Gene Expression and Bacterial Culture Conditions

High quality (>95% purity) and quantity (>10 mg) of purified proteins are mandatory prerequisites for the investigation of relationships between protein structure and function. Recombinant expression systems and the development of a variety of fusion tags have dramatically facilitated purification. Nevertheless, the choice of the right purification strategy is still a matter of trial and error and has to be elaborated for each individual protein.

To optimize the synthesis of the protein of interest in *E. coli*, various culture conditions should be examined, including the IPTG concentration as the inducer of the *lac*-promoter-controlled gene expression, the optical density (OD_{600}) at the time of induction, and the culture temperature and expression time post induction. Culture condition tests varying these parameters should be performed in small-scale studies prior to upscaling the cultures for preparative protein expression. To improve a maximal yield of the desired protein, we alternatively use, besides the *E. coli* strain BL21 (DE3), strains containing additional plasmids, such as pLysS (to improve bacterial lysis and for the expression of toxic proteins) and Codon plus RIL or Rosetta (to improve the codon usage).

1. Grow 30–250 mL of precultures of the desired *E. coli* strain in TB medium in a 150–1,000-mL flask overnight at 37°C (see Note 1).

2. Fill 5-L Erlenmeyer flasks with 2.5 L of TB medium. Inoculate each flask with 25 mL of an overnight preculture (see Note 2). Place the inoculated culture flasks in an environmental shaker and let them grow at 37°C with shaking at 160 rpm.

3. When the logarithmic growth phase is reached (OD 0.4–0.8), lower the temperature to the previously optimized expression condition (usually, 18–30°C), add IPTG (usually, 0.05–0.5 mM) (see Note 3), and incubate the culture usually overnight and in rare cases for only 3–6 h.

4. Transfer the cells to 1,000-mL centrifuge bottles and harvest the cells by centrifugation at $5,000 \times g$ for 15 min at 4°C using a 6-L rotor if available. Repeat this step several times if the culture volume exceeds the capacity of the available rotor.

5. Wash the bacterial pellet in each rotor bottle with 20 mL of wash buffer. Combine the resuspended cell pellets into a smaller rotor bucket and centrifuge again at $5,000 \times g$ for 20 min at 4°C (see Note 4).

6. Discard the supernatant and determine the weight of the bacterial pellets (the tare of the centrifuge beaker should be known before). Resuspend the pellets in wash buffer (3 mL/g bacterial pellet) or any other appropriate buffer that is able to solubilize and stabilize the desired protein and distribute in aliquots in 50-mL plastic tubes.

7. Store aliquots at –20°C (see Note 5).

3.2. Bacterial Lysis

The efficient bacterial lysis is an important prerequisite for the complete recovery of the recombinant protein. Cell walls of bacteria must be disrupted in order to allow access to intracellular components. Different methods have been evolved to achieve this goal, which vary considerably in the severity of the disruption process, reagents needed, and the equipment available. Besides enzymatic methods, e.g., lysozyme treatment, which is suitable for analytical scales and not always reproducible, there are several mechanical methods available, including bead milling with glass beads, the "cell disruption bomb," high shear mechanical methods like the "French press" and the "microfluidizer," or sonication with ultrasound to gently disrupt bacterial cell walls. We commonly use the latter two methods, which are efficient and fairly quick.

*3.2.1. Bacterial Lysis
by Sonication*

1. Equip the cell sonicator with a titanium horn of 3- to 19-mm diameter (depending on the culture volume to be disrupted) and clean it before use with 70% ethanol.

2. Transfer the defrosted bacterial suspension from all aliquots to a beaker of suitable size and place it on ice.

3. Place the sonicator horn about 0.5–1 cm immersed into the suspension and stir the suspension on a magnetic stirrer.

4. Start the sonication procedure by increasing the output control (5–10 W each) in 10-s intervals starting with low, 30 W, to reach finally 95 W. Repeat this procedure several times (8–12) and always wait for 30 s in between to prevent overheating of the sample. For the latter reason, the beaker with the bacterial solution also needs to be stored on ice during the whole procedure (see Notes 6–8).

*3.2.2. Bacterial Lysis Using
a Microfluidizer*

The microfluidizer is an instrument that uses high pressure to squeeze the bacterial solution through an interaction chamber containing a narrow channel, thereby generating high shear forces that pull the cells apart. The system permits controlled cell breakage and does not need addition of detergent or higher ionic strength. Since heat is generated during this process, an interaction chamber needs to be cooled.

1. Wash the instrument extensively with water. For a final wash step, use the standard buffer used for the protein purification (see Note 9). Pour the defrosted bacterial suspension into the instrument's reservoir and turn on the instrument (see Note 10). Direct the flow on the instrument's outlet toward the wall of a beaker to prevent foam formation.

2. Prevent intake of air on the inlet as this also produces foam and lead to protein denaturation. For this, turn off the instrument before air enters the instrument's inlet. Wash with a small volume of standard buffer and switch the instrument on again. Stop again before air enters the inlet. By repeating this step 2–3 times, nearly all of the bacterial suspension is processed.

3. If necessary, flush the bacterial suspension 2–3 times through the instrument until a color change from milky to slightly more translucent is observed.

4. Wash the instrument extensively with water and finally with 2-propanol. Store it in this alcohol.

**3.3. Protein
Purification**

*3.3.1. Purification
Steps as GST Fusion*

1. Centrifuge the bacterial lysate typically derived from 5 to 15 L of cell culture to sediment insoluble components at 35,000–100,000×g at 4°C for 40 min. If possible, centrifuge at 100,000×g to remove insoluble cell fragments quantitatively.

If a high-speed rotor/centrifuge is not available, a minimal force of 35,000×g might also be sufficient.

2. Equilibrate a GSH sepharose column (10–25 mL bed volume) with approximately 3–4-column volumes of standard buffer until a stable baseline absorption monitored at 280 nm is achieved (see Note 11).

3. After centrifugation, apply the cleared bacterial lysate on the GSH sepharose column (4 mL/min, if using fast flow material). After all lysate is applied, wash with standard buffer until the baseline at 280 nm is reached again. Wash with 100–200 mL of high-salt ATP buffer (see Note 12).

4. Wash with at least 100 mL of standard buffer for removal of KCl and ATP until the original baseline level is achieved.

5. Elute GST fusion proteins from the GSH sepharose column with 100–150 mL of GSH elution buffer and collect the eluting GST fusion protein in 5–10-mL fractions (see Note 13).

6. Pool the GST fusion protein-containing fractions after analysis by SDS-polyacrylamide gel electrophoresis (SDS-PAGE) and Coomassie staining following standard procedures. Regenerate the column with 50 mL of aqueous 6 M guanidinium hydrochloride and wash with 100–150 mL of standard buffer afterward.

3.3.2. Removal of the GST Tag

The GST tag that helps to purify a recombinant protein from crude cell extracts should be removed when the protein shall be used for structural or biochemical analysis. Usually, expression vectors have protease-specific cleavage sites inserted between the coding sequence for the fusion tag and the multiple cloning site. The corresponding fusion protein, thus, can be processed and cut with the appropriate protease and finally the fusion tag can be removed by further chromatographic purification steps.

1. Cleave fusion proteins (≥1 mg/mL) in batch by applying 1–2 U of the appropriate protease (thrombin, TEV, or PreScission depending on the available cleavage site in the vector) per mg of GST fusion protein and incubate for 4–20 h at 4°C. Take a sample of 10 μL from the reaction batch after 4 h and after overnight incubation and analyze by SDS-PAGE and Coomassie blue staining using standard methods to monitor progress of the cleavage.

2. Further purification and removal of protein impurities or small components including reduced GSH is achieved by size-exclusion chromatography (gel filtration) on the scale of 16/60 or 26/60 columns using Superdex 75 or Superdex 200 material (see Note 14). Equilibrate the column with at least one-column volume (130 mL for 16/60 Superdex or 340 mL for

26/60 Superdex) of standard buffer. Load a 1–5-mL sample of the concentrated protein (≤20 mg/mL).

3. Collect in 3–5-mL fractions and withdraw 10-μL samples for analysis by SDS-PAGE and Coomassie staining. Identify the desired protein by its molecular mass and pool the corresponding fractions.

4. In cases where the desired protein (after protease cleavage of the GST tag) has a similar molecular mass as GST, the desired protein cannot be simply removed by gel filtration. Then, the tag has to be removed by a second chromatography on a GSH sepharose column. For that, apply the cleavage reaction on a GSH sepharose column equilibrated with standard buffer and collect the flow through containing the desired protein in 5–10-mL fractions, and then apply standard buffer until the absorbance at 280 nm reaches the baseline level (see Note 15). To regenerate the column, wash with GSH elution buffer to elute the bound GST tag. Withdraw 10-μL samples of the flow-through fractions and analyze by SDS-PAGE and Coomassie staining.

5. Pool and concentrate the desired protein to 10–20 mg/mL.

6. Snap freeze purified proteins in 50–500-μL aliquots in liquid nitrogen and store at –20°C or preferably at –80°C (see Note 16).

3.4. Preparation of Nucleotide-Free Forms of Rho GTPases

Preparation of nucleotide-free GTPase is carried out in two steps: (1) The GTPase-bound GDP is degraded by agarose bead-coupled alkaline phosphatase and replaced by $Gpp(CH_2)p$ (a nonhydrolysable GTP analog, which is resistant to degradation by alkaline phosphatase but sensitive to phosphodiesterase). (2) After the GDP is completely degraded, phosphodiesterase from snake venom is added to the solution of the $Gpp(CH_2)p$-bound GTPase to cleave this nucleotide to GMP and P_i.

1. Add a 1.5 molar excess of $Gpp(CH_2)p$ to 1 mg of GDP-bound GTPase in standard buffer, apply to the reaction batch the 10× exchange mix, thereby diluting the latter to a 1× concentration, mix rapidly, and withdraw a sample for isocratic ion-pair reversed-phase HPLC analysis on a C_{18} column of the GDP and $Gpp(CH_2)p$ content using HPLC buffer containing 7.5% acetonitrile; for this, dilute the withdrawn sample to 20–100 μM GTPase.

2. Add 0.5–1 U of agarose bead-coupled alkaline phosphatase to the reaction setup, mix rapidly, and incubate at 4°C for 2–16 h (depending on the GTPase used). Analyze the GDP degradation regularly by HPLC determination of the GDP content as described in step 1.

3. After the GDP is completely degraded, centrifuge the suspension at $1,500 \times g$, 4°C, for 2 min to remove the bead-coupled

alkaline phosphatase. Repeat this process two to three times to remove quantitatively all traces of alkaline phosphatase-coupled beads (see Note 17).

4. Add 0.002 U of snake venom phosphodiesterase per mg of GTPase to cleave $Gpp(CH_2)p$ to GMP, guanosine, and P_i. Monitor $Gpp(CH_2)p$ degradation by HPLC as described in step 1.

5. When degradation of $Gpp(CH_2)p$ is complete, inactivate phosphodiesterase by snap freezing in liquid nitrogen. Quickly defrost by warming up the vials in the hands and freeze again in liquid nitrogen; repeat these steps two times and store the protein solution in aliquots at –80°C (see Note 18).

3.5. Preparation of mantGDP-Bound GTPases

Loading of nucleotide-free forms of GTPases with fluorescently labeled nucleotides can be achieved by simply mixing both components followed by a small-scale size-exclusion chromatography on a desalting column. The steps described below are usually necessary for the preparation of GTPases bound to fluorescent GDP analogs.

1. Equilibrate an NAP5 column with 2–3-column volumes of standard buffer.

2. Mix 0.5 mg of a nucleotide-free GTPase (e.g., 50 µL from a 0.5 mM solution) with a 1.5-fold molar excess of mantGDP (e.g., 3.75 µL from a 10 mM stock solution).

3. Apply the complete sample volume on the NAP5 column and allow the sample to enter the gel bed completely.

4. Apply standard buffer to achieve a total applied volume of 500 µL and allow the buffer to enter the gel bed completely (e.g., for the upper example, add 446.25 µL).

5. Add 1 mL of standard buffer and collect fractions (2 drops per fraction).

6. Analyze the protein content of the fractions by dotting 2 µL from each fraction on a nitrocellulose membrane and subsequently staining with Ponceau S. This qualitative test is just to determine which fractions do contain the protein of interest for subsequent pooling.

7. Pool protein-containing fractions and determine the concentration of mantGDP bound to the protein by HPLC using an HPLC buffer containing 20–25% acetonitrile.

8. Store the protein in aliquots at –80°C.

3.6. Preparation of GppNHp-Bound and mantGppNHp-Bound GTPases

GppNHp-bound and mantGppNHp-bound GTPases are prepared using soluble alkaline phosphatase, which degrades the bound GDP as described in Subheading 3.4, steps 1–3, and Subheading 3.5, steps 7–8.

3.7. Intrinsic and GEF-Catalyzed Nucleotide Exchange Reactions

Different procedures are available for the investigation of the nucleotide exchange on small GTPases. The dissociation of a protein-bound nucleotide can be easily determined in real time by fluorescence spectroscopy using a fluorescent GDP analog. Usually, mant derivatives of guanosine nucleotides, which are coupled at the 2′(3′) hydroxyl group of the ribose, are used.

In principle, each nucleotide-binding protein has a defined intrinsic rate of GDP release, which is often too slow to be physiologically relevant. Thus, GEFs operate on these small GTPases and catalyze the generation of the active GTP-bound state from the inactive GDP-bound form. This process is often a result of the GEFs themselves being activated or recruited to the vicinity of the corresponding GTPase in response to extracellular signaling events.

3.7.1. Measurement of Slow Reactions

Specificity and activity of GEFs can be analyzed qualitatively by comparison of intrinsic and GEF-stimulated fluorescence measurements. Usually, this is performed in a fluorescence spectrometer, since the timescale of these reactions is slow (>1,000 s). Here, the bacterially expressed and highly pure (>90% homogeneity) recombinant mantGDP-loaded GTPases are mixed with an excess of a pure nonfluorescent nucleotide solution in a cuvette. The decrease of the mant-fluorescence signal is monitored with a fluorescence spectrometer. GEF and also GAP assays do not need posttranslationally modified GTPases. Thus, proteins and protein domains produced in *E. coli* can be used (see Note 19). Cleared cell lysate is not suitable in this assay for several reasons: (1) the protein concentration may not be sufficient; (2) the protein of interest may exist in complex with other proteins and may thus not be freely accessible; (3) the activity of other regulators may interfere with the assay. The latter aspect should also be considered with solid-phase-enriched proteins (using GST- or His-tagged proteins) or subcellular fractionated samples.

1. Preincubate the solution of 0.1 μM mantGDP-bound GTPase (see Subheading 3.5) in a fluorescence cuvette (see Note 20) in degassed GEF buffer in a final volume of 600 μL and at 25°C for at least 5 min.

2. Record the mant-fluorescence signal in a fluorescence spectrometer applying an excitation wavelength of 366 nm and an emission wavelength of 450 nm, an integration time of at least 2 s, and a recording time for each data point of 20 s.

3. If the fluorescence signal is stable, add 1.2 μL of a 10 mM nonfluorescent GDP solution (20 μM final GDP concentration) and mix rapidly with a pipette to start the reaction.

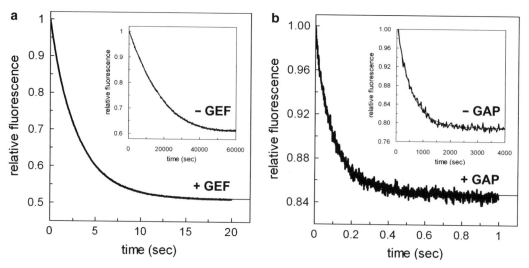

Fig. 2. Monitoring the nucleotide exchange and hydrolysis reactions. (**a**) A sample containing 0.1 μM RhoA·mantGDP was rapidly mixed with 20 μM GDP and 10 μM PDZ-RhoGEF in a stopped-flow instrument to monitor the GEF-catalyzed mantGDP dissociation from RhoA in real time. The intrinsic mantGDP dissociation (in the absence of the GEF protein) was measured under the same conditions in a fluorescence spectrometer (inset). The data are fitted as single-exponential decay of the curve and the observed rate constants (k_{obs}) obtained were 0.000069 s^{-1} for the intrinsic reaction and 0.34 s^{-1} for the GEF-catalyzed reaction that indicate a highly efficient catalysis of about 4,928 fold. (**b**) A sample containing 0.2 μM Cdc42·tamraGTP was rapidly mixed with 2 μM p50Cdc42GAP in a stopped-flow instrument to follow the GAP-stimulated tamraGTP hydrolysis reaction in real time. The intrinsic tamraGTP hydrolysis (in the absence of the GAP protein) was measured under the same conditions in a fluorescence spectrometer (inset). The data was fitted as single-exponential decay to obtain the k_{obs} values of 0.0018 s^{-1} and 9.66 s^{-1} for the intrinsic and the GAP-stimulated reaction, respectively, indicating a rather efficient catalysis of about 5,367 fold.

4. Normally, an exponential decrease in fluorescence occurs and can be monitored over the time course of the reaction (2–72 h), which is due to the mantGDP release into the aqueous solution.

5. When there is no further change in fluorescence, add 24 μL of a 0.5 M EDTA solution to adjust a final concentration of 20 mM and monitor the reaction for additional 10 min. This reveals whether the nucleotide dissociation reaction is completed at all.

6. Fit the data single exponentially with, for example, the Grafit program to provide the dissociation (k_{off}) rates, which are in case of small GTPases usually around 10^{-3} to 10^{-5} s^{-1} (Fig. 2a, inset).

3.7.2. Measurement of Fast Reactions

For fast GEF-catalyzed nucleotide dissociation reactions, the time resolution of a fluorescence spectrometer is insufficient. Instead, a stopped-flow instrument is routinely used for the analysis of rapid kinetics as obtained by quantitative GEF-stimulated nucleotide exchange reactions. Here, equal volumes of two different twofold concentrated samples are automatically shot into a mixing chamber, where the fluorescence changes can be detected directly after the rapid mixing (death time around 2–5 ms) (see Note 21). Five to eleven identical measurements are recorded and averaged in order to obtain a higher accuracy.

1. Wash the drive syringes of the instrument several times with 5–10 mL of GEF buffer and adjust the temperature to 25°C.

2. Prepare two different samples in degassed GEF buffer at room temperature (~25°C) and a final volume of 1,000 µL: (1) one sample contains 0.2 µM of mantGDP-bound GTPase; (2) the other sample contains the GEF protein (at varying concentrations of 2–1,000 µM depending on the activity and affinity of the GEF for the respective GTPase) plus 40 µM GDP (200-fold excess above mantGDP) (see Note 22).

3. Load each sample into one of the two drive syringes of the stopped-flow instrument.

4. Set the excitation wavelength for the mant nucleotides to 366 nm and detect the fluorescence with a cutoff filter mounted in front of a photomultiplier (408 nm for mant nucleotides).

5. Start the measurement with the supplied stopped-flow software which initiates the pushing of the two syringes containing the samples to the sample cell. At this stage, both samples join together and rapidly mix to a final volume of about 50–75 µL. Repeat the mixing and fluorescence recording event up to 11 times until all volume in the drive syringe reservoir is consumed.

6. Evaluate obtained data by single-exponential fitting with scientific software, e.g., the Grafit program, to obtain the observed rate constant (k_{obs}) for the respective concentration of the GEF protein (see Fig. 2a).

3.8. Intrinsic and GAP-Stimulated GTP Hydrolysis Reaction

The intrinsic and GAP-stimulated GTP hydrolysis reaction of GTPases can be measured by different methods. A generally useful and accurate method is HPLC, by which concentrations of GDP and GTP can be determined from the area of the elution peaks. The relative GTP content determined as the ratio [GTP]/([GTP]+[GDP]) is used to describe the reaction progress as described (14). Measured by HPLC, intrinsic GTP hydrolysis rates of small GTPases of the Rho family are about fivefold faster than that of Ras proteins (e.g., 0.028 min^{-1} for H-Ras) (16, 17). A different approach, which is less material and time consuming, is the real-time measurement of tryptophane fluorescence with an excitation wavelength of 295 nm and an emission wavelength of 350 nm, by which the GTPase reaction rates can be conveniently measured. The Ras(Y32W) mutant provided a large increase in fluorescence signal upon hydrolysis of GTP to GDP, which has been used to study the mechanism of the intrinsic GTPase reaction (17).

Whether this approach is useful for the members of the Rho family remains to be investigated.

Unlike mant, tamra is a powerful fluorescence reporter group to study in real time the GTP hydrolysis of Rho GTPases in the presence and absence of their GAPs (15). The intrinsic and GAP-stimulated tamraGTP hydrolysis reaction of RhoGTPases can be detected by conventional fluorescence spectrometric and stopped-flow measurements, showing a significant decrease in the fluorescence signal.

3.8.1. Measurement of Slow Reactions

In contrast to other fluorescent nucleotide derivatives, including the mant nucleotides, tamraGTP (a ribose hydroxyl-substituted tetramethylrhodamine derivative of GTP) enables us to measure the intrinsic and GAP-stimulated GTPase reactions of Rho and Ras proteins using fluorescence spectroscopy (15). Besides much lower consumption of proteins and nucleotides as compared to the HPLC-based assay, tamraGTP hydrolysis assay allows to monitor the real-time kinetics of the hydrolysis reaction of the Ras and Rho families.

1. A solution of 0.1 µM tamraGTP (stock solution of >1 mM) is preincubated in a fluorescence cuvette in GAP buffer at a final volume of 600 µL and at 25°C for at least 5 min.

2. Set fluorescence spectrometer at an excitation wavelength of 546 nm and an emission wavelength of 583 nm, with an integration time of at least 2 s and a recording time for each data point of 20 s.

3. Add 1.8 µL from a 50 µM stock solution of the nucleotide-free GTPase (0.15 µM final concentration) to observe complex formation with the nucleotide through a fast and strong increase in fluorescence.

4. After this initial phase of nucleotide association, monitor the significant fluorescence decay as a result of GTP hydrolysis, which lasts between 0.5 and 6 h, depending on the GTPase variant used (15).

5. Continue the measurement until no further decrease in fluorescence can be observed.

6. Evaluate obtained data by single-exponential fitting with scientific software, e.g., the Grafit program, to obtain the observed rate constant (k_{obs}) for the respective concentration of the GAP protein (see Fig. 2b, inset).

3.8.2. Measurement of Fast Reactions

Measure GAP-stimulated tamraGTP hydrolysis by a stopped-flow instrument (see Subheading 3.7.2), but use appropriate settings to detect the tamra fluorescence.

1. Wash the drive syringes of the instrument several times with 5–10 mL of GAP buffer and adjust the temperature to 25°C.

2. Prepare two different samples in GAP buffer at room temperature (25°C) and a final volume of 1,000 μL: (1) one sample contains 0.6 μM nucleotide-free GTPase and 0.4 μM tamraGTP; (2) the other sample contains the GAP protein (at varying concentrations of 0.2–200 μM that depend on the activity of the GAP for the respective GTPase) (see Note 23).

3. Load each sample into one of the two drive syringes of the stopped-flow instrument.

4. Set the excitation wavelength for the tamra nucleotides to 546 nm and detect the fluorescence with a cutoff filter mounted in front of a photomultiplier (570 nm for tamra nucleotides).

5. Start the measurement with the supplied stopped-flow software which initiates the pushing of the two syringes containing the samples to the sample cell. At this stage, both samples join together and rapidly mix to a final volume of about 50–75 μL. Repeat the mixing and fluorescence recording event up to 11 times until all volume in the drive syringe reservoir is consumed.

6. Fit the data single exponentially with, for example, the Grafit program, to provide the hydrolysis rates. GAP proteins of Rho GTPases vary in their activity in stimulating hydrolysis from 10^1 to 10^{-3} s^{-1} (see Fig. 2b).

3.9. GTPase–Effector Interaction

Fluorescence-based measurement of the interaction between mantGppNHp-bound GTPases and their effectors can be investigated in different ways. A direct method for the time-resolved detection and quantification of interactions is the first step of analysis. However, some GTPase/effector interactions cannot be monitored by direct fluorescence measurements. In these cases, two alternative approaches can be utilized, the guanine nucleotide dissociation inhibition (GDI) assay and equilibrium fluorescence polarization.

3.9.1. Kinetic Measurements (Direct)

To study GTPase–effector interaction, mant-labeled nonhydrolyzable GTP analogs, such as mantGppNHp in complex with the GTPase, can lead to a large change of the fluorescence intensity, like in the case of the Cdc42/WASp interaction (18). In such a kinetic approach, the association and dissociation rates of the effector interaction with mantGppNHp-bound GTPase can be measured using a stopped-flow instrument as described in Subheading 3.7.2.

The association rate constant k_{on} can be measured by using the stopped-flow setup for fast kinetics as follows:

1. Wash the drive syringes of the instrument several times with 5–10 mL effector buffer and adjust the temperature to 25°C.

2. Prepare two different samples in effector buffer at room temperature (~25°C) and a final volume of 1,000 μL: (1) one sample contains 0.2 μM of mantGppNHp-bound GTPase; (2)

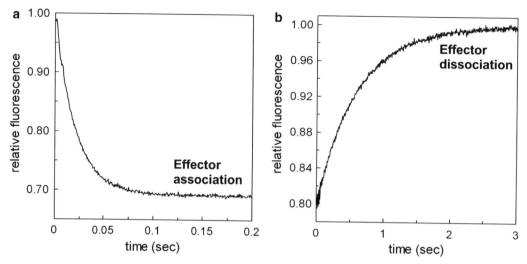

Fig. 3. Association and dissociation reactions for the GTPase–effector interaction. (**a**) A sample containing 0.1 μM Cdc42·mantGppNHp was rapidly mixed with 2 μM WASP in a stopped-flow instrument to monitor the WASP–Cdc42 association in real time. The data was fitted as single-exponential decay to obtain an extremely fast k_{obs} value for association of 48.4 s^{-1}. The association rate constant (k_{on}) can be measured and evaluated by varying the effector concentrations. (**b**) A sample containing 0.1 μM Cdc42·mantGppNHp and 2 μM WASP was rapidly mixed with 10 μM Cdc42·GppNHp in a stopped-flow instrument to monitor the WASP displacement from its complex with Cdc42·mantGppNHp to measure the dissociation rate constant (k_{off}) 1.65 s^{-1}. The ratio of k_{off} divided by k_{on} gives the dissociation constant (K_d) for this bimolecular interaction.

the other sample contains the effector protein (at varying concentrations of 0.2–100 μM depending on the affinity of the effector for the respective GTPase).

3. Next steps are similar to procedures described in Subheading 3.7.2, steps 3–5.

4. Analyze the data by fitting the data for each effector protein concentration with a monoexponential function (see Fig. 3a). Plot the resulting k_{obs} values against the effector concentration and determine the slope of the resulting line, which is k_{on}.

The dissociation rate constant k_{off} can be measured as follows:

1. Mix mantGppNHp-bound GTPase (0.1 μM for high-affinity or 0.5 μM for low-affinity binders in effector buffer) and the effector (0.4 μM for high-affinity or 2 μM for low-affinity binders in effector buffer) with unlabeled GppNHp-bound GTPase (10 μM for low-affinity or 100 μM for high affinity binders) in the stopped-flow instrument to obtain a single-exponential fluorescence change.

2. Next steps are similar to procedures described in Subheading 3.7.2, steps 3–5.

3. Fit the curve single exponentially to obtain the k_{off} value (Fig. 3b). The dissociation constant K_d can now be calculated from the ratio k_{off}/k_{on}.

3.9.2. Equilibrium Measurements Using Fluorescence Titration (Direct)

The investigations described above (see Subheading 3.9.1) can also be carried out using fluorescence equilibrium titration measurements particularly if there is no stopped-flow instrument or if only small amounts of protein are available. In this type of analysis, which has been successfully used for Cdc42/PAK (19), the fluorescence level is measured in dependence of the concentration of the unlabeled reaction partner.

1. Incubate 0.1 μM mantGppNHp-bound GTPase in effector buffer in a quartz cuvette (volume 600 μL) at 25°C in a fluorescence spectrometer and follow fluorescence emission at 450 nm (excitation 366 nm) until the signal is stable.

2. Titrate increasing concentrations of the effector (0.01–100 μM), and wait after each titration step until you reach a stable emission signal level indicating equilibrium.

3. Continue the titration steps until there is no further change of the fluorescence signal indicating that the system is saturated.

4. Plot the fluorescence intensities against the respective effector concentration and fit according to ref. 13 to obtain the equilibrium dissociation constant eK_d.

3.9.3. Measurements of Nucleotide Dissociation Inhibition Through Effector Binding (Indirect)

Some GTPase–effector interactions cannot be monitored by direct fluorescence measurements, but the large fluorescence decay on dissociation of bound mantGppNHp can be utilized to determine indirectly the binding affinity of effector domains for their GTPases as described for Rac/PAK (20, 21), Cdc42/WASp (20), and Rho/Rhotekin, Rho/PKN, and Rho/Rho kinase interactions (22). This method is based on the observation that guanine nucleotide dissociation from the GTPase is inhibited by interaction with effectors (GDI effect), as the effectors predominantly bind close to the nucleotide-binding region of the GTPase. According to the used effector concentration, a fraction of the GTPase is bound by the effector, thus slowing down the nucleotide dissociation. The observed nucleotide dissociation rate is a combination of the dissociation rates of free and effector-bound GTPase.

1. Add 0.2 μM mantGppNHp-bound GTPase together with different concentrations of the full length effector or the GTPase-binding domain (0.2–2.0 μM for high-affinity interactions and up to 100 μM for low-affinity binders) into a quartz cuvette containing effector buffer, equilibrate at 25°C, and monitor the fluorescence emission at 450 nm (excitation at 366 nm).

2. Add 20 μM unlabeled GppNHp into the solution and mix to start the reaction. For every experimental setup, four reactions can be measured simultaneously if the fluorescence spectrometer is equipped with a four-position cuvette holder.

3. For better comparability of the different setups, the intrinsic nucleotide dissociation in the absence of the effector should be determined in one of the four reactions.

4. Evaluate obtained data by single-exponential fitting with scientific software, e.g., the Grafit program, to obtain the equilibrium dissociation constant (eK_d) as a consequence of the GDI effect.

3.9.4. Equilibrium Measurements Using Fluorescence Polarization Titration (Direct)

Fluorescence polarization can also be used to determine the equilibrium dissociation constant (eK_d), and works best if the effector protein is larger than 40 kDa. In cases where the effector proteins are to small, like in the isolated GTPase-binding domains, a GST- or maltose-binding protein fusion of the effector can be used to obtain sufficiently large polarization signal changes (23, 24).

1. Incubate 0.2 μM mantGppNHp-bound GTPase in effector buffer (see reagents and buffers) in a fluorescence cuvette (volume 600 μL) at 25°C and monitor the fluorescence polarization in a fluorescence spectrometer (emission at 450 nm, excitation 366 nm) until the signal is stable.

2. Titrate increasing concentrations of the effector (0.01–100 μM), and wait after each titration step until a stable polarization signal level is reached, indicating equilibrium.

3. Continue the titration steps until there is no further change of the polarization signal, indicating that the system is saturated.

4. Fit the concentration-dependent binding curve using a quadratic ligand-binding equation (24) to obtain the equilibrium dissociation constant (eK_d) for the respective GTPase–effector interaction.

3.10. Anticipated Results

The elucidation of the molecular switch mechanism of the GTPases and particularly their specificities and affinities for regulators and effectors requires the dissection of such interactions at the molecular level by utilizing sensitive biochemical assays. Fluorescence spectroscopic methods provide researchers with a number of tools for studying the intercommunication of a GTPase with nucleotides and binding partners. Compared to qualitative assays (e.g., the filter-binding assay or thin-layer chromatography, which contain between 3 and 6 data points), the fluorescence methods described in this chapter allow to monitor the activity of GEFs and GAPs in real time as well as the interaction with effectors at which every single measurement consists of at least 400 data points per reaction trace. These assays are highly sensitive and, in principle, reproducible provided that the proteins and reagents are carefully prepared from high-purity materials and tested for quality. Thus, optimal gene expression and protein purification as well as the quality of fluorescent nucleotide-bound GTPases and other components, including GEFs, GAPs, effector proteins, and nucleotide derivatives, are prerequisites for reliable and reproducible measurements. In all assays described in this chapter, a decrease in fluorescence should mostly be monitored in a time-dependent manner; in other cases, an increase in fluorescence may be observed. However, another important aspect to be considered is the fluorescence offset, which represents the final fluorescence. This should be relatively similar for all experiments (1) under the same concentrations of the fluorescent nucleotides in the complex with the GTPase and (2) independent of the GEF, GAP, and effector concentrations.

4. Notes

1. Remember to add the required antibiotics to the TB medium to maintain transformed plasmids. In case of using BL21 (DE3) Codon plus RIL, BL21 (DE3) pLysS, or Rosetta (DE3) strains, chloramphenicol (25 mg/L) needs to be added.

2. Cultures are carried out usually in 2.5–20-L scale (depending on the expression level and yield of the particular protein).

3. The small GTPases as well as GEF and GAP proteins, including their catalytic domains, are usually expressed at an OD_{600} of 0.6–0.8 with 0.1 mM IPTG and at 18–25°C overnight.

4. This step is carried out in order to remove residual medium.

5. Cryo preservation through storage at –20°C also helps to improve the efficiency of bacterial lysis.

6. Optionally, the wave duty cycle function of the ultrasonic instrument can be used to reduce heat production and free radical formation.

7. A color change from very milky to slightly more translucent should be observed and can be used as an indicator for cell disruption.

8. This method permits cell disruption in smaller samples (≥200 μL and ≤200 mL).

9. This step removes all traces of alcohol in which the instrument usually is stored to prevent microbial growth.

10. The microfluidizer system provides a convenient and efficient method for cell lysis of larger cell suspensions (≥5 mL to several liters).

11. When purifying guanine nucleotide-binding proteins, it is mandatory to add 0.1 mM GDP during the first affinity chromatography on GSH sepharose and magnesium ions to the standard buffer, which are essential especially in the case of low-affinity GDP/GTP-binding proteins. Always determine the

GDP concentration by HPLC in addition to determining protein concentration.

12. ATP is used for removing chaperons which might be associated with the desired protein and the high concentration of KCl to remove proteins which unspecifically bind to the column or to the GST fusion protein of interest. Note that the absorbance at 280 nm is not reaching the previous baseline level due to the absorbance of ATP present in the buffer applied.

13. Be aware that one has to readjust the pH of the elution buffer with NaOH due to the acidity of GSH.

14. The size and, therefore, the choice of the column size depend on the amount of purified protein and its molecular mass: 16/60 for protein amounts of ≤30 mg or 26/60 for 30–100 mg. If the protein amount to be purified exceeds 100 mg, divide the sample into several portions of ≤100 mg and perform several consecutive runs.

15. After cleaving the protein with protease, the GST tag is not attached to the desired protein anymore. Therefore, while passing again on a GSH sepharose column, the desired protein passes without binding or retention through the column and elute as flow through.

16. Freezing and thawing of protein solutions is a very critical step and has a large impact on protein stability. It is absolutely mandatory to freeze a protein solution in liquid nitrogen and to store it afterward at –20°C or even –80°C. For longer storage periods, the latter is recommended. Before freezing a protein, a small-scale test whether the protein can be frozen in plain standard buffer is recommended, and addition of supplements like glycerol or sucrose might help to prevent protein denaturation during freezing. Thawing of protein solutions should be fast by warming up the vials in the hands. For each protein, the optimal strategy has to be elaborated.

17. Residual alkaline phosphatase-coupled beads might interfere with subsequent spectrofluorometric assays performed with the nucleotide-free GTPase.

18. Releasing GDP (or GTP in the case of the constitutive mutants) from the GTPase and reloading with fluorescent nucleotides, as described above, are prerequisites to perform fluorescence measurements. The incubation time for preparing the nucleotide-free proteins varies among GDP/GTP-binding proteins and has to be established for every GTPase. GppNHp is resistant to phosphodiesterase and thus should not be used in place of Gpp(CH$_2$)p.

19. In order to obtain reliable and reproducible kinetic data, the protein and nucleotide quality needs to be high. Thus, nucleotides

with more than 90% purity should be used and, if necessary, additional purification steps should be carried out to obtain nucleotides and proteins of high purity.

20. Spectroscopic measurements like the fluorescence assays described in this chapter require the use of very clean quartz cuvettes, filtered and degassed buffers, as well as protein solutions without precipitate or any other solid material. Otherwise, light dispersion will take place and the signal-to-noise ratio will be poor. It is, therefore, very important to centrifuge the protein solution immediately before using.

21. Because the samples are mixed 1:1, all stock solutions for components of the samples should be 2× and are described in this section of the protocol as 2× concentrated.

22. Example: Dilute 2 μL of a 100 μM mantGDP-bound GTPase solution in 998 μL of GEF buffer to obtain a 0.2 μM solution of the respective mantGDP-bound protein. A tenfold excess of the GEF protein usually is the first choice to determine the activity of the GEF protein. For that, mix 20 μL of a 100 μM GEF solution (20 μM final concentration) and 4 μL of a 10 mM GDP solution (40 μM final concentration) in 976 μL of GEF buffer. Accordingly, both solutions have a final volume of 1 mL.

23. Example: Dilute 12 μL from a 50 μM solution of nucleotide-free GTPase (0.6 μM final concentration) in 968 μL of GAP buffer and add 20 μL of a 20 μM tamraGTP solution (0.2 μM final concentration) just prior to loading the drive syringes of the stopped flow with your sample. In a double mixing stopped-flow system, both components (tamraGTP and GTPase) can be premixed for a defined short time before the GAP reaction is started.

24. The amounts of proteins and (fluorescent) nucleotides required are rather dependent on the assay used. For the determination of intrinsic nucleotide dissociation or intrinsic GTP hydrolysis in a cuvette (at a final volume of 600 μL), between about 10 and 20 μg nucleotide-bound GTPase is required for three identical experiments. At least 60 μg of catalytic domains of GEF or GAP proteins (with 250–300 amino acid residues) is needed for one experiment to measure the specificity and activity of these regulatory proteins (19). A stopped-flow experiment requires 3–6 μg of GTPase and 15–200 μg of GEFs, GAPs, or effectors but provides an averaged value obtained from 5 to 7 identical measurements. However, between 2 and 10 mg of binding proteins are required to quantitatively analyze the GEF or GAP activities or the effector-binding affinity.

References

1. Ridley, A.J. (2001) Rho family proteins: coordinating cell responses. *Trends Cell Biol* 11, 471–477.

2. Etienne-Manneville, S., and Hall, A. (2002) Rho GTPases in cell biology. *Nature* 420, 629–635.

3. Burridge, K., and Wennerberg, K. (2004) Rho and Rac take center stage. *Cell* 116, 167–179.

4. Raftopoulou, M., and Hall, A. (2004) Cell migration: Rho GTPases lead the way. *Develop Biol* 265, 23–32.

5. Vetter, I.R., and Wittinghofer, A. (2001) The guanine nucleotide-binding switch in three dimensions. *Science* 294, 1299–1304.

6. Bishop, A.L., and Hall, A. (2000) Rho GTPases and their effector proteins. *Biochem J* 348, 241–255.

7. Herrmann, C. (2003) Ras-effector interactions: after one decade. *Curr Opin Struct Biol* 13, 122–129.

8. Dvorsky, R., Blumenstein, L., Vetter, I.R., and Ahmadian, M.R. (2004) Structural insights into the interaction of ROCKI with the switch regions of RhoA. *J Biol Chem* 279, 7098–7104.

9. Moon, S.Y., and Zheng, Y. (2003) Rho GTPase-activating proteins in cell regulation. *Trends Cell Biol* 13, 13–22.

10. Rossman, K.L., Der, C.J., and Sondek J. (2005) GEF means go: turning on RHO GTPases with guanine nucleotide-exchange factors. *Nat Revs Mol Cell Biol* 6, 167–180.

11. Dvorsky, R., and Ahmadian, M.R. (2004) Always look on the bright site of Rho – structural implications for a conserved intermolecular interface. *EMBO Rep* 5, 1130–1136.

12. Ahmadian, M.R., Wittinghofer, A., and Herrmann, C. (2002) Fluorescence methods in the study of small GTP-binding proteins. *Methods Mol Biol* 189, 45–63.

13. Hemsath, L., and Ahmadian, M.R. (2005) Fluorescence approaches for monitoring interactions of RhoGTPases with nucleotides, regulators and effectors. *Methods* 37, 173–182.

14. Eberth, A., and Ahmadian, M.R. (2009) In vitro GEF and GAP assays. In: Bonifacino, J.S, Dasso, M., Harford, J.B., Lippincott-Schwartz, J. and Yamada, K.M. (eds) Current Protocols in Cell Biology, Unit 14.9. Wiley, New York.

15. Eberth, A., Dvorsky, R., Becker, C.F., Beste, A., Goody, R.S., and Ahmadian, M.R. (2005) Monitoring the real-time kinetics of the hydrolysis reaction of guanine nucleotide-binding proteins. *Biol Chem* 386, 1105–1114.

16. Eberth, A., Lundmark, R., Gremer, L., Dvorsky, R., Koessmeier, K.T., McMahon, H.T,. and Ahmadian, M.R. (2009) A BAR domain-mediated autoinhibitory mechanism for RhoGAPs of the GRAF family. *Biochem J* 417, 371–377.

17. Ahmadian, M.R., Zor, T., Vogt, D., Kabsch, W., Selinger, Z., Wittinghofer, A., and Scheffzek, K. (1999) Guanosine triphosphatase stimulation of oncogenic Ras mutants. *Proc Nat Acad Sci USA* 96, 7065–7070.

18. Hemsath, L., Dvorsky, R., Fiegen, D., Carlier, M.F., and Ahmadian, M.R. (2005) An electrostatic steering mechanism of Cdc42 recognition by Wiskott-Aldrich syndrome proteins. *Mol Cell* 20, 313–324.

19. Stevens, W.K., Vranken, W., Goudreau, N., Xiang, H., Xu, P., and Ni, F. (1999) Conformation of a Cdc42/Rac interactive binding peptide in complex with Cdc42 and analysis of the binding interface. *Biochem* 38, 5968–5975.

20. Haeusler, L.C., Blumenstein, L., Stege, P., Dvorsky, R., and Ahmadian, M.R. (2003) Comparative functional analysis of the Rac GTPases. *FEBS Lett* 555, 556–560.

21. Fiegen, D., Haeusler, L.C., Blumenstein, L., Herbrand, U., Dvorsky, R., Vetter, I.R. and Ahmadian, M.R. (2004) Alternative splicing of Rac1 creates a self-activating GTPase. *J Biol Chem* 279, 4743–4749.

22. Blumenstein, L., and Ahmadian, M.R. (2004) Models of the cooperative mechanism for Rho-effector recognition: Implications for RhoA-mediated effector activation. *J Biol Chem* 279, 53419–53426.

23. Gremer, L., Merbitz-Zahradnik, T., Dvorsky, R., Cirstea, I.C., Kratz, C.P., Zenker, M., Wittinghofer, A., and Ahmadian, M.R. (2010) Impacts of germline mutations in KRAS on the GTPase cycle and effector interaction. *Hum Mutat* 32, 33–43.

24. Gremer, L., De Luca, A., Merbitz-Zahradnik, T., Dallapiccola, B., Morlot, S., Tartaglia, M., Kutsche, K., Ahmadian M.R., and Rosenberger, G. (2010) Duplication of Glu37 in the switch I region of HRAS impairs effector/GAP binding and underlies Costello syndrome by promoting enhanced growth factor-dependent MAPK and AKT activation *Hum Mol Genet* 19, 790–802.

Chapter 4

How to Analyze Bacterial Toxins Targeting Rho GTPases

Heike Bielek and Gudula Schmidt

Abstract

Bacterial pathogens developed several strategies to overcome defense systems of eukaryotic hosts. Within the infection process, they need to attach to and cross through epithelial layers, escape from the innate and adaptive immune response, and find a physiological niche to survive. One target to modulate the host–pathogen interaction in order to deceit pathogen resistance is the actin cytoskeleton and its regulators: the family of Rho GTPases. Some bacterial toxins catalyze a covalent modification of Rho GTPases to keep these molecular switches in a constitutive active or inactive state. This leads to rearrangement of the actin cytoskeleton. Toxin-treated cells show typical morphological changes depending on substrate specificity and action of the toxins. In this chapter, we discuss the classes of bacterial toxins based on their mode of action, their recombinant expression (specifically CNF1), intoxication and subsequent morphological changes of the actin cytoskeleton, and cell shape.

Key words: Bacterial toxins, Rho GTPases, Toxin B, C2, C3, CNF1, CNFY, Actin

1. Introduction

Many pathogenic microorganisms have developed strategies to survive in the hostile environment. Some of them produce effectors/toxins which are capable of killing eukaryotic cells by disruption of the actin cytoskeleton or at least inducing extensive changes in cell morphology by controlling the family of Rho GTPases. As previously described, Rho GTPases are crucial molecular switches in actin signaling that are controlled within the GTPase cycle by diverse regulators, such as guanine nucleotide exchange factors (GEFs), GTPase-activating proteins (GAPs), and guanine nucleotide dissociation inhibitors (GDIs). Within this cycle, Rho GTPases localize at the membrane but also bind downstream effectors in the cytosol. Bacterial toxins can affect all phases of Rho GTPase

Francisco Rivero (ed.), *Rho GTPases: Methods and Protocols*, Methods in Molecular Biology, vol. 827,
DOI 10.1007/978-1-61779-442-1_4, © Springer Science+Business Media, LLC 2012

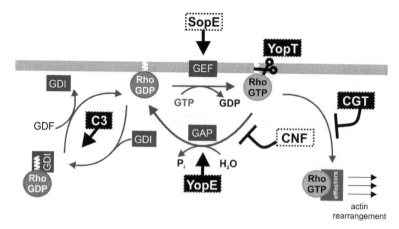

Fig. 1. Overview of bacterial toxins/effectors acting on the Rho GTPase cycle. Rho GTPases are molecular switches shuttling between a cytosolic, inactive, GDP-bound state in complex with GDI and an active GTP-bound state capable of binding to downstream effectors. GDP to GTP exchange is induced by GEFs at the membrane when GDI is released through, e.g. GDFs. Rho GTPases signaling is terminated by GTP hydrolysis through GAPs (21). Bacterial toxins interfere with this Rho GTPase cycle at multiple steps by inactivation (*black boxes*) or activation (*white boxes*). The most efficient inactivating toxins are the CGTs like toxin A/B which glucosylate Thr37/35 (RhoA/Rac, Cdc42) and thereby block interaction with GDIs, GEFs, GAPs and most importantly interaction with effectors. Consequently, downstream signaling, e.g., to the actin cytoskeleton is inhibited. ADP ribosylation of RhoA at Asn41 by C3 exoenzymes leads to increased GDI affinity and inhibition of GEF interaction. The cysteine protease YopT cleaves off the C-terminal isoprenylated cysteine, causing release of Rho proteins from membranes. YopE mimics GAP activity to enhance GTP hydrolysis. Activating effectors like SopE function as GEFs to facilitate GDP/GTP exchange and activate Rho GTPases. Deamidation and transglutamination of Gln63/61 (Rho/Rac, Cdc42) by CNFs and DNT constitutively activate Rho proteins (1). Abbreviations: *GDI* guanine nucleotide dissociation inhibitor, *GEF* nucleotide exchange factor, *GDF* GDI displacement factor, *GAP* GTPase-activating protein, *CGT* clostridial glucosylating toxin, *CNF* cytotoxic necrotizing factor, *YopT/E* Yersinia outer protein E/T, *SopE* Salmonella outer protein E (1).

cycling, leading to an imbalanced pool of active and inactive GTPase that finally results in dramatic changes of downstream signaling events as depicted in Fig. 1. This is achieved either by direct or indirect modulation (1).

1.1. Inactivating Toxins

Modifications of Rho GTPases through glucosylation, ribosylation, and proteolytic cleavage are caused by the class of irreversible, inactivating toxins/effectors (see Table 1). The very potent large clostridial glucosylating cytotoxins (CGTs), e.g. toxin A/B from *Clostridium difficile*, modify Rho GTPases by glucosylation of a crucial threonine residue, Thr37 in RhoA and Thr35 in Rac/Cdc42, which is located within the effector-binding region (switch I). This glucosylation blocks binding to regulators of the GTPase cycle (GEFs, GDIs, GAPs), but most importantly it blocks Rho effector-binding and downstream signaling, which consequently leads to cell rounding (2). Similar effects are seen by intoxication

Table 1
A selection of inactivating toxins

Inactivating effectors/toxins	Pathogen	Enzymatic activity	Substrate	Refs.
Clostridial glucosylating toxins (e.g., *Toxin A/B*)	*Clostridium* species (*C. difficile*)	Glucosylation of Thr35/37	Rho, Rac, Cdc42	(2)
Yersinia outer protein T (YopT)	*Yersinia enterocolitica*	Proteolytic cleavage of C-terminal isoprenylated cysteine	Rho, Rac, Cdc42	(3, 4)
C3-like toxins (e.g., C3bot)	*Clostridium botulinum*	ADP ribosylation of Asn41	RhoA, RhoB, RhoC	(5, 6)
Yersinia outer protein E (YopE)	*Y. enterocolitica*	Rho GTPase-activating protein RhoGAP	Rho, Rac, Cdc42	(7–9)

with *Yersinia* effector YopT. As a cysteineprotease, YopT cleaves off the C-terminal isoprenylated cysteine that is essential for binding of the GTPase to membranes and GDI. Thereby, Rho GTPases detach from the membrane and are incapable of interacting with membrane-bound GEFs: the GTPase cycle is blocked. As seen for the glucosylating cytotoxins, cells display similar morphological effects: actin disruption and cell rounding (3, 4).

Another inactivating toxin is the exoenzyme C3 from *Clostridium botulinum*, a member of the C3-like toxin family. It is a ribosyltransferase that modifies Rho (A, B, C) at Asn41. This modification enhances binding affinity to RhoGDI and leads to Rho inactivation and subsequent disappearance of RhoA-induced stress fiber formation (5, 6). Reversible inhibition is achieved through the mimicry of GAP-like activity, e.g., by *Salmonella* effector YopE, promoting GTP hydrolysis to its GDP-bound inactive state (7–9).

1.2. Activating Toxins

Cytotoxic necrotizing factors (CNFs) from *Escherichia coli* and *Yersinia* as well as dermonecrotizing toxin (DNT) from *Bordetella* species belong to the class of activating toxins. They deamidate and/or transglutaminate a specific glutamine (Gln63 in RhoA, Gln61 in Cdc42/Rac) that is crucial for GTP hydrolysis, since it is involved in the coordination of the water molecule. This modification leads to constitutive activation of the Rho GTPases and to disarrangement of the actin cytoskeleton (10–13). Due to uncontrolled RhoA activity, regulation of cytokinesis is disturbed; multinucleated cells are observed with CNF toxins (14). As an example for reversible activating effectors, SopE from *Salmonella* species

Table 2
A selection of activating toxins

Activating effectors/toxins	Pathogen	Enzymatic activity	Substrate	Refs.
Cytotoxic necrotizing factors (CNF1, 2, 3, Y)	*Escherichia coli, Yersinia pseudotuberculosis*	Deamidation/ transglutamination of Gln61/63	Rho, Rac, Cdc42	(10–12)
Dermonecrotizing toxin (DNT)	*Bordetella* sp.	Transglutamination/ deamidation of Gln61/63	RhoA, Rac, Cdc42	(13)
Salmonella outer protein E (SopE)	*Salmonella* sp.	GEF activity	Rac, Cdc42	(15)

imitates host GEF activity in the same manner as GAP-like effectors and thereby enhances GTP exchange (15). A summary of activating effectors is presented in Table 2.

1.3. Toxin Uptake

There are several ways how toxins enter the host cell: e.g., receptor-mediated endocytosis or injection via Type III Secretion Systems (T3SS).

The class of autotransporter toxins usually harbors similar structural elements: a receptor-binding, a translocation, (a protease), and an enzymatic domain within a one-chain protein. The receptor-binding domain binds to a cell surface receptor and leads to endocytosis. Upon endosome acidification, the toxin refolds with its hydrophobic translocation domain inserting into the endosomal membrane. By proteolytic cleavage, the release of the enzymatic domain into the cytosol is caused. This was shown for CNFs (see Fig. 2a) and large CGTs (2, 16).

Binary AB toxins as in Anthrax consist of two nonlinked components: the A, receptor-binding component protective antigen (PA), and the B, enzymatic components lethal toxin (LT) and edema factor (EF). LT and EF bind to the activated PA heptamer which is attached to a receptor and is taken up into the endosome. After acidification, the PA heptamer inserts into the endosomal membrane and EF and/or LT are released into the cytosol. Another example is C2 from *C. botulinum*. C2 binary toxin consists of the A enzymatic part, C2I, and the B receptor-binding precursor, C2II. After cleavage, C2IIa (activated) forms a heptamer which interacts with its receptor, binds, and releases C2I into the host cytosol via the endosomal route, similar to Anthrax toxin (17). Other effectors like YopT only consist of the enzymatic component and are released directly into the host cytosol via, e.g., the T3SS, a bacterial needle that crosses bacterium and

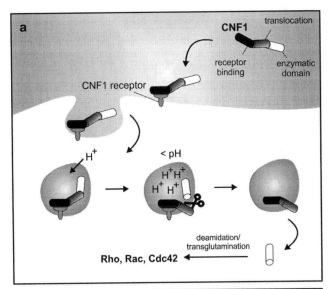

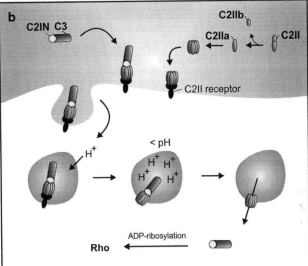

Fig. 2. Toxin uptake: Cell delivery system for the fusion protein C2IN-C3 and CNF1. (**a**) CNF1 inherits an N-terminal receptor-binding component and a C-terminal enzymatic component that are connected via the putative hydrophobic translocation domain. Receptor binding through the N terminus leads to endocytosis. Decrease of pH results in insertion of the translocation domain into the membrane and cleavage of the enzymatic part. After release into the cytosol, the CNF1 enzymatic domain modifies Rho GTPases by deamidation and/or transglutamination. (**b**) C2II precursor is activated by proteolytic cleavage to C2IIa. This activated B-component forms a heptamer which binds a specific cell surface receptor. The A component (C2IN–C3 fusion protein) docks with the cell-bound C2IIa heptamer, and the complex is then taken up via receptor-mediated endocytosis into early endosomes. These endosomes become acidified by proton pumps, resulting in insertion of the heptamer into the endosomal membrane. The C2IN–C3 chimera is translocated through the pore into the host cytosol. There, C3 exoenzyme ADP ribosylates Rho proteins (16, 18).

host cell membranes (4, 6). This mechanism requires the pathogen for full toxicity while microinjection only allows intoxication of single cells.

Uptake efficiency can also vary depending on the cell type. Some cells may lack or display only low expression levels of specific receptors. Therefore, it is an advantage to construct toxins as fusion proteins: the enzyme domain combined with the receptor-binding part of autotransporters known to intoxicate cells of interest. Via these chimeras, it is possible to have higher yields of intoxication for biochemical approaches. Thus, it is a great advantage to adopt the uptake mechanism of autotransporter toxins for large-scale intoxication. As an example (see Fig. 2b), the Rho-ADP-ribosylating C3 exoenzyme was fused to the receptor-binding domain of C2 toxin as the chimera C2IN-C3. For the intoxication process, the pore-forming component C2IIa needs to be added to the chimera construct (18). Also, translocation by transactivator of transcription (TAT) from HIV has been described to function as shuttle for toxins, as in DNT-TAT (19, 20).

Interference with Rho GTPase regulation by toxins can be investigated by direct biochemical approaches to measure Rho GTPase activity, e.g., with effector pull-downs. RT (rhotekin) pull-down is used for RhoA activity, whereas PAK (p21-activated kinase) pull-down is a readout for Rac and Cdc42 activity (see Chapter 3).

1.4. Methods Outline

Here, we exemplify the method of recombinant expression with the activating AB-type toxin CNF1 from *E. coli*, purification via its GST tag, and intoxication of cells. Furthermore, we describe the analysis of morphological changes elicited by Toxin B, C2, C3, CNF1, and CNFY treatment using light/fluorescence microscopy of living and fixed (rhodamine-phalloidin/DAPI-stained) HeLa cells.

2. Materials

2.1. Buffers

1. Lysis buffer: 20 mM Tris–HCl, pH 7.3, 10 mM NaCl, 5 mM $MgCl_2$.

2. Lysis buffer +: 20 mM Tris–HCl, pH 7.3, 10 mM NaCl, 5 mM $MgCl_2$, 1% Triton X-100.

3. Lysis buffer +++: 20 mM Tris–HCl, pH 7.3, 10 mM NaCl, 5 mM $MgCl_2$, 1% Triton X-100, 5 mM dithiothreitol (DTT), 1 mM phenylmethanesulfonyl fluoride.

4. High-salt buffer: 50 mM Tris–HCl, pH 7.5, 150 mM NaCl, 5 mM $MgCl_2$.

5. Glutathione buffer: 10 mM reduced glutathione, 50 mM Tris–HCl, pH 9.0.

6. Thrombin cleavage buffer (pH 7.4): 50 mM triethanolamine, 100 mM NaCl, 2.5 mM MgCl$_2$.

7. 5× Laemmli buffer: 100 mM Tris–HCl, pH 6.8, 10% sodium dodecylsulfate, 50% glycerol, 100 mM DTT, 0.01% bromophenol blue.

8. Phosphate-buffered saline (PBS): 10 mM Na$_2$HPO$_4$, 140 mM NaCl, 2.7 mM KCl, 1.8 mM KH$_2$PO$_4$, pH 7.4.

2.2. Protein Expression and Purification

1. pGEX2T vectors (GE Healthcare) containing the respective gene of interest are transformed into TG-1 *E. coli* strains; 500 μL of bacterial suspension is mixed with 500 μL of 70% glycerol and stored at –80°C as a bacterial plasmid stock.

2. Final 1× LB medium for *E. coli* cultivation is prepared from 10× LB Luria/Miller concentrate (Roth) with sterile water and stored at 4°C until use.

3. 100 mg/mL ampicillin stock solution in 50% ethanol (stored at –20°C) is added to the LB medium at a dilution of 1:1,000 just before use.

4. 1 M isopropyl β-d-1-thiogalactopyranoside (IPTG) stock in water. Store at –20°C.

5. Empty PD-10 columns (Amersham Pharmacia).

6. Sonicator: SonoPuls HD60 (Bandelin).

7. DNAseI (Roche). Prepare a 10 mg/mL stock and store at –20°C.

8. Thrombin (Sigma).

9. Glutathione-sepharose 4B beads (GE Healthcare).

10. Benzamidine-sepharose 6B beads (Amersham).

2.3. Cell Biology

1. HeLa medium: Dulbecco's modified Eagle's medium (DMEM, Biochrome), 10% fetal calf serum (Biochrome), 1% nonessential amino acids (100×, PAA), 1% penicillin and streptomycin (100×, PAA).

2. Trypsin/EDTA (0.25/0.02%) (PAN biotech).

3. Light microscope Axiovert 25 (Zeiss).

4. Camera AxioCam HRC (Zeiss).

5. Fixing solution: 3.7% formaldehyde, 0.1% Triton X-100 in PBS.

6. Phalloidin–tetramethylrhodamine isothiocyanate (TRITC) B (Sigma): Prepare a 40 μg/mL stock in methanol.

7. Mounting solution: ProLong® Gold antifade reagent with DAPI (Molecular Probes/Invitrogen).

3. Methods

3.1. Protein Expression and Purification

Recombinant expression of CNF1 toxin is performed in *E. coli* TG-1 cells, an expression system close to uropathogenic *E. coli* that harbors *cnf1*. This protocol (see Fig. 3) is optimized for 4 L of main culture to yield about 1 mg of protein.

1. Overnight culture: Prepare 400 mL of LB medium with 400 μL of ampicillin (100 mg/mL stock), inoculate from glycerol stock of pGEX2T-CNF1, and incubate for 24 h at 37°C with shaking.

2. Main culture: Inoculate 4 L of LB medium (with ampicillin) with 1:10 overnight culture, and incubate at 37°C with shaking until an OD_{600} 0.7–0.8 is attained. Control: Withdraw 500 μL of bacterial suspension as noninduced sample, centrifuge down for 20 s at $15,000 \times g$, resuspend pellet with 30 μL of 5× Laemmli buffer, and boil at 95°C for 5 min.

3. Induce gene expression of GST-CNF1 with 100 μM IPTG. Incubate for another 2–3 h at 37°C with shaking.

4. Centrifuge the culture for 10 min at $4,000 \times g$ at 4°C. Decant supernatant. Pellet can be stored at –20°C until lysis. Before lysis, thaw pellet on ice (see Note 1).

5. With the pellet on ice, add 10–15 mL of cold lysis buffer+++ (see Note 2) and resuspend by vortexing and/or pipetting up and down.

6. Sonicate the suspension three times for 1 min on ice (70% power/ cycle 70%) (see Note 3). Add 2 μL of DNAseI (10 mg/mL stock) if pellet becomes viscous. DNAseI treatment can be performed on ice for 1 h or preferably with rotation at 4°C.

7. Centrifuge lysate for 30 min at $15,000 \times g$ at 4°C to separate cell debris from cytosol.

8. Transfer supernatant to a 15- or 50-mL conical tube (see Note 4) containing 200–300 μL of 50% prewashed glutathione sepharose beads (see Note 5) in 5 mL of lysis buffer+.

9. Incubate for 60–90 min at 4°C tilting upside down.

10. Always working in the cold, transfer the sample to an empty PD-10 column (prewashed with water) (see Note 6), open the valve, and allow the lysate to run through the column by gravity flow (see Note 7).

    If CNF1 without the GST tag is the main goal, continue with step 11 for thrombin cleavage. In order to obtain GST-CNF1, continue with glutathione elution at step 19.

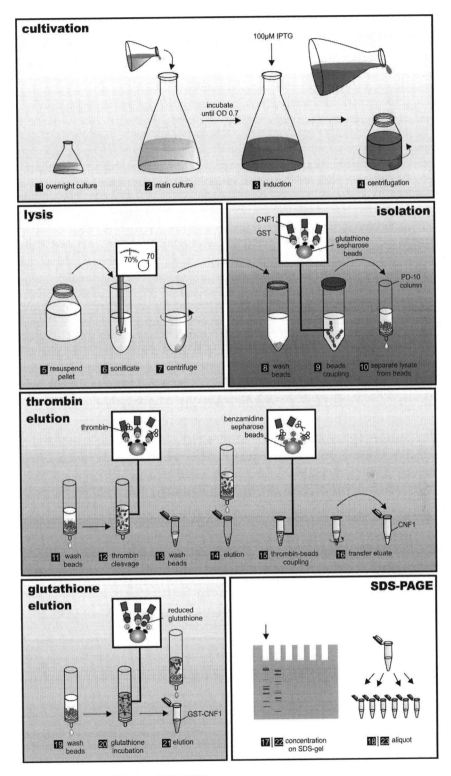

Fig. 3. Protein expression and purification of GST-CNF1.

11. Perform the following five washing steps at 4°C, every time letting the buffer run through by gravity flow:

    (a) 10 mL of lysis buffer+++
    (b) 10 mL of lysis buffer+
    (c) 10 mL of lysis buffer without any supplements
    (d) 10 mL of high-salt buffer
    (e) 5 mL of thrombin cleavage buffer

    Beads control: Withdraw 10 μL of beads, mix with 30 μL of 5× Laemmli buffer, and boil at 95°C for 5 min.

12. Elute CNF1 by applying 200–300 μL of thrombin cleavage buffer and 2 μL of thrombin (10 NIH-units/) to the column. Incubate at room temperature for 15 min with shaking (see Note 8).

13. Equilibrate benzamidine-sepharose beads twice with thrombin cleavage buffer (see Note 9).

14. Open valve and collect eluate in a 1.5-mL precooled tube placed on ice. Repeat steps 12 and 14 three times (E_{1-3}). For the last elution step, remove the column from the valve so that the residual volume can be collected. Thrombin cleavage elution control: Withdraw 10 μL of beads out of the column, mix with 30 μL of 5× Laemmli buffer, and boil at 95°C for 5 min.

15. Add E_1 onto the prepared benzamidine-sepharose beads. Incubate for 15 min at 4°C, tilting upside down.

16. Centrifuge for 1 min at 15,000×g at 4°C and transfer supernatant to a fresh, precooled 1.5-mL tube placed on ice. Repeat for E_2 and E_3 with the same benzamidine-sepharose beads. Benzamidine-beads control: Withdraw 10 μL of beads out of the column, mix with 30 μL of 5× Laemmli buffer, and boil at 95°C for 5 min.

17. Check for protein concentrations with SDS-PAGE followed by Coomassie staining (see Note 10). If concentrations are similar, pool eluates (E_1, E_2, E_3).

18. Store 20-μL aliquots of CNF1 at –20°C (see Note 11).

19. Perform the following five washing steps at 4°C, every time letting the buffer run through by gravity flow:

    (a) 10 mL of lysis buffer+++
    (b) 10 mL of lysis buffer+
    (c) 10 mL of lysis buffer without any supplements
    (d) 10 mL of high salt buffer
    (e) 5 mL of 50 mM Tris, pH 7.4

    Beads control: Withdraw 10 μL of beads from the column with a 20-μL pipette (cut the tip), mix with 30 μL of 5× Laemmli buffer, and boil at 95°C for 5 min.

20. Elute GST-CNF1 by applying 200–300 µL of glutathione buffer to the column. Incubate at room temperature for 15 min with shaking (see Note 8).

21. Open valve and collect eluate in a 1.5-mL precooled tube placed on ice. Repeat steps 20 and 21 three times (E_{1-3}). For the last elution step, remove the column from the valve so that the residual volume can be collected. Glutathione elution control: Withdraw 10 µL of beads out of the column, mix with 30 µL of 5× Laemmli buffer, and boil at 95°C for 5 min.

22. Check for protein concentrations on SDS-PAGE followed by Coomassie staining (see Note 10). If concentrations are similar, pool eluates (E_1, E_2, E_3).

23. Store 20-µL aliquots of GST-CNF1 (see Note 11) at –20°C (see Note 12).

3.2. Intoxication of Eukaryotic Cells

As a model cell line, HeLa cells are used to examine the effect of intoxication on Rho GTPases and downstream actin rearrangements. HeLa cells, isolated from cervical cancer, constitute a well-defined spreading cell type for morphological studies concerning the actin cytoskeleton.

1. In a 12-well plate under a sterile hood, place one coverslip (stored in 70% ethanol) into each well. Wash with 2 mL of sterile PBS. Add 1 mL of HeLa medium into each well.

2. Trypsinize HeLa cells (see Note 13) with 2 mL of trypsin and incubate at 37°C until cells detach (~1–2 min). Resuspend in 10 mL of preheated (37°C) HeLa medium (final volume of 12 mL). Add 1 mL of cell suspension into each well of the 12-well plate. Incubate overnight at 37°C, 5% CO_2, until cells are ~50% confluent.

3. Remove medium and add 2 mL of fresh preheated medium to each well (see Note 14).

4. With a 2-µL pipette, dispense 200–400 ng of toxin per mL into the medium (see Note 15).

5. Incubate for 2–24 h at 37°C, 5% CO_2 (see Note 16).

6. Analyze using light microscopy. Then, continue with actin staining (see Note 17).

7. Remove medium and add 1 mL of fixing solution to each well. Incubate for at least 10 min at room temperature (see Note 18).

8. Wash three times with 1 mL of PBS.

9. Prepare master mix of TRITC–phalloidin staining solution diluting the stock 1:200 in PBS containing 1% BSA, e.g., 25 µL for each coverslip, 270 µL for 12 coverslips. Prepare 300 µL of PBS–BSA and add 1.5 µL of TRITC–phalloidin stock solution (40 µg/mL).

10. Dispense of 25-µL drops of staining solution on marked spots (e.g., numbers) on parafilm and transfer each coverslip with a forceps from its well onto a drop, taking care that the side with the cells faces the TRITC–phalloidin solution (see Note 19).

11. Incubate for 30 min–1 h at room temperature in the dark (see Note 20).

12. Turn coverslip around and add PBS dropwise to the cell surface. Incubate for 5 min. Remove PBS with a pipette from the edge of the coverslip. Repeat the PBS wash three times (see Note 21).

13. Prepare glass slides. Apply one drop of mounting solution for each coverslip on a glass slide. Avoid making bubbles.

14. Take coverslip with forceps and dip in a beaker filled with distilled water, dry the edge of the coverslip on a paper towel. Repeat washing five times (see Note 22).

15. Place the coverslip upside down onto mounting solution (side with the cells toward the drop). Incubate at room temperature overnight in the dark before samples are ready for fluorescence microscopy.

3.3. Anticipated Results

In our experiments, we use untreated HeLa cells and 2–24-h intoxications with toxin B, C2 (C2IIa + C2I), C3 (C2IIa + C2IN-C3), CNF1, and CNFY. Analysis of intoxication is done with living cells for light microscopy and with fixed cells stained for actin for fluorescence microscopy.

Figure 4 shows light microscopic and fluorescent pictures of DAPI- and TRITC–phalloidin-stained cells. For some toxins, morphological changes of the actin cytoskeleton can be observed by light microscopy; other toxins reveal their effects more obviously by distinct actin staining. While HeLa control cells show an attached, cell phenotype with some basal stress fiber formation, C2 intoxicated cells round up within 1 h due to the loss of actin filament formation, which can be monitored by light microscopy. In the same manner, cells become round through treatment with toxin B. It also affects the actin cytoskeleton by inactivating its regulators: Rho, Rac, and Cdc42. On the contrary, C3 inactivation of RhoA does not result in a rounded cell phenotype, but in the loss of stress fibers. This effect is clearly apparent by actin staining.

The activating toxins, CNF1 and CNFY, differ in their substrate specificity. CNF1 activates Rho, Rac, and Cdc42, and CNFY is specific for RhoA. Thereby, CNF1 elicits a flattened phenotype after 2 h with ruffling, lamellipodia, and stress fibers. CNFY shows a more contracted phenotype and extensive stress fiber formation throughout the cell body. Cells also display impairment of cytokinesis as an effect of uncontrolled RhoA activity and are incapable

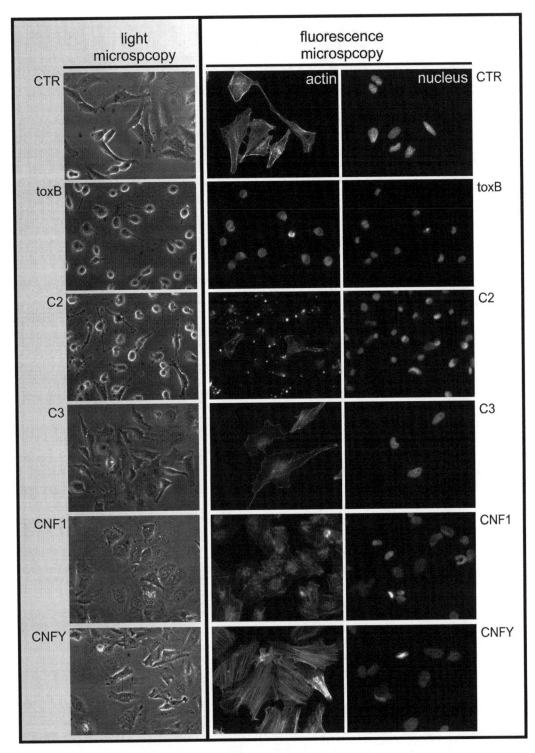

Fig. 4. HeLa intoxication with toxin B, C2, C3, CNF1/Y, and untreated (CTR). *Left panels*: Light microscopic pictures at 32× magnification. *Right panels*: Fluorescent microscopic pictures with 40× magnification; TRITC–phalloidin for actin and DAPI for nucleus staining.

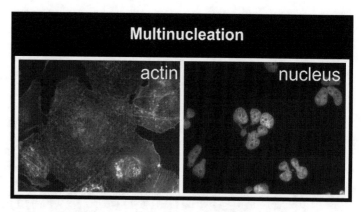

Fig. 5. HeLa cells with multiple nuclei after a 24-h intoxication with CNF1. *Left panel*: TRITC–phalloidin-stained actin. *Right panel*: DAPI-stained nuclei. Nuclei are not divided symmetrically. 40× magnification

of dividing. This results in an expanded cell body with multiple nuclei which is detectable after long-time exposure (at least 24 h) to CNF1 or CNFY (see Fig. 5).

4. Notes

1. Due to protein instability, higher yields of protein are obtained when the pellet is not frozen but processed immediately.

2. Add protease inhibitors just before lysis since they are instable. Alternatively, there are also protease inhibitor cocktails commercially available, e.g., protease inhibitor cocktail without EDTA as tablets (Roche).

3. Since sonication heats up the lysate, it is recommended to use an adequate volume of at least 5 mL. It is preferable to sonicate in several rounds of 20 s with cooling intervals of 20 s to keep the temperature low.

4. Do not decant the supernatant into a fresh tube, use a pipette to transfer it, and avoid contamination with any cell debris.

5. Beads are removed from the stock solution with a 1-mL pipette (the point of the tip is cut off to decrease the amount of remaining beads). Glutathione-sepharose beads are washed three times in lysis buffer. Be careful to remove only supernatant without touching the beads with the tip; it is recommended to leave a layer liquid on the bead pellet.

6. Emptied PD-10 columns can be reused; rinse with water after use.

7. Washings can also be done by centrifugation of the pellet. The supernatant can be removed with a pipette, but may also remove some beads. PD-10 columns prevent the loss of beads.

8. Attach the column, e.g., to a thermomixer and shake at 300 rpm.

9. Benzamidine is a thrombin inhibitor. Thrombin is removed by benzamidine beads.

10. Carry out SDS-PAGE followed by Coomassie staining using conventional methods. If concentrations are similar, pool eluates (E_1, E_2, E_3).

11. GST-CNF1 can be used for cell intoxication because the GST tag does not disturb cellular effects when focusing on signal transduction. Working on the toxin itself, it might be recommended to cleave off the GST tag, although the yield is much less.

12. Recommendation: Shock freeze in liquid nitrogen before storing the aliquots at –20°C. Expected yield is around 1.5 μg/μL.

13. Use HeLa cells grown in a T75 flask of 50% confluency. If counting the cells is preferred, add 10^5 cells per well.

14. Cells should be 50% confluent, single cells with little cell–cell contact. If there is still a lot of space between cells, leave for another day.

15. Here, we use 400 ng/mL CNF1/Y, 200 pM toxin B, 400 ng/mL C2IIa+200 ng/mL C2I, 400 ng/mL C2IIa+200 ng/mL C2IN-C3, and untreated cells.

16. Very potent toxins, like the inactivating toxins toxin A and B, already provoke morphological changes observable by light microscopy after 30 min. Multinucleation by CNF intoxication can efficiently be seen after overnight intoxication, most likely with DAPI staining of the nuclei.

17. Some actin rearrangements can only be observed by TRITC–phalloidin staining.

18. Cells are permeabilized and proteins denatured. From this step, handling is done under nonsterile conditions. Coverslips can also be stored in fixing solution at 4°C for a couple of days.

19. This can also be done in the 12-well plate. But to reduce volume and expenses for TRITC–phalloidin solution or antibody dilutions, we reduce the volume to 25 μL per sample. In order to prevent sample mix-up, write numbers on the bottom of a 15-cm dish and place parafilm on it. Coverslips are each assigned to a number.

20. Use the lid of the used dish and close it. Wrap up the complete dish to protect from light.

21. Be careful when adding PBS to the coverslip; this side is covered with cells, so try not to damage them!

22. To check the cell side, one can use a small tip and scratch a little bit at the very edge of the coverslip on each side. Little smear reveals the cell side.

References

1. Aktories, K., and Barbieri, J. T. (2005) Bacterial cytotoxins: targeting eukaryotic switches. *Nat Rev Microbiol* **3**, 397–410.

2. Jank, T., and Aktories, K. (2008) Structure and mode of action of clostridial glucosylating toxins: the ABCD model. *Trends Microbiol.* **16**, 222–229.

3. Sorg, I., Goehring, U. M., Aktories, K., and Schmidt, G. (2001) Recombinant *Yersinia* YopT leads to uncoupling of RhoA-effector interaction. *Infect Immun* **69**, 7535–7543.

4. Aepfelbacher, M., Zumbihl, R., and Heesemann, J. (2005) Modulation of Rho GTPases and the actin cytoskeleton by YopT of *Yersinia*. *Curr Top Microbiol Immunol* **291**, 167–175.

5. Genth, H., Gerhard, R., Maeda, A., Amano, M., Kaibuchi, K., Aktories, K., and Just, I. (2003) Entrapment of Rho ADP-ribosylated by *Clostridium botulinum* C3 exoenzyme in the Rho-guanine nucleotide dissociation inhibitor-1 complex. *J Biol Chem* **278**, 28523–28527.

6. Aktories, K., Wilde, C., and Vogelsgesang, M. (2004) Rho-modifying C3-like ADP-ribosyltransferases. *Rev Physiol Biochem Pharmacol* **152**, 1–22.

7. Aili, M., Telepnev, M., Hallberg, B., Wolf-Watz, H., and Rosqvist, R. (2003) In vitro GAP activity towards RhoA, Rac1 and Cdc42 is not a prerequisite for YopE induced HeLa cell cytotoxicity. *Microb Pathog* **34**, 297–308.

8. Andor, A., Trulzsch, K., Essler, M., Roggenkamp, A., Wiedemann, A., Heesemann, J., and Aepfelbacher, M. (2001) YopE of *Yersinia*, a GAP for Rho GTPases, selectively modulates Rac-dependent actin structures in endothelial cells. *Cell Microbiol* **3**, 301–310.

9. Von Pawel-Rammingen, U., Telepnev, M. V., Schmidt, G., Aktories, K., Wolf-Watz, H., and Rosqvist, R. (2000) GAP activity of the *Yersinia* YopE cytotoxin specifically targets the Rho pathway: a mechanism for disruption of actin microfilament structure. *Mol Microbiol* **36**, 737–748.

10. Lerm, M., Selzer, J., Hoffmeyer, A., Rapp, U. R., Aktories, K., and Schmidt, G. (1999) Deamidation of Cdc42 and Rac by *Escherichia coli* cytotoxic necrotizing factor 1: activation of c-Jun N-terminal kinase in HeLa cells. *Infect Immun* **67**, 496–503.

11. Schmidt, G., Sehr, P., Wilm, M., Selzer, J., Mann, M., and Aktories, K. (1997) Gln 63 of Rho is deamidated by *Escherichia coli* cytotoxic necrotizing factor-1. *Nature* **387**, 725–729.

12. Hoffmann, C., Pop, M., Leemhuis, J., Schirmer, J., Aktories, K., and Schmidt, G. (2004) The *Yersinia pseudotuberculosis* cytotoxic necrotizing factor (CNFY) selectively activates RhoA. *J Biol Chem* **279**, 16026–16032.

13. Horiguchi, Y., Inoue, N., Masuda, M., Kashimoto, T., Katahira, J., Sugimoto, N., and Matsuda, M. (1997) *Bordetella bronchiseptica* dermonecrotizing toxin induces reorganization of actin stress fibers through deamidation of Gln-63 of the GTP-binding protein Rho. *Proc Natl Acad Sci USA* **94**, 11623–11626.

14. Huelsenbeck, S. C., May, M., Schmidt, G., and Genth, H. (2009) Inhibition of cytokinesis by *Clostridium difficile* toxin B and cytotoxic necrotizing factors--reinforcing the critical role of RhoA in cytokinesis. *Cell Motil Cytoskeleton* **66**, 967–975.

15. Friebel, A., Ilchmann, H., Aepfelbacher, M., Ehrbar, K., Machleidt, W., and Hardt, W. D. (2001) SopE and SopE2 from *Salmonella typhimurium* activate different sets of Rho GTPases of the host cell. *J Biol Chem* **276**, 34035–34040.

16. Knust, Z., Blumenthal, B., Aktories, K., and Schmidt, G. (2009) Cleavage of *Escherichia coli* cytotoxic necrotizing factor 1 is required for

full biologic activity. *Infect Immun* **77**, 1835–1841.

17. Barth, H., Aktories, K., Popoff, M. R., and Stiles, B. G. (2004) Binary bacterial toxins: biochemistry, biology, and applications of common *Clostridium* and *Bacillus* proteins. *Microbiol Mol Biol Rev* **68**, 373–402.

18. Barth, H., Blocker, D., and Aktories, K. (2002) The uptake machinery of clostridial actin ADP-ribosylating toxins--a cell delivery system for fusion proteins and polypeptide drugs. *Naunyn Schmiedebergs Arch Pharmacol* **366**, 501–512.

19. Stratmann, H., Schwan, C., Orth, J. H., Schmidt, G., and Aktories, K. (2010) Pleiotropic role of Rac in mast cell activation revealed by a cell permeable *Bordetella* dermonecrotic fusion toxin. *Cell Signal* **22**, 1124–1131.

20. Berks, B. C., Sargent, F., and Palmer, T. (2000) The Tat protein export pathway. *Mol Microbiol* **35**, 260–274.

21. DerMardirossian, C. and Bokoch, G. M. (2005) GDIs: central regulatory molecules in Rho GTPase activation. *Trends Cell Biol* **15**, 356–363.

Chapter 5

Assessing Ubiquitylation of Rho GTPases in Mammalian Cells

Anne Doye, Amel Mettouchi, and Emmanuel Lemichez

Abstract

Rho GTPases including RhoA, Cdc42, and Rac1 are master regulators of cell cytoskeleton dynamic, thus controlling essential cellular processes notably cell polarity, migration and cytokinesis. These GTPases undergo a spatiotemporal regulation primarily controlled by cellular factors inducing both the exchange of GDP for GTP and the hydrolysis of GTP into GDP. Recent findings have unveiled another layer of complexity in the regulation of Rho proteins consisting in their ubiquitylation followed by their proteasomal degradation. Here, we describe how to assess the level of ubiquitylation of Rho proteins in cells, taking Rac1 as an example.

Key words: GTPases, Rac1, Cdc42, RhoA, Ubiquitin, Proteasome, CNF1, Smurf1, BACURD

1. Introduction

Small GTPases of the Rho family, notably Rac1, are master regulators of cell cytoskeleton and direct targets of numerous virulence factors of pathogenic bacteria (1–4).

Rho GTPases undergo a spatiotemporal regulation primarily controlled by three families of factors comprising guanine nucleotide exchange factors (GEFs), GTPase activating proteins (GAPs) and guanine nucleotide dissociation inhibitors (GDIs) (1, 2). GEF and GAP factors regulate the switch between the inactive GDP-bound form and the active GTP-bound form of Rho proteins at the membrane. This GTP-based core regulation of Rho proteins is modulated by diverse post-translational modifications, notably ubiquitylation (4).

Francisco Rivero (ed.), *Rho GTPases: Methods and Protocols*, Methods in Molecular Biology, vol. 827,
DOI 10.1007/978-1-61779-442-1_5, © Springer Science+Business Media, LLC 2012

Protein modification by ubiquitylation involves the covalent attachment of ubiquitin, an 8-kDa polypeptide, to lysine residues on the target (5). Conjugation of ubiquitin to cellular targets is achieved through a sequential cascade of transfer reactions between ubiquitin carrier proteins. Additional molecules of ubiquitin can be subsequently attached to one of the seven lysines of the previously conjugated ubiquitin molecule, producing several types of poly-ubiquitin chains. Modification of proteins by mono, multi (several mono-ubiquitylations), or by the assembly of poly-ubiquitin chains determines the fate of modified proteins by conferring them the capacity to bind to specific ubiquitin-binding proteins. The modification by assembly of a K48-polyubiquitin chain is a proteasome targeting signal for protein degradation.

The covalent attachment of ubiquitins to its targets can be assessed in cells expressing hexameric-histidine-tagged ubiquitin upon affinity purification in denaturing conditions. This method has enabled to establish that the activation of Rho proteins, notably Rac1, by the cytotoxic necrotizing factor-1 (CNF1) toxin of uropathogenic *Escherichia coli*, triggers their ubiquitin-mediated proteasomal degradation (6, 7). This regulation of Rho proteins is implicated in a growing number of signaling pathways notably in serotonin, TGF-beta, and c-MET signaling (4, 8). Ubiquitylation and degradation of Rac1 triggers epithelial cell disjunctions and motility (6, 9). Caveolin-1 controls Rac1 ubiquitylation and degradation by a mechanism that remains to be determined (10). The ubiquitylation and proteasomal degradation of activated RhoA is catalyzed by the E3 ubiquitin-ligase Smurf1 (11, 12). This regulation controls the epithelial-mesenchymal transition induced by TGF-beta, as well as CNF1 cytotoxicity (11–13). In addition, it has been recently described that the multimeric E3-ubiquitin ligase complex SCF-BACURD regulates cellular levels of RhoA by ubiquitin-mediated proteasomal degradation of GDP-bound RhoA to control cell movement (14). All these studies point to the importance of the regulation of Rho proteins by ubiquitylation in the control of tissue cohesion and cell motility.

2. Materials

2.1. Cells and Culture Medium

Hamster ovary epithelial cells (CHO) are obtained from ATCC (CCL-61). Cells are grown in "DSG" medium, composed of DMEM/F12 (Invitrogen) supplemented with 10% (v/v) fetal bovine serum (EU Approved Origin, Invitrogen) and 50 μg/mL gentamicin.

2.2. Cell Transfection

1. Electroporator permitting a pulse at 300 V and 450 μF.

2. 4-mm gap electroporation cuvettes.

3. Sterile 2-mL transfer pipets.

4. 15-mL and 50-mL sterile conical tubes.

5. 10-cm sterile cell culture dishes.

6. Hemocytometer or cell counter.

7. 0.05% Trypsin-EDTA (Invitrogen).

8. Plasmids: expression vectors for His-tagged ubiquitin, HA-tagged Rac1 wild type and Q61L mutant (vectors are described in ref. 6).

9. Cold DMEM/F12 medium.

10. DSG medium (see Subheading 2.1) prewarmed at 37°C.

11. Sterile phosphate buffer saline (PBS): 137 mM NaCl, 2.7 mM KCl, 10 mM Na_2HPO_4, 1.7 mM KH_2PO_4, pH 7.4 (also commercially available).

2.3. Metal Affinity Precipitation

1. BU buffer: 20 mM Tris–HCl pH 7.5, 200 mM NaCl, 10 mM imidazole, 0.1% Triton X-100, 8 M urea, 100 µM PR-619 (ubiquitin isopeptidase inhibitor, LifeSensors) (see Note 1). Prepare this buffer extemporaneously. Calculate sufficient amount to have 6 mL of buffer for each experimental condition. Dissolve the components in a water bath at 37°C (around 10–15 min). Store this buffer at room temperature (see Note 2).

2. Cobalt chelated resin (Talon metal affinity resin, Clontech) (see Note 3).

3. 2 mg/mL bovine serum albumin (BSA).

4. Rubber policeman.

5. 1.5-mL microtubes.

6. Rotating shaker.

2.4. Sodium Dodecyl Sulfate-Polyacrylamide Gel Electrophoresis

1. Resolving gel buffer: Dissolve 90.75 g of Tris in 300 mL of distilled water. Add 5 mL of 20% SDS. Mix and adjust the pH to 8.8 with HCl. Make up to 500 mL with distilled water. Store at 4°C.

2. 40% acrylamide/Bis solution (29:1), commercially available, e.g., QBiogene.

3. Stacking gel buffer: 76 mL of 1 M Tris–HCl, pH 6.8, 100 mL of 40% acrylamide/Bis solution, and 3 mL of 20% SDS. Make up to 600 mL with distilled water. Store at 4°C.

4. 10% (w/v) ammonium persulfate (APS).

5. N,N,N,N'-tetramethyl-ethylenediamine (TEMED).

6. 10× SDS-PAGE running buffer: 25 mM Tris–HCl pH 8.3, 190 mM glycine and 0.1% SDS in distilled water.

7. SDS lysis buffer (2× Laemmli buffer): 100 mM Tris–HCl pH 6.8, 0.2% bromophenol blue, 20% glycerol, 4% SDS and 1% β-mercaptoethanol in distilled water.

8. Prestained protein ladder (10–170 kDa) (Euromedex).

9. Vertical minigel electrophoresis system.

2.5. Immunoblotting

1. Hybond-P polyvinylidene fluoride (PVDF) membrane (Amersham).

2. Methanol.

3. 10× Western blot transfer buffer: 12.5 mM $NaHCO_3$, 3.75 mM Na_2CO_3. Adjust pH to 9.9.

4. 1× Western blot transfer buffer: 100 mL of 10× Western blot transfer buffer, 200 mL of ethanol. Make up to 1 L with distilled water. Store at 4°C.

5. Amidoblack stain: 0.1% (w/v) Amidoblack, 2% acetic acid and 10% ethanol in distilled water.

6. Destaining solution: 20% ethanol and 5% acetic acid in distilled water.

7. 10× Tris buffered saline (TBS): 1.5 M NaCl, 0.03 M KCl, 0.25 M Tris–HCl, pH 7.4.

8. TBST: 1× TBS (prepare from 10× TBS) containing 0.05% Tween-20.

9. Blocking solution: 5% (w/v) skimmed milk powder in TBST.

10. Diluent solution: 2% BSA in TBS.

11. Mini TransBlot Cell system (Biorad).

12. Whatman no. 3 filter paper.

13. Two foam pads per membrane.

14. Enhanced chemoluminescence (ECL) horseradish peroxidase (HRP) substrate (Immobilon Western, Millipore), composed of buffers A and B.

15. Signals can be detected with a light-sensitive camera, such as Fuji Las4000 (Fujifilm) or using X-ray films and a developing machine.

2.6. Antibodies and Conjugates

1. Primary antibodies: purified mouse monoclonal antibody to HA tag (HA 16B12, Covance) and purified mouse monoclonal antibody to penta-His (Qiagen) for the detection of recombinant 6× His tagged proteins.

2. Secondary antibody: polyclonal goat anti-mouse immunoglobulin HRP-conjugated (DAKO).

3. Methods

3.1. Cell Culture

Grow cells in 15-cm dishes filled with 20 mL of DSG medium until reaching 80% confluence in an incubator at 37°C, 5% CO_2.

3.2. Cell Transfection

1. Before performing the electroporation, prepare 15-mL conical tubes (one for each electroporation condition) filled with 5 mL of warm DSG medium. Place a sterile transfer pipet in each tube and aspirate about 1 mL of medium.

2. Cell detachment: aspirate the culture medium, rinse cells twice with 5 mL of PBS and incubate cells in 5 mL of 0.05% Trypsin-EDTA. Let cells detach for 5 min at 37°C. Add 5 mL of DSG medium (final volume: 10 mL), mix gently with the pipet to disrupt cell clusters. Transfer to a 15-mL conical tube. Count cells using a hemocytometer or cell counter.

3. For each condition of transfection (n), use 5×10^6 cells. Centrifuge the required total amount of cells at $1,000 \times g$ for 2 min at room temperature. Discard the supernatant. Add $n \times 300$ μL of cold DMEM to the cell pellet. Resuspend gently with a pipet (see Note 4).

4. Dispense up to 15 μg of each plasmid DNA at the bottom of a 4-mm gap electroporation cuvette and add 300 μL of cell suspension, mix gently (see Note 5). Place the cuvette in the electroporator and apply one pulse at 300 V and 450 μF (see Note 6).

5. Immediately resuspend the electroporated cells in DSG medium, using the transfer pipet containing 1 mL of medium prepared in step 1 and transfer the cell suspension to the corresponding 15-mL conical tube. Aspirate and flush three times to homogenously disperse the cells (see Note 7).

6. Plate each electroporated cell suspension in a 10-cm dish and place the dish in the incubator at 37°C, 5% CO_2. Let cells adhere (about 3 h), then replace the medium with 10 mL of fresh DSG medium (see Note 8).

7. Wait for plasmids expression for 24 h before performing the metal affinity precipitation.

3.3. Metal Affinity Precipitation

Carry out all procedures at room temperature.

3.3.1. Preparation of the Cobalt Resin

1. Resuspend very well the resin by vortexing. For each assay use 60 μL of resin slurry. Pipet the required total amount of resin in a microtube.

2. Wash the beads with 1 mL of BU buffer, vortex, centrifuge briefly in a microcentrifuge, and remove the supernatant.

3. Add 1 mL of BU and 50 µL of 2 mg/mL BSA solution (final concentration 100 µg/mL) (see Note 9). Place on a rotating shaker for 1 h.

4. Wash the beads twice with 1 mL of BU, as in step 2. Adjust the volume of beads to 100 µL of slurry per assay.

3.3.2. Processing of Samples for Metal Affinity Precipitation

1. Prepare 3 microtubes per assay. The first tube is for collecting the entire lysate, the second for the "total proteins" and the third for the "ubiquitin precipitation".

2. Preparation of cell lysate: Rinse cells with 5 mL of PBS (see Note 10). Add 1 mL of BU, scrap cells with a rubber policeman, transfer the cell suspension to the first series of microtubes, and centrifuge 10 min at $15,000 \times g$.

3. Transfer 50 µL of the supernatant to the second series of microtubes ("total proteins") and add 50 µL of 2× Laemmli buffer. Heat the samples for 5 min at 95°C and centrifuge briefly.

4. Place the remaining 950 µL of supernatant in the third series of microtubes ("ubiquitin precipitation") and add 100 µL of cobalt beads slurry (see Subheading 3.3.1). Let the affinity precipitation proceed for 1 h on a rotating shaker.

5. Washing of beads: Centrifuge samples briefly in a microcentrifuge, discard the supernatant. Wash the beads with 1 mL of BU buffer, vortex briefly. Repeat these washing steps four times.

6. After the last wash, resuspend the beads in 30 µL of 2× Laemmli buffer. Heat the samples for 5 min at 95°C and centrifuge briefly.

3.4. SDS-PAGE Sample Resolution

Process 30 µL of the "total proteins" sample by SDS-PAGE followed by immunoblotting with anti-HA antibody to verify the level of expression of HA-Rac mutants. Resolve the "ubiquitin precipitation" samples (30 µL) on the same SDS-PAGE gel to compare signal intensities, if possible. In addition, process 20 µL of "total proteins" for immunoblotting with anti-His antibody in order to verify expression of His-Ubiquitin (Fig. 1).

1. Prepare a 12% resolving gel by mixing 2 mL of distilled water, 5 mL of resolving gel buffer and 3 mL of acrylamide/Bis in a 50-mL conical tube. Add 150 µL of APS and 20 µL of TEMED. Cast gel within a 10×8 cm glass plate and 1.5 mm spacers (about 6.5 mL of gel). Allow space for the stacking gel and gently overlay with 70% ethanol. Let polymerize for about 20 min.

2. Prepare the stacking gel by mixing 3 mL of stacking gel buffer with 100 µL of APS and 20 µL of TEMED. Insert a 10-well

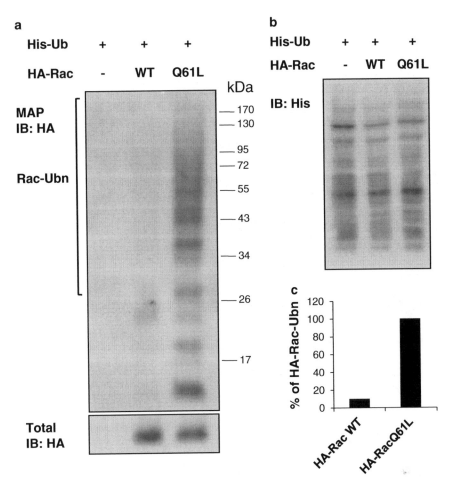

Fig. 1. Increased sensitivity of Rac-Q61L to ubiquitylation. (**a**) Ubiquitylation profiles show the increase of sensitivity of Rac-Q61L to ubiquitylation. Transfected CHO cells expressing His-tagged ubiquitin (His-Ub) together with wild-type Rac (HA-Rac WT) or the dominant-active form of Rac (HA-Rac-Q61L) were subjected to metal affinity purification (MAP) for ubiquitylated-Rac purification (Rac-Ubn: n for 1 to several ubiquitin molecules) and visualized by anti-HA immunoblotting (IB:HA). The HA-immunoblot in the *lower panel* shows levels of HA-Rac engaged in the assay. Note that Rac1 migrates at around 21 kDa and that ubiquitin is around 8 kDa. Bands under 26 kDa likely correspond to degraded forms of poly-ubiquitylated Rac, which are not detected in the absence of HA-Rac expression. (**b**) Immunoblot anti-His showing the level of His-Ub expression in the transfected cells (Total proteins). (**c**) Quantification of Rac ubiquitylation efficiencies corresponding to the ratio of poly-ubiquitylated Rac (Rac-Ubn) to total Rac, taking Rac-Q61L as 100%.

comb immediately without introducing air bubbles. Let polymerize for about 5 min.

3. Remove gently the comb and place the glass plate sandwich in the migration tank, filled with SDS-PAGE running buffer. Load 5 μL of prestained protein standard, 30 μL of "total protein" and 30 μL of "ubiquitin precipitation" sample.

4. Electrophoresis is carried at 140 V until the samples have entered the resolving gel and then continued at 200 V until the dye front reaches the bottom of the gel.

3.5. Protein Transfer

1. Immediately following SDS-PAGE, separate the gel plates and remove the stacking gel. Transfer carefully the gel to a container filled with western blot transfer buffer.

2. Cut two pieces of Whatman paper and one piece of PVDF membrane to the size of the gel. Immerse the PVDF membrane in methanol for 2 min and rinse in transfer buffer. Soak the pieces of Whatman paper in transfer buffer.

3. Assemble a "sandwich" with one layer of foam pad, one layer of Whatman paper, the gel, the PVDF membrane, one layer of Whatman paper, and one layer of foam pad.

4. Place the sandwich in the gel holder cassette and insert the cassette in the transfer tank with the side of the PVDF membrane facing the anode of the transfer apparatus. Fill the transfer tank with Western blot transfer buffer and transfer overnight at 4°C, 30 V.

3.6. Immunoblotting

1. Stain the membrane in Amidoblack for 5 min.

2. Destain the membrane with the destaining solution for 10 min.

3. Rinse the membrane in TBST.

4. Block the membrane with blocking solution for 2 h.

5. Add primary antibodies anti-HA (dilution 1/3,000) or anti-His (dilution 1/1,000) diluted in TBS-2% BSA and incubate for 2 h.

6. Wash membrane three times with TBST, 10 min each time.

7. Add secondary antibody diluted (1/3,000) in blocking solution and incubate for 1 h.

8. Wash as in step 6.

9. Incubate the membrane in ECL reagent following the manufacturer's instructions. Drain off excess reagent and wrap the membrane with plastic foil.

10. Signals can be detected with a light sensitive camera or upon exposure of the foil covered membrane to X-ray films.

4. Notes

1. Addition of 10 mM imidazole is important to prevent nonspecific protein binding to the resin. All samples after His precipitation can be stored at −20°C if not directly processed for SDS-PAGE. Warm the sample before loading to resuspend urea precipitates. The use of urea is preferred to that of guanidinium because addition of urea does not add salt, which strongly interferes with protein migration in SDS-PAGE.

2. Given that dissolution of urea is endothermic it is more rapid at 37°C. Also avoid incubating the buffer at 4°C otherwise urea will precipitate.

3. This type of resin is preferred to Nickel resin because it gives no background in these conditions.

4. It is important to avoid vortexing the cells.

5. To compare ubiquitylation efficiency between Rac mutants, or different conditions, it is crucial to adjust quantities of each DNA to obtain similar expression levels of HA-Rac constructs. Indeed, the levels of ubiquitylated-Rac recovered directly correlate with both levels of Rac expression and the degree of activation of Rac.

6. This should result in the formation of a white cluster of cells.

7. It is important to disrupt the white cluster gently for optimal cell separation.

8. It is recommended to remove dead cells for a better recovery of transfected cells. Approximately 70% of cells will survive to the electroporation.

9. BSA is used to saturate beads and prevent unspecific protein binding.

10. Aspirate correctly all the PBS to avoid BU buffer dilution.

Acknowledgments

This work was supported by an institutional funding from the INSERM, from the Agence Nationale de la Recherche (ANR-07-MIME-007 and ANR-07-BLAN-0046), and from the Association pour la Recherche sur le Cancer (ARC 3800).

References

1. Heasman SJ, Ridley AJ (2008) Mammalian Rho GTPases: new insights into their functions from *in vivo* studies. *Nat Rev Mol Cell Biol* 9, 690–701.

2. Jaffe AB, Hall A (2005) RHO GTPases: Biochemistry and Biology. *Annu Rev Cell Dev Biol* 21, 247–269.

3. Aktories, K., and Barbieri J.T. (2005) Bacterial cytotoxins: targeting eukaryotic switches. *Nat Rev Microbiol* 3, 397–410.

4. Visvikis, O., Maddugoda, M.P., and Lemichez, E. (2010) Direct modifications of Rho proteins: deconstructing GTPase regulation. *Biol Cell* 102, 377–389.

5. Kerscher, O., Felberbaum, R., and Hochstrasser, M. (2006) Modification of proteins by ubiquitin and ubiquitin-like proteins. *Annu Rev Cell Dev Biol* 22, 159–180.

6. Doye, A., Mettouchi, A., Bossis, G., Clément, R., Buisson-Touati, C., Flateau, G., Gagnoux, L., Piechaczyk, M., Boquet, P., and Lemichez, (2002) CNF1 exploits the ubiquitin-proteasome machinery to restrict Rho GTPase activation for bacterial host cell invasion. *Cell* 111, 553–564.

7. Lemonnier, M., Landraud, L., and Lemichez, E. (2007) Rho GTPase-activating bacterial toxins: from bacterial virulence regulation to eukaryotic cell biology. FEMS Microbiol Rev 31, 515–534.

8. Walther, D.J., Peter, J.U., Winter S, Höltje, M., Paulmann, N., Grohmann, M., Vowinckel, J., Alamo-Bethencourt, V., Wilhelm, C.S., Ahnert-Hilger, G., and Bader, M. (2003) Serotonylation of small GTPases is a signal transduction pathway that triggers platelet alpha-granule release. *Cell* 115, 851–862.

9. Lynch, E.A., Stall, J., Schmidt, G., Chavrier, P., and D'Souza-Schorey, C. (2006) Proteasome-mediated degradation of Rac1-GTP during epithelial cell scattering. *Mol Biol Cell* 17, 2236–2242.

10. Nethe, M., Anthony, E.C, Fernandez-Borja, M., Dee, R., Geerts, D., Hensbergen, P.J., Deelder, A.M., Schmidt, G., and Hordijk, P.L. (2010) Focal-adhesion targeting links caveolin-1 to a Rac1-degradation pathway. *J Cell Sci* 123, 1948–1958.

11. Ozdamar, B., Bose, R., Barrios-Rodiles, M., Wang, H.R., Zhang, Y., and Wrana, J.L. (2005) Regulation of the polarity protein Par6 by TGFbeta receptors controls epithelial cell plasticity. *Science* 307, 1603–1609.

12. Wang, H.R., Zhang, Y., Ozdamar, B., Ogunjimi, A.A., Alexandrova, E., Thomsen, G.H., and Wrana, J.L. (2003) Regulation of cell polarity and protrusion formation by targeting RhoA for degradation. *Science* 302, 1775–1779.

13. Boyer, L., Turchi, L., Desnues, B., Doye, A., Ponzio, G., Mege, J.L., Yamashita, M., Zhang, Y.E., Bertoglio, J., Flatau, G., Boquet, P., and Lemichez, E. (2006) CNF1-induced ubiquitylation and proteasome destruction of activated RhoA is impaired in Smurf1−/− cells. *Mol Biol Cell* 17, 2489–2497.

14. Chen, Y., Yang, Z., Meng, M., Zhao, Y., Dong, N., Yan, H., Liu, L., Ding, M., Peng, H.B., and Shao, F. (2009) Cullin mediates degradation of RhoA through evolutionarily conserved BTB adaptors to control actin cytoskeleton structure and cell movement. *Mol Cell* 35, 841–855.

Chapter 6

Posttranslational Lipid Modification of Rho Family Small GTPases

Natalia Mitin, Patrick J. Roberts, Emily J. Chenette, and Channing J. Der

Abstract

The Rho family comprises a major branch of the Ras superfamily of small GTPases. A majority of Rho GTPases are synthesized as inactive, cytosolic proteins. They then undergo posttranslational modification by isoprenoid or fatty acid lipids, and together with additional carboxyl-terminal sequences target Rho GTPases to specific membrane and subcellular compartments essential for function. We summarize the use of biochemical and cellular assays and pharmacologic inhibitors instrumental for the study of the role of posttranslational lipid modifications and processing in Rho GTPase biology.

Key words: CAAX motif, Farnesylation, Geranylgeranylation, Palmitoylation, Ras-converting enzyme 1, Isoprenylcysteine carboxyl methyltransferase

1. Introduction

Rho proteins are members of the Ras superfamily of small GTPases and function as GDP/GTP-regulated switches (1, 2). Rho GDP/GTP cycling is regulated by Rho-specific guanine nucleotide exchange factors (RhoGEFs) that promote the formation of the active GTP-bound form (3) and GTPase-activating proteins (RhoGAPs) that catalyze the intrinsic GTPase activity and promote the formation of inactive GDP-bound Rho (4). Active, GTP-bound Rho GTPases bind preferentially to downstream effectors, stimulating diverse cytoplasmic signaling cascades that control actin reorganization and regulate cell shape, polarity, motility, adhesion, and membrane trafficking (5).

The majority of Rho family GTPases undergoes a series of posttranslational modifications that promote proper subcellular

Francisco Rivero (ed.), *Rho GTPases: Methods and Protocols*, Methods in Molecular Biology, vol. 827,
DOI 10.1007/978-1-61779-442-1_6, © Springer Science+Business Media, LLC 2012

localization to the plasma membrane and/or endomembranes which is required for biological activity (Fig. 1). These series of modifications are initiated by the recognition of a carboxyl-terminal CAAX tetrapeptide motif (C = cysteine, A = aliphatic amino acid, and X = terminal amino acid; dictates prenyltransferase specificity), which is found on 16 of 20 Rho GTPases. Canonical CAAX motifs are not present in Wrch-1, Chp/Wrch-2, RhoBTB1, or RhoBTB2. Depending on the X residue of the CAAX sequence, either farnesyltransferase (FTase; e.g., Rnd3) or geranylgeranyltransferase type I (GGTase-I; e.g., RhoA) acts on the GTPase to covalently add a farnesyl or geranylgeranyl isoprenoid lipid, respectively, to the cysteine residue of the CAAX sequence (6). Then, the AAX peptide is cleaved from the carboxyl terminus by the Ras-converting enzyme 1 (Rce1) endoprotease (7, 8). Finally, isoprenylcysteine-O-carboxyl methyltransferase (Icmt) catalyzes the addition of a methyl group to the prenylated cysteine residue (9, 10). Together, these modifications increase protein hydrophobicity and facilitate membrane association. Mutation of the cysteine residue (C to S; SAAX mutant), deletion of the AAX residues, or pharmacologic inhibition of prenyltransferase activity prevents all three modifications and renders Rho GTPases inactive due to mislocalization to the cytosol (11). Additionally, CAAX-less GTPases Wrch-1 and Chp/Wrch-2 and some CAAX-terminating Rho GTPases have C-terminal sequences that are modified by palmitate addition (12–14).

Substrate specificity of FTase and GGTase-I toward small GTPases is determined primarily by the sequence of the *CAAX* tetrapeptide motif. Biochemical and structural studies of *CAAX* peptides in complex with FTase and GGTase-I have defined rules that govern substrate selectivity (6). In addition, even though Rho GTPases are known to undergo Rce1- and Icmt-dependent processing, the contribution of Rce1- and Icmt-catalyzed modifications to their cellular functions is beginning to be defined. Given the essential function of Rho family GTPases in normal cell physiology and their aberrant activation in oncogenesis, establishing the sensitivity of Rho GTPases to FTI and GGTI inhibitors and the contribution of Rce1- and Icmt-catalyzed modifications to their

---

Fig. 1. (continued) (**b**) Summary of posttranslational lipid modifications reported for Rho GTPase C-terminal membrane-targeting sequences. Whereas most GTPases are either F- or GG-modified, Rif is naturally prenylated by both FTase and GGTase-I and exists in independent F- and GG-modified pools. Whereas normally farnesyl-modified, RhoH and RhoB become alternatively prenylated (GG-modified) when FTase activity is inhibited by the FTI treatment. Wrch-1 terminates with a CAAX-like motif but is not prenylated, and instead is modified by palmitoylation. Wrch-2/Chp lacks a CAAX motif, as it terminates with a CXX sequence, and is modified by palmitoylation. At least two additional carboxyl-terminal sequence elements are required for Chp subcellular membrane localization, polybasic amino acids and an invariant tryptophan residue. RhoBTB1/2 proteins lack any known carboxyl-terminal lipid modifications, and instead terminate with tandem Bric-a-brac, Tramtrack, Broad-complex (BTB) domains.

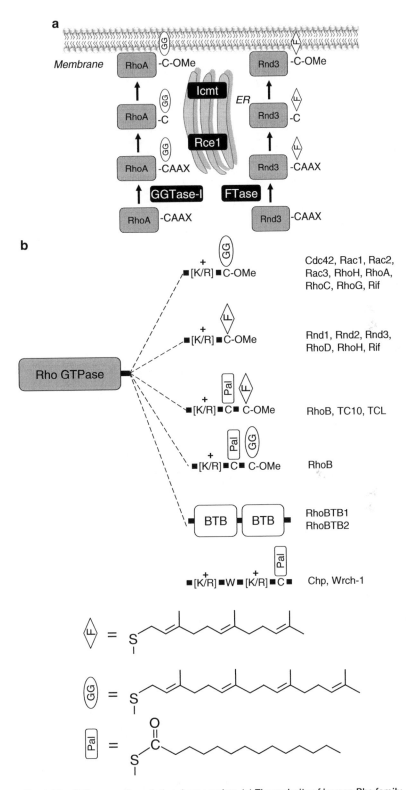

Fig. 1. Rho GTPase posttranslational processing. (**a**) The majority of human Rho family GTPases terminate with CAAX motifs that signal for modification by farnesyl (F) and/or geranylgeranyl (GG) isoprenoids, AAX proteolysis, and carboxylmethylation (OMe).

cellular functions is critical to the successful development of inhibitors of C*AAX*-signaled modifications. Therefore, here, we describe pharmacologic and genetic approaches that can be used to characterize the role of C*AAX*-signaled modifications on the functional and biological properties of Rho GTPases.

2. Materials

2.1. Expression Constructs

pEGFP mammalian expression vectors that encode full-length human Rho GTPase (RhoA, RhoB, RhoC, Rnd1, Rnd2, Rnd3, RhoH, TC10, TCL, Rif, RhoD, and placental and brain isoforms of Cdc42) fusion proteins with N-terminal green fluorescent protein (GFP) tags were constructed and characterized as described previously (14, 15). All constructs were sequence verified, with cloning details available upon request, and have been deposited with Addgene (http://www.addgene.org).

2.2. Cell Lines

1. HEK 293T cells are obtained from ATCC and maintained in Dulbecco's modified minimum essential medium supplemented with 10% fetal calf serum (Sigma), 100 units/mL penicillin, and 100 μg/mL streptomycin (Gibco).

2. NIH 3T3 mouse fibroblasts (obtained originally from Geoffrey M. Cooper, Boston University) are maintained in Dulbecco's modified minimum essential medium supplemented with 10% calf serum (Sigma), 100 units/mL penicillin, and 100 μg/mL streptomycin.

3. Spontaneously immortalized mouse embryonic fibroblasts (MEFs) were originally prepared from $Icmt^{-/-}$ and $Rce1^{-/-}$ mouse embryos, along with control fibroblasts ($Icmt^{+/+}$ and $Rce1^{+/+}$) from littermate embryos (16) and were kindly provided by Stephen G. Young (University of California, Los Angeles). MEF cultures are maintained in Dulbecco's modified minimum essential medium supplemented with 15% calf serum (Colorado Serum, Denver, CO), nonessential amino acids, and L-glutamine (GIBCO) (see Note 1).

2.3. Chemicals, Antibodies, and Pharmacological Inhibitors

1. FTase inhibitor (FTI-2153; provided by Saïd Sebti, Moffitt Cancer Center and Andrew Hamilton, Yale) (17): 10 mM solution in dimethyl sulfoxide (DMSO). A similar FTase-selective inhibitor, FTI-277, is available from commercial sources (18) (Sigma, EMD4Biosciences, Tocris Bioscience).

2. GGTase-I inhibitor (GGTI-2417; provided by Saïd Sebti, Moffitt Cancer Center and Andrew Hamilton, Yale) (19): 8.5 mM solution in DMSO. A similar GGTase-I-selective inhibitor, GGTI-298, is available from commercial sources (18) (Sigma, EMD4Biosciences, Tocris Bioscience).

3. Phosphate-buffered saline (PBS): 137 mM NaCl, 2.7 mM KCl, 4.3 mM Na_2HPO_4, 1.47 mM KH_2PO_4, pH 7.4.

4. 2-Bromopalmitate (2-BP) (Sigma) (13): Prepare a 100 mM solution in DMSO.

5. MatTek tissue culture dishes: 35-mm culture dishes that have a 14-mm cutout at the bottom sealed with a No 1.5 glass coverslip (MatTek, Ashland, MA). Available uncoated and poly-D-lysine coated.

6. Lipofectamine Plus reagent is used for transient transfection of mammalian cells according to manufacturers' instructions.

7. Inverted laser-scanning confocal microscope and temperature-controlled stage. We use a Zeiss LSM 510 microscope with an oil immersion ×63 numerical aperture (NA) 1.4 objective.

8. HEPES-buffered saline (HBS; 1×): 140 mM NaCl, 50 mM HEPES, 1.5 mM Na_2HPO_4. Adjust to pH 7.1 with 1 N NaOH, filter sterilize, and store at room temperature.

9. Calcium chloride (1.25 M): Dissolve $CaCl_2$ in distilled water and filter sterilize. Store at room temperature.

10. N-ethylmaleimide (NEM): 200 mM of NEM in ultrapure water. Prepare immediately before use to prevent hydrolysis of the maleimide group (see Note 2).

11. Anti-GFP mouse monoclonal antibody (clone JL-8; Clontech): 1 mg/mL.

12. Protein G-Agarose (Santa Cruz Biotechnology).

13. Hydroxylamine•HCl: 1 M solution in ultrapure distilled water. After dissolving, adjust pH to 7.4. Prepare immediately before use.

14. 1-biotinamido-4-[4′-(maleimidomethyl)cyclohexanecarbox-amido] butane (biotin–BMCC): 20 mM solution in DMSO (Pierce). To make 1 μM biotin–BMCC working stock, dilute in 0.5 M Tris–HCl, pH 7.0.

15. Biotin–BMCC lysis buffer: 150 mM NaCl, 5 mM EDTA, 50 mM Tris–HCl, pH 7.4, 0.02% sodium azide, 2% Triton X-100, protease inhibitor cocktail (Roche).

16. Polyvinylidene difluoride (PVDF) membrane (Millipore).

17. 2× sample loading buffer: 100 mM Tris–HCl, pH 6.8, 10% glycerol, 4% SDS, 72 mM 2-mercaptoethanol, 0.2% bromophenol blue.

18. TBS-Tween: 20 mM Tris–HCl, pH 7.4, 500 mM NaCl, 0.05% Tween 20.

19. Protein gel electrophoresis and transfer chamber. We use a Mini-PROTEAN system (BioRad).

20. Streptavidin horseradish peroxidase (Pierce). Dissolve 1 mg of powder in 400 μL of distilled water. Store aliquots at –20°C.

21. Enhanced Chemiluminescent (ECL) Western Blotting Substrate (Pierce).

3. Methods

3.1. Transfection and Microscopy

Although localization of GFP-tagged Rho GTPases can be observed in transiently transfected, fixed cells, studying the localization of these proteins in live cells offers tremendous advantages. These advantages include increased brightness, higher resolution, and the ability to observe filopodia formation, membrane trafficking, and organelle movement in real time. In addition, localization of small GTPases in fixed cells can be artifactual (20). Therefore, we recommend performing live-cell imaging analysis whenever possible.

1. Cells are plated, transfected, and visualized in the same MatTek dish. Uncoated dishes are used for NIH 3T3 cells, and poly-D-lysine-coated dishes are used for immortalized MEFs (see Note 1). Cells are plated at 1×10^5 cells per 35-mm MatTek dish and transfected with 0.5 μg of GFP-Rho GTPase construct using Lipofectamine Plus reagent according to the manufacturer's instructions.

2. Three hours after transfection, cells are washed with PBS and refed with fresh growth medium. In the inhibitor assays, transfected cells are washed and refed with growth medium supplemented with 10 μM FTI-2153, 10 μM GGTI-2417, or 100 μM 2-BP. Control cultures are treated with ethanol or DMSO (vehicle control).

3. Cells are imaged 18–22 h after transfection and only cells expressing low to medium levels of GFP-tagged constructs are chosen for the analysis (see Note 3).

4. For high-resolution images, cells are examined with an inverted laser-scanning confocal microscope using an oil immersion ×63 numerical aperture 1.4 objective. Images are captured by scanning with the 488-nm spectral line of an argon-ion laser using the LP 505 emission filter. Rho GTPases reside on both plasma membrane and endomembrane compartments (Fig. 2); therefore, we collect 0.4-μm confocal z-sections that show both nuclear and membrane/cytosolic localization of GFP fusion proteins to ensure comprehensiveness of the analysis. The GFP peptide alone, in the absence of other targeting sequences, is nuclear localized.

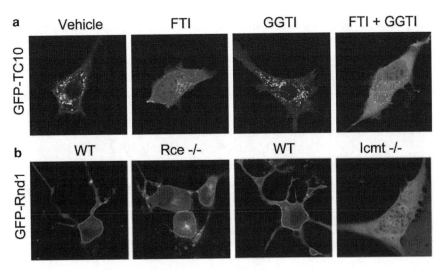

Fig. 2. Analyses of CAAX motif-signaled posttranslational modifications. (**a**) TC10 subcellular localization is dependent on farnesylation. NIH 3T3 cells transfected with an expression construct for GFP-tagged human TC10 were treated with FTI-2153, GGTI-2417, both, or DMSO (vehicle). Treatment with FTI-2153 and not GGTI-2417 mislocalized GFP-TC10 to the cytoplasm. (**b**) Differential requirements for Rce1- and Icmt-mediated processing in the subcellular localization and function of Rho family proteins. Wild-type (WT), *Rce1*⁻/⁻, and *Icmt*⁻/⁻ MEFs were transiently transfected with expression constructs for GFP-tagged Rnd1. GFP-Rnd1 localization to plasma membrane was completely disrupted in *Icmt*⁻/⁻ and only slightly perturbed in *Rce1*⁻/⁻ cells, suggesting that methylation of the terminal cysteine is essential for Rnd1 localization to the plasma membrane.

3.2. Biotin–BMCC Labeling

We use biotin–BMCC labeling (21) for direct analyses of palmitoylation of Rho GTPases expressed in mammalian cells (13). It is important to note that this assay does not detect palmitate groups added to serine and threonine residues by means of oxyester linkages because these bonds are not hydrolyzed by hydroxylamine. This assay involves blocking all available free cysteines in a protein with NEM (see Note 2), which alkylates thiol groups. Cysteines with bound palmitate are not sensitive to NEM-induced alkylation; hence, after the thioester bond holding palmitate to a cysteine is cleaved with hydroxylamine, only the free thiol groups at the sites of palmitate linkage are recognized and bound by the biotin–BMCC compound. This method, therefore, provides a sensitive and efficient way of detecting palmitoylation of cysteines, although a definitive demonstration of protein modification by palmitoylation requires analysis by mass spectroscopy.

1. Seed 10^6 293T cells in 100-mm dishes and transfect the next day with 7 μg of GFP-tagged Rho GTPase construct using a calcium phosphate transfection technique. For this, dilute plasmid DNA in 900 μL of 1× HBS, then add 100 μL of 1.25 M CaCl₂, and mix by vortexing gently. After a 5-min incubation, add the DNA–calcium solution directly to the cells.

2. Twenty-four hours after transfection, feed cells with fresh growth medium. Forty-eight hours after transfection, lyse cells in 700 µL of biotin–BMCC lysis buffer, spin down and incubate supernatant with 5 µg of anti-GFP monoclonal antibody at 4°C for 1–2 h, then add 20 µL of protein G-agarose, and incubate at 4°C for 1 h.

3. Wash bound protein twice with lysis buffer and incubate with 1 mL of lysis buffer containing 50 mM NEM for 48 h at 4°C. Change NEM solution once during incubation (see Note 2).

4. Wash bound proteins once with cold lysis buffer and resuspend in 1 M hydroxylamine, pH 7.4, to cleave thioester bonds for 1 h at 25°C with gentle rocking.

5. Wash again with cold lysis buffer and resuspend in 1 µM biotin–BMCC. Gently rock the beads for 2 h at 25°C.

6. Wash bound protein two or three times with lysis buffer, resuspend in 50 µL of 2× sample loading buffer, and resolve by 12% SDS-PAGE using standard procedures.

7. Transfer proteins to PVDF membrane using standard procedures. Detect biotin-labeled protein by incubation with streptavidin horseradish peroxidase (1:3,000) (diluted in TBS-Tween) for 2 h at room temperature, wash three times with TBS-Tween, and develop using ECL reagents and exposing to X-ray film.

8. To verify the presence of each GFP-tagged protein in whole cell lysates, resolve 20 µg of lysate by SDS-PAGE and detect by Western blotting with anti-GFP antibody.

4. Notes

1. We have noticed that later passages of MEFs isolated from $Icmt^{-/-}$ mice have decreased proliferation rate. This property makes it difficult to synchronize them with the $Icmt^{+/+}$ control MEFs. However, we noticed that plating cells at half the density 48 h before the transfection (instead of 24 h) allows $Icmt^{-/-}$ to recover and produce cell densities comparable to the control.

2. NEM is a labile compound, so it is especially important to make fresh NEM solution for each experiment and to keep all solutions containing NEM cold and protected from light. We also find that ordering fresh NEM compound every 3–4 months helps keep background low.

3. Using Lipofectamine Plus transfection method, we routinely obtain 50–70% transfection efficiencies for NIH 3T3 cells and about 30% efficiencies for the MEFs.

Acknowledgments

This work was supported, in whole or in part, by National Institutes of Health Grants CA063071, CA67771, and CA92240 to C.J.D.

References

1. Colicelli, J. (2004) Human RAS superfamily proteins and related GTPases. *Sci STKE* 2004(250), RE13.

2. Wennerberg, K., Rossman, K.L., and Der, C.J. (2005) The Ras superfamily at a glance. *J Cell Sci* 118, 843–846.

3. Rossman, K.L., Der, C.J., and Sondek, J. (2005) GEF means go: turning on RHO GTPases with guanine nucleotide-exchange factors. *Nat Rev Mol Cell Biol* 6, 167–180.

4. Lamarche, N., and Hall, A. (1994) GAPs for rho-related GTPases. *Trends Genet* 10, 436–440.

5. Etienne-Manneville, S., and Hall, A. (2002) Rho GTPases in cell biology. *Nature* 420, 629–635.

6. Reid, T.S., Terry, K.L., Casey, P.J., and Beese, L.S. (2004) Crystallographic analysis of CaaX prenyltransferases complexed with substrates defines rules of protein substrate selectivity. *J Mol Biol* 343, 417–433.

7. Ashby, M.N. (1998) CaaX converting enzymes. *Curr Opin Lipidol* 9, 99–102.

8. Boyartchuk, V.L., Ashby, M.N., and Rine, J. (1997) Modulation of Ras and a-factor function by carboxyl-terminal proteolysis. *Science* 275, 1796–1800.

9. Hrycyna, C.A., Sapperstein, S.K., Clarke, S., and Michaelis, S. (1991) The *Saccharomyces cerevisiae* STE14 gene encodes a methyltransferase that mediates C-terminal methylation of a-factor and RAS proteins. *EMBO J* 10, 1699–1709.

10. Sebti, S.M., and Der, C.J. (2003) Opinion: Searching for the elusive targets of farnesyltransferase inhibitors. *Nat Rev Cancer* 3, 945–951.

11. Winter-Vann, A.M., and Casey, P.J. (2005) Post-prenylation-processing enzymes as new targets in oncogenesis. *Nat Rev Cancer* 5, 405–412.

12. Michaelson, D., Silletti, J., Murphy, G., Eustachio, P., Rush, and M., Philips, M.R. (2001) Differential localization of Rho GTPases in live cells: regulation by hypervariable regions and RhoGDI binding. *J Cell Biol* 152, 111–126.

13. Chenette, E.J., Abo, A., and Der, C.J. (2005) Critical and distinct roles of amino- and carboxyl-terminal sequences in regulation of the biological activity of the Chp atypical Rho GTPase. *J Biol Chem* 280, 13784–13792.

14. Berzat, A.C., Buss, J.E., Chenette, E.J., Weinbaum, C.A., Shutes, A., Der, C.J., Minden, A., Cox, A.D. (2005) Transforming activity of the Rho family GTPase, Wrch-1, a Wnt-regulated Cdc42 homolog, is dependent on a novel carboxyl-terminal palmitoylation motif. *J Biol Chem* 280, 33055–33065.

15. Roberts, P.J., Mitin, N., Keller, P.J., Chenette, E.J., Madigan, J.P., Currin, R.O., Cox, A.D., Wilson, O., Kirschmeier, P., and Der, C.J. (2008) Rho Family GTPase modification and dependence on CAAX motif-signaled posttranslational modification. *J Biol Chem* 283, 25150–25163.

16. Bergo, M.O., Leung, G.K., Ambroziak, P., Otto, J.C., Casey, P.J., Gomes, A.Q., Seabra, M.C., and Young, S.G. (2001) Isoprenylcysteine carboxyl methyltransferase deficiency in mice. *J Biol Chem* 276, 5841–5845.

17. Sun, J., Blaskovich, M.A., Knowles, D., Qian, Y., Ohkanda, J., Bailey, R.D., Hamilton, A.D., and Sebti, S.M. (1999) Antitumor efficacy of a novel class of non-thiol-containing peptidomimetic inhibitors of farnesyltransferase and geranylgeranyltransferase I: combination therapy with the cytotoxic agents cisplatin, Taxol, and gemcitabine. *Cancer Res* 59, 4919–4926.

18. McGuire, T.F., Qian, Y., Vogt, A., Hamilton, A.D., and Sebti, S.M. (1996) Platelet-derived growth factor receptor tyrosine phosphorylation requires protein geranylgeranylation but not farnesylation. *J Biol Chem* 271, 27402–27407.

19. Falsetti, S.C., Wang, D.A., Peng, H., Carrico, D., Cox, A.D., Der, C.J., Hamilton, A.D., and Sebti,.S.M. (2007) Geranylgeranyltransferase I inhibitors target RalB to inhibit anchorage-dependent growth and induce apoptosis and RalA to inhibit anchorage-independent growth. *Mol Cell Biol* 27, 8003–8014.

20. Bivona, T.G., Wiener, H.H., Ahearn, I.M., Silletti, J., Chiu, V.K., and Philips, M.R. (2004) Rap1 up-regulation and activation on plasma membrane regulates T cell adhesion. *J Cell Biol* 164, 461–470.

21. Drisdel, R.C., and Green, W.N. (2004) Labeling and quantifying sites of protein palmitoylation. *Biotechniques* 36, 276–285.

Chapter 7

Analysis of the Role of RhoGDI1 and Isoprenylation in the Degradation of RhoGTPases

Etienne Boulter and Rafael Garcia-Mata

Abstract

RhoGDI1 is one of the three major regulators of the Rho switch along with RhoGEFs and RhoGAPs. RhoGDI1 extracts prenylated Rho proteins from lipid membranes, sequesters them in the cytosol, and prevents nucleotide exchange or hydrolysis. In addition, RhoGDI1 protects prenylated Rho proteins from degradation. Here, we describe techniques to monitor Rho proteins degradation upon depletion of RhoGDI1 and their dependence upon prenylation for degradation.

Key words: RhoA, RhoGDI, Prenylation, YopT, Degradation

1. Introduction

Rho proteins act as molecular switches by cycling between an active (GTP bound) and an inactive (GDP bound) state. The activation of Rho proteins is mediated by specific guanine nucleotide exchange factors (GEFs), which catalyze the exchange of GDP for GTP. In their active state, GTPases interact with one of several downstream effectors to modulate their activity and localization. The signal is terminated by hydrolysis of GTP to GDP, a reaction that is stimulated by GTPase-activating proteins (GAPs) (1). An additional layer of regulation for the RhoGTPases is mediated by the Rho guanine nucleotide dissociation inhibitor (GDI) family of proteins. RhoGDIs were initially characterized based on their ability to inhibit the dissociation of the bound nucleotide (usually, GDP) from the RhoGTPases (2–4). It was subsequently shown that RhoGDI can also interact with the GTP-bound form of the RhoGTPases preventing both intrinsic and GAP-mediated GTP hydrolysis, as well as effector interaction (5–7). One of the main functions of RhoGDIs

Francisco Rivero (ed.), *Rho GTPases: Methods and Protocols*, Methods in Molecular Biology, vol. 827,
DOI 10.1007/978-1-61779-442-1_7, © Springer Science+Business Media, LLC 2012

is to modulate the cycling of RhoGTPases between the cytosol and cellular membranes (8). RhoGDIs form a high-affinity interaction with RhoGTPases, promoting their extraction from membranes and sequestering them in an inactive state in the cytosol.

RhoGDIs' interaction with Rho proteins requires the GTPases to be prenylated. RhoGTPases contain a conserved CAAX motif at their carboxyl terminus, which is posttranslationally modified by isoprenylation at the cysteine residue (9). In the case of RhoA, Rac1, and Cdc42, a 20-carbon geranylgeranyl group is added. Upon release from RhoGDI, this isoprenyl group can be inserted into the lipid bilayer and anchors the Rho proteins to cellular membranes. Membrane association is essential for the function of RhoGTPases. When the Rho proteins are extracted from the membrane, the lipid moiety is transferred into a hydrophobic pocket in the RhoGDI molecule that prevents its exposure to the solvent (7). Structural studies showed that the interaction between RhoGTPases and RhoGDI requires the association of the N-terminal domain of RhoGDI with the switch region of the GTPase, which inhibits nucleotide release, and the insertion of the prenyl group of the GTPase into the hydrophobic C-terminal pocket of the GDI (10).

There are three RhoGDIs in the human genome: RhoGDI1 (GDIα), RhoGDI2 (Ly/D4GDI or GDIβ), and RhoGDI3 (GDIγ) (8). RhoGDI1 is ubiquitously expressed while GDI2 and 3 expression is restricted to certain tissues (hematopoietic for GDI2 and lung, brain, testis for GDI3) (11–14). RhoGDI1 has orthologues in *Saccharomyces cerevisiae* (RDI1), *Caenorhabditis elegans* (rhi-1), and *Drosophila melanogaster* (RhoGDI)(15).

Recently, our understanding of RhoGDI's function has progressed as we showed that it protects prenylated Rho proteins from misfolding and degradation (16). Prenylation is a major feature of Rho proteins which generates a biological dilemma: isoprenylation of Rho proteins is absolutely required for their proper subcellular localization and signaling (17), but simultaneously, this lipid moiety disturbs Rho protein folding in solution and triggers their degradation (16). Prenylation of RhoGTPases can be inhibited using general inhibitors of the cholesterol/isoprenoid biosynthetic pathway or specific inhibitors for the geranyl-geranyl-transferase enzyme (GGTaseI). 3-hydroxy-3-methyl-glutaryl (HMG)-CoA reductase inhibitors, such as statins, inhibit the conversion of HMG-CoA to mevalonate in the biosynthesis pathway for both isoprenoids and cholesterol (18). Alternatively, the isoprenoid group can be cleaved using YopT, a cystein protease expressed by pathogenic species of *Yersinia*. YopT cleaves N-terminal to the prenylated cysteine in RhoA, Rac, and Cdc42. This cleavage results in the irreversible removal of the lipid modification from the GTPases and their subsequent membrane detachment (19). YopT cleaves GTP- and GDP-bound forms of RhoA equally, suggesting that the cleavage does not depend upon the conformation status of the GTPases (19).

Here, we describe techniques to monitor the degradation of Rho proteins upon depletion of RhoGDI1 and to characterize the role of prenylation in the RhoGDI-mediated degradation of Rho-GTPases.

2. Materials

Prepare all solutions using deionized water unless specified otherwise. Similarly, prepare and store all reagents at room temperature unless indicated otherwise.

2.1. SiRNA and RNA/DNA Ttransfection

1. We routinely order siRNAs as desalted dsRNA oligos from Sigma-Genosys. No further purification seems to be required for efficient silencing. SiRNAs are resuspended in ultrapure sterile water at a stock concentration of 20 μM.

2. We design siRNAs using the neural network technology-based designer BIOPREDsi at http://www.biopredsi.org (20) which is now part of Qiagen.

3. The sequence of the siRNA against human RhoGDI1 is 5′-UCAAUCUUGACGCCUUUCCTT-3′ and the mismatch containing control siRNA is 5′-UCACUCGUGCCGCAUUU CCTT-3′ (16).

4. 2× HBSP buffer: 280 mM NaCl, 10 mM KCl, 1.5 mM Na_2HPO_4, 50 mM Hepes, pH 7.05, 12 mM glucose. Filter on a 0.45-μm pore-size filter. Store at room temperature.

5. 2.5 M $CaCl_2$. Filter on a 0.45-μm pore-size filter.

2.2. Cell Culture

Growth medium is Dulbecco's Modified Eagle Medium (DMEM, Invitrogen) containing 10% fetal bovine serum and antibiotics: penicillin 100 U/mL and streptomycin 100 μg/mL (Invitrogen).

2.3. Lysis and Western Blotting

1. 10× PBS: Dissolve 80 g of NaCl, 2 g of KCl, 14.4 g of NaH_2PO_4 $2H_2O$, and 2 g of KH_2PO_4 in 900 mL of water. Complete to 1 L with water. Filter on 0.8-μm pore-size membrane. Store at room temperature.

2. Laemmli lysis buffer: 100 mM Tris–HCl, pH 6.8, 2.5% SDS, 10% glycerol. Beta-mercaptoethanol (5% final concentration) and bromophenol blue (0.0025% final concentration) are added after cell lysis.

3. Sonicator: Branson Sonifier S150D.

4. 2× resolving gel buffer: 750 mM Tris–HCl, pH 8.8, 0.2% SDS. Dissolve 90.8 g of Tris in 800 mL of water. Adjust pH to 8.8. Complete to 1 L with water. Add 2 g of SDS. Filter on a 0.8-μm pore-size membrane. Store at room temperature.

5. 2× stacking gel buffer: 250 mM Tris–HCl, pH 6.8, 0.2% SDS. Dissolve 30.3 g of Tris in 800 mL of water. Adjust pH to 6.8. Complete with water to 1 L. Add 2 g of SDS. Filter on a 0.8-μm pore-size membrane. Store at room temperature.

6. Stacking gel: 125 mM Tris–HCl, pH 6.8, 0.1% SDS, 4% acrylamide–bisacrylamide (29:1) (Biorad). Mix 50 mL of 2× stacking gel buffer, 20 mL of 30% acrylamide–bisacrylamide (29:1) solution, and 30 mL of water. Store at 4°C.

7. Ammonium persulfate (APS): 10% solution (w/v) in water.

8. N,N,N,N'-tetramethyl-ethyldiamine (TEMED). Store at 4°C.

9. 10× SDS-PAGE running buffer: 250 mM Tris–HCL, pH 8.5, 2.2 M glycine, 1% SDS. Dissolve 30.5 g of Tris and 164 g of glycine in 800 mL of water. Check that pH is between 8.3 and 8.8. Dissolve 10 g of SDS. Complete to 1 L with water. Store at room temperature.

10. 10× transfer buffer: 250 mM Tris, pH 8.5, 1.92 M glycine. Dissolve 30 g of Tris and 144 g of glycine in 800 mL of water. Complete to 1 L with water. Store at room temperature.

11. Polyvinylidene fluoride (PVDF) membrane.

12. Ponceau red solution: 0.5% Ponceau red, 1% acetic acid in water.

13. Blocking buffer: PBS supplemented with 5% milk powder.

14. Antibodies: Anti-RhoA monoclonal antibody 26C4 (Santa Cruz Biotechnology). Anti-RhoGDI1 polyclonal antibody A20 (Santa Cruz Biotechnology). Mouse monoclonal anti-HA antibody (clone 16B12) (Covance). Secondary horseradish peroxidase (HRP)-coupled anti-mouse or anti-rabbit antibodies (Jackson Immunoresearch).

15. Immobilon Western chemiluminescent HRP substrate (Millipore).

16. Protein gel electrophoresis and transfer equipment: Mini-PROTEAN tetracell (Biorad).

2.4. Prenylation Inhibition

1. Lovastatin (Axxora): Resuspend in ethanol at a concentration of 2.5 mM.

2. HA-tagged YopT: pPTuner IRES2 HA-YopT was engineered in our laboratory. (16) (avaliable from the authors upon request).

3. Methods

3.1. Transfection and Lovastatin Treatment

1. Plate HeLa cells at a density of 50% for RNA/DNA transfection or 80% for pharmacological inhibitor treatment in a 100-mm culture dish with 5 mL of growth medium. Perform transfection the next morning.

2. Mix 12.5 µL of siRNA (20 µM stock) with 225 µL of sterile ultrapure water. Add 25 µL of 2.5 M CaCl$_2$ (see Note 1). Vortex and add 250 µL of 2× HBSP buffer starting from the bottom of the tube (see Note 2). Wait for 30–60 s and dispense the mix on the cells. This should give a final concentration of siRNA of around 50 nM.

3. Approximately 8 h after transfection, aspirate the medium, wash once with PBS, and add 5 mL of growth medium (see Note 3).

4. Perform transfection again on the next day according to the same protocol and then wait for 24–48 h before assessing RhoGDI1 silencing.

5. Alternatively, it is possible to cotransfect YopT cDNA by adding 2 µg of HA-YopT plasmid to the transfection mix either during the first or second transfection. Single transfection of YopT can also be achieved by mixing 2 µg of HA-YopT cDNA with 225 µL of sterile ultrapure water and 25 µL of 2.5 M CaCl$_2$. Like previously (step 2), add 250 µL of 2× HBSP buffer and add to the cells for 8 h.

6. Alternatively, cells can be treated with lovastatin, which inhibits HMG-CoA reductase resulting in impaired prenylation. Add 10 µL of 2.5 mM lovastatin to 10 mL of growth medium (final concentration of lovastatin is 2.5 µM) for 24–48 h maximum (see Note 4).

3.2. Cell Lysis

1. Discard growth medium and wash cells with ice-cold PBS (see Note 5).

2. For cell lysis, add 1 mL of Laemmli buffer to a 100-mm dish (see Note 6). Incubate Laemmli buffer on cells for 5 min at room temperature (see Note 7) and then harvest cell lysate.

3. Sonicate cell lysate for 5–10 s (see Note 8). Add β-mercaptoethanol to a final concentration of 5% and bromophenol blue to a final concentration of 0.0025%.

3.3. SDS-PAGE and Western Blotting

1. Cast a 15% mini-gel by mixing 5 mL of 2× resolving gel buffer with 5 mL of acrylamide–bisacrylamide solution. Add 100 µL of 10% APS solution and 10 µL of TEMED. Mix by inverting the tube. Dispense the gel immediately between the glass plates (see Note 9), leaving 1–1.5 cm of empty space on top of the gel and carefully layer it with 100–200 µL of water (see Note 10). The gel usually polymerizes within 5–10 min.

2. When the resolving gel is polymerized, remove the top layer of water. Add 30 µL of 10% APS and 3 µL of TEMED to 3 mL of stacking gel solution. Dispense on top of the resolving gel and insert the comb.

3. Remove the comb and rinse the wells with distilled water to remove any excess of nonpolymerized acrylamide. Assemble

the gel electrophoresis unit and add SDS-PAGE running buffer on top of the gel and in the wells.

4. Heat the samples at 95°C for 5 min. Load the samples (15–25 μlL/lane) according to your loading scheme and run the gel at 140 V for approximately 1 h 40 min (until the blue dye exits the gel).

5. Soak the PVDF membrane in ethanol 95% to rehydrate it. Equilibrate the PVDF membrane and the pieces of Whatman paper in 1× transfer buffer with 20% ethanol. Disassemble the gel electrophoresis unit and the glass plates to recover the gel. Assemble the transfer sandwich as follows: Starting from the cathode-facing side of the sandwich, place two pieces of Whatman paper, one piece of PVDF membrane, the gel, and two pieces of Whatman paper (see Note 11). Transfer the gel in 1× transfer buffer with 20% ethanol 95% at 100 mA for 2 h. Stain the membrane to check transfer efficiency with Ponceau red solution (optional).

6. Block the membrane in blocking buffer for 10 min and incubate overnight at 4°C with the appropriate antibodies. We use the 26C4 antibody at a dilution of 1/1,000, anti-GDI1 antibody at 1/10,000, and anti-HA antibody at 1/5,000 in blocking buffer.

7. Wash the membrane three times in PBS for approximately 10 min each. Incubate with the secondary antibody at the dilution (in blocking buffer) recommended by the manufacturer for 1 h (see Note 12).

8. Wash the membrane three times in PBS for approximately 10 min each, and develop using the Immobilon Western chemiluminescent HRP substrate following the manufacturer's instructions (see Note 13). Expose the membrane to an X-ray film.

A typical result from such an experiment is shown in Fig. 1. By western blot, the RhoA band is observed around 21 kDa while RhoGDI1 is slightly higher at around 28–30 kDa. HA-YopT can be seen much higher on the membrane. RhoGDI1 silencing triggers degradation of RhoA (Fig. 1a). This can be rescued either by expression of YopT which cleaves prenylated RhoA C-terminal tail (Fig. 1b) or by treating cells with lovastatin (Fig. 1c).

4. Notes

1. The $CaCl_2$ solution has a higher density than the water/DNA mix; it sinks in the water/DNA solution and requires thorough mixing.

2. The HBSP solution is less dense than the $DNA/CaCl_2$ mix; therefore, it is better to add it starting at the bottom of the

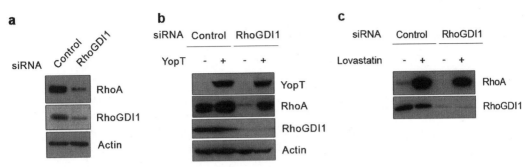

Fig. 1. (a) RhoGDI1 depletion triggers degradation of Rho proteins. Control or RhoGDI1 siRNA-transfected HeLa cells were analyzed by western blotting. Upon depletion of RhoGDI1, many Rho proteins including RhoA are degraded. (b) Prenylation removal rescues Rho protein degradation in the absence of RhoGDI1. HeLa cells were cotransfected with control or RhoGDI1 siRNA and a cDNA encoding the bacterial protease YopT. Cell lysates were resolved by SDS-PAGE and analyzed by western blotting (reproduced with permission from Nature Publishing Group). (c) Inhibition of RhoA prenylation prevents its degradation in the absence of RhoGDI1. HeLa cells were transfected with control or RhoGDI1 siRNA and treated with lovastatin 2.5 μM for 24 h. Cell lysates were resolved by SDS-PAGE and analyzed by western blotting. Reproduced with permission from Nature Publishing Group (16).

tube with a circular motion going up in the tube. Additional mixing can be achieved by generating bubbles from the bottom of the tube.

3. At this point, the medium may look like the cells are contaminated with bacteria. Indeed, the calcium phosphate method is based on the generation of a precipitate of calcium phosphate which traps DNA/RNA molecules. This precipitate looks like contamination although it is not. Usually, the remaining precipitate after washing is finally engulfed by cells during the following 10–12-h period.

4. It can be noticed that during treatment with lovastatin cells tend to round up after 24 h. Cells need to be carefully monitored since they finally detach after rounding up. A similar effect is observed upon expression of YopT cDNA.

5. Washing the cells with ice-cold PBS washes out proteins from the growth medium, such as bovine serum albumin, and cools down the cells to inhibit any signaling during lysis.

6. The volume of lysis buffer is to be modified according to the size of the culture vessel. We usually use 1 mL for 100-mm dish and 300 μL for each well of a 6-well plate.

7. We perform cell lysis with this high SDS concentration buffer at room temperature to avoid SDS precipitation. It is worth noting that cell lysis with such a high concentration of SDS results in an almost immediate cell lysis and denaturation of proteins blocking all enzymatic activities.

8. Lysis with SDS disrupts the nuclear envelope and releases genomic DNA which results in an extremely viscous lysate almost impossible to pipet. A simple way to reduce viscosity of

the lysate is to sonicate it for a few seconds at lowest settings in order to shear the genomic DNA.

9. Any gel thickness can be used, but the present protocol refers to 1.5-mm-thick gels.

10. During polymerization, polyacrylamide gels have a tendency to retract at the interface between the gel and surrounding air, leaving an irregular interface. In order to avoid this, it is common to create an artificial liquid interface. This can be done using water-saturated butanol (which tends to be avoided since it is toxic), a viscous solution of SDS, or just plain water. Water is the simplest way to get a straight interface but requires extra caution when dispensing over the gel.

11. Cut pieces of Whatman paper and PVDF membrane slightly larger than the gel itself. When placing the gel on the PVDF membrane, make sure to soak the gel with 1× transfer buffer to avoid tearing it as it may stick to the gloves when drying. Use a test tube or pipet as a rolling pin and roll over the membrane carefully in both directions to remove any air bubbles trapped between the gel and the membrane. Air bubbles block the current and impair protein transfer at the site of the bubble.

12. The recommended dilution for secondary antibodies to start off with is 1/10,000. Alternatively, if the background is too high, primary and secondary antibodies can be washed using TBS or PBS with 0.5% Tween 20. For these antibodies, a 1-h incubation at RT is usually enough to get a good signal.

13. The Immobilon HRP substrate can be used to amplify very weak signals; therefore, since blotting for RhoA or RhoGDI1 leads to very strong signals, we usually dilute the HRP substrate to 1:4 in water.

References

1. Jaffe, A.B., and Hall, A. (2005). Rho GTPases: biochemistry and biology. *Annu Rev Cell Dev Biol* 21, 247–269.

2. Fukumoto, Y., Kaibuchi, K., Hori, Y., Fujioka, H., Araki, S., Ueda, T., Kikuchi, A., and Takai, Y. (1990). Molecular cloning and characterization of a novel type of regulatory protein (GDI) for the rho proteins, ras p21-like small GTP-binding proteins. *Oncogene* 5, 1321–1328.

3. Ueda, T., Kikuchi, A., Ohga, N., Yamamoto, J., and Takai, Y. (1990). Purification and characterization from bovine brain cytosol of a novel regulatory protein inhibiting the dissociation of GDP from and the subsequent binding of GTP to rhoB p20, a ras p21-like GTP-binding protein. *J Biol Chem* 265, 9373–9380.

4. Leonard, D., Hart, M.J., Platko, J.V., Eva, A., Henzel, W., Evans, T., and Cerione, R.A. (1992). The identification and characterization of a GDP-dissociation inhibitor (GDI) for the CDC42Hs protein. *J Biol Chem* 267, 22860–22868.

5. Hart, M.J., Maru, Y., Leonard, D., Witte, O.N., Evans, T., and Cerione, R.A. (1992). A GDP dissociation inhibitor that serves as a GTPase inhibitor for the Ras-like protein CDC42Hs. *Science* 258, 812–815.

6. Chuang, T.H., Xu, X., Knaus, U.G., Hart, M.J., and Bokoch, G.M. (1993). GDP dissociation inhibitor prevents intrinsic and GTPase activating protein-stimulated GTP hydrolysis by the Rac GTP-binding protein. *J Biol Chem* 268, 775–778.

7. Nomanbhoy, T.K., and Cerione, R. (1996). Characterization of the interaction between RhoGDI and Cdc42Hs using fluorescence spectroscopy. *J Biol Chem* 271, 10004–10009.

8. DerMardirossian, C., and Bokoch, G.M. (2005). GDIs: central regulatory molecules in Rho GTPase activation. *Trends Cell Biol 15*, 356–363.

9. Casey, P.J., and Seabra, M.C. (1996). Protein prenyltransferases. *J Biol Chem* 271, 5289–5292.

10. Hoffman, G.R., Nassar, N., and Cerione, R.A. (2000). Structure of the Rho family GTP-binding protein Cdc42 in complex with the multifunctional regulator RhoGDI. *Cell* 100, 345–356.

11. Adra, C.N., Manor, D., Ko, J.L., Zhu, S., Horiuchi, T., Van Aelst, L., Cerione, R.A., and Lim, B. (1997). RhoGDIgamma: a GDP-dissociation inhibitor for Rho proteins with preferential expression in brain and pancreas. *Proc Natl Acad Sci USA* 94, 4279–4284.

12. Lelias, J.M., Adra, C.N., Wulf, G.M., Guillemot, J.C., Khagad, M., Caput, D., and Lim, B. (1993). cDNA cloning of a human mRNA preferentially expressed in hematopoietic cells and with homology to a GDP-dissociation inhibitor for the rho GTP-binding proteins. *Proc Natl Acad Sci USA* 90, 1479–1483.

13. Scherle, P., Behrens, T., and Staudt, L.M. (1993). Ly-GDI, a GDP-dissociation inhibitor of the RhoA GTP-binding protein, is expressed preferentially in lymphocytes. *Proc Natl Acad Sci USA* 90, 7568–7572.

14. Zalcman, G., Closson, V., Camonis, J., Honore, N., Rousseau-Merck, M.-F., Tavitian, A., and Olofsson, B. (1996). RhoGDI-3 is a new GDP dissociation inhibitor (GDI). Identification of a non-cytosolic GDI protein interacting with the small GTP-binding proteins RhoB AND RhoG. *J. Biol. Chem.* 271, 30366–30374.

15. Dovas, A., and Couchman, J.R. (2005). RhoGDI: multiple functions in the regulation of Rho family GTPase activities. *Biochem J* 390, 1–9.

16. Boulter, E., Garcia-Mata, R., Guilluy, C., Dubash, A., Rossi, G., Brennwald, P.J., and Burridge, K. (2010). Regulation of Rho GTPase crosstalk, degradation and activity by RhoGDI1. *Nat Cell Biol* 12, 477–483.

17. Cox, A.D., and Der, C.J. (1992). Protein prenylation: more than just glue? *Curr Opin Cell Biol* 4, 1008–1016.

18. Tobert, J.A. (2003). Lovastatin and beyond: the history of the HMG-CoA reductase inhibitors. *Nat Rev Drug Discov* 2, 517–526.

19. Shao, F., Vacratsis, P.O., Bao, Z., Bowers, K.E., Fierke, C.A., and Dixon, J.E. (2003). Biochemical characterization of the *Yersinia* YopT protease: Cleavage site and recognition elements in Rho GTPases. *Proc Natl Acad Sci USA 100*, 904–909.

20. Huesken, D., Lange, J., Mickanin, C., Weiler, J., Asselbergs, F., Warner, J., Meloon, B., Engel, S., Rosenberg, A., Cohen, D., et al. (2005). Design of a genome-wide siRNA library using an artificial neural network. *Nat Biotechnol* 23, 995–1001.

Chapter 8

A Quantitative Fluorometric Approach for Measuring the Interaction of RhoGDI with Membranes and Rho GTPases

Jared Johnson, Richard A. Cerione, and Jon W. Erickson

Abstract

Tight regulation of Rho GTPase-signaling functions requires the proper localization of proteins to the membrane and cytosolic compartments, which can themselves undergo reconfiguration in response to signaling events. The importance of lipid-mediated membrane signal transduction continues to emerge as a critical event in many Rho GTPase-signaling pathways. Here we describe methods for the reconstitution of lipid-modified Rho GTPases with defined lipid vesicles and how this system can be used as a real-time assay for monitoring protein–membrane interactions.

Key words: Cdc42, Rho, Fluorescence resonance energy transfer, Synthetic lipid vesicles, Guanine nucleotide dissociation inhibitor

1. Introduction

Membrane localization of signaling proteins and their interaction with constituent lipid headgroups underlies the recruitment of reactive partners to the membrane surface. It has been well documented that proteins often harbor binding domains such as those that bind to specific lipid headgroups (e.g., pleckstrin homology (PH) domains exhibit general specificity for phosphoinositide lipids) (1–6). Green fluorescent protein probes fused to lipid-specific domains have been used successfully to demonstrate the asymmetric distribution of different lipids in living cells and to follow their accumulation and redistribution during cellular stimulation (7–9). In order to study membrane–protein interactions more quantitatively, we have been developing reconstituted lipid/protein systems

Francisco Rivero (ed.), *Rho GTPases: Methods and Protocols*, Methods in Molecular Biology, vol. 827,
DOI 10.1007/978-1-61779-442-1_8, © Springer Science+Business Media, LLC 2012

where one can measure in real time the association–dissociation kinetics of Rho GTPases (10–12) interacting with membranes and with one of their key regulatory proteins, RhoGDI (for *Rho-Guanine* nucleotide *Dissociation Inhibitor*). RhoGDI forms a cytosolic complex with the isoprenylated forms of the Rho GTPases, having the overall effect of stabilizing a soluble form of the Rho GTPase and affecting its translocation from the membrane to the solution compartments (13, 14).

The X-ray crystal structure of the complex formed between Cdc42 and RhoGDI provides a structural basis for the stabilization of the cytosolic form of Cdc42, as it shows the C-terminal lipid modification of Cdc42 (geranylgeranyl) nestled in a hydrophobic pocket in the immunoglobulin-like domain of RhoGDI (15). The membrane-cytosolic partitioning of Rho GTPases is dictated by the relative affinity of the lipid surface or RhoGDI, respectively, for the GTPase. How do different lipid headgroups, nucleotide-bound states, and protein–protein interactions other than those involving RhoGDI influence the reactivity and distribution of Rho GTPases and their signaling properties? In order to address these questions, we have used a reconstitution approach, which uses fluorescence resonance energy transfer (FRET) as a spectroscopic method for monitoring the degree of Rho GTPase association with membranes. Our approach uses the fluorescent guanine nucleotide analog 2′-(or-3′)-*O*-(*N*-methylanthraniloyl)guanosine 5′-diphosphate (MANT-GDP) or 2′-(or-3′)-*O*-(*N*-methylanthraniloyl)-β:γ-imidoguanosine 5′-triphosphate (MANT-GMPPNP) as the fluorescence donor and hexadecanoylaminofluorescein incorporated into synthetic vesicles of defined composition as the fluorescence resonance energy acceptor (16, 17). Monitoring changes in the fluorescence of the MANT probes bound to Cdc42 provides an extremely sensitive assay for studying the kinetics of Cdc42–membrane interactions that has allowed for the delineation of discreet mechanistic steps in the membrane-to-cytosolic transition of this GTPase and how RhoGDI plays an important role in regulating these events.

2. Materials

2.1. Buffers, Reagents, and Equipment

1. *Spodoptera frugiperda* (Sf21) cells for adherent cultures (Invitrogen). Our contract supplier, Kinnakeet Biotechnology (Midlothian, VA), performs large-scale infections on suspension cultures of Sf9 cells with the viruses we provided.

2. Bacterial strain *E. coli* BL21(DE3) (Novagen).

3. Parental bacterial expression vectors pET28a (Invitrogen) and pGEX-KG (ATCC).

4. Hypotonic buffer: 20 mM sodium borate, pH 10.2, 5 mM $MgCl_2$, 200 µM phenylmethylsulfonyl fluoride (PMSF), and 1 µg/mL each of aprotinin and leupeptin.

5. Tris buffered saline (TBSM): 50 mM Tris–HCl, pH 7.5, 150 mM NaCl and 5 mM $MgCl_2$. For solubilizing membranes, this buffer is supplemented with 1% Triton X-100. For eluting GST-RhoGDI this buffer is supplemented with 10 mM glutathione.

6. High salt wash buffer: 50 mM Tris–HCl, pH 7.5, 700 mM NaCl, 5 mM $MgCl_2$, 0.1% CHAPS, and 20 mM imidazole, for immobilized metal ion affinity chromatography (IMAC).

7. Elution buffer for IMAC: 50 mM Tris–HCl, pH 7.5, 150 mM NaCl, 5 mM $MgCl_2$, 0.1% CHAPS, and 500 mM imidazole TEDA buffer: 20 mM Tris–HCl, pH 8.0, 1 mM EDTA, 1 mM dithiothreitol, and 1 mM sodium azide.

8. Chelating Sepharose beads charged with Ni^{2+} (Qiagen).

9. Glutathione Sepharose beads (GE Healthcare).

10. 40-mL glass homogenizer (Wheaton).

11. Sonicator: Sonic Dismembrator Fisher Scientific 550.

12. Fritted glass columns, 10×1 cm bore (BioRad).

13. Amicon Ultra 10 K concentrators (Millipore).

14. Colloidal Blue (Boston Scientific).

15. Polyacrylamide gel electrophoresis is carried out using a Novex Mini-Cell (Invitrogen) with precast 4–20% Tris–glycine gradient gels (1.5 mm × 10 well).

16. Lipids are obtained as lyophilized powders (Avanti Polar Lipids) or as reconstituted samples in chloroform (Sigma Aldrich). Between uses, the aliquots are stored under argon at –20°C. Cholesterol is from Nu-Chek Prep (see Fig. 1).

17. Hexadecanoylaminofluorescein (HAF), MANT-GDP, and Mant-GMPPNP (Invitrogen, formerly Molecular Probes). Store the 5 mM stock in a readily accessible dark location at room temperature.

18. Lipid vesicles for fluorescence spectroscopy are prepared using an Avanti mini-extruder (Avanti Polar Lipids) that comes with easy to follow instructions. Use 1-µm diameter pore polycarbonate membranes (Avanti, Cat. No. 610010).

19. 1 M $MgCl_2$ solution.

20. Fluorescence spectrometer Varian Cary Eclipse (Varian, Inc.) operated in the counting mode. To be used with a 1-mL quartz cuvette (Hellma Analytics) and a magnetic stirrer.

Liposome Stock Preparation (10 mg/mL)

Components:

Avanti (http://avantilipids.com/)
- •840032C   L-α-phosphatidylserine (PS) 25mg
- •840026C   L-α-phosphatidylethanolamine (PE) 25mg
- •840042C   L-α-phosphatidylinositol (PI) 10mg

Nu Chek Prep (http://www.nu-chekprep.com/)
- •Ch-800   Cholesterol 1g (dissolved in ultrapure (>99.99%) chloroform)

Lipids (concentration in chloroform)	Molecular Weight (g/mol)	Final Molar Percentage	Mass needed for 1 mole of lipid mix [a]	Fraction of the total in the previous column
Phosphatidylethanolamine 10 mg/mL	768	40	307.2 g	0.47
Phsophatidylserine 10 mg/mL	812	20	162.4 g	0.25
Phosphatidylinositol 10 mg/mL	909	5	135.5 g	0.07
Cholesterol 10 mg/mL	387	35	45.5 g	0.21

[a] Molecular Weight (g/mole) x molar percentage (1/100) x 1 mole

Determine Limiting Reagent *

Lipid	Amount available	Mass Fraction	Limiting Reagent Column 2 ÷Column 3	Multiply fractional masses by the conversion factor (53)	Volume needed for a 10 mg/mL stock
PE	25 mg	0.47	53 *	25 mg	2.5 mL
PS	25 mg	0.25	100	13.25 mg	1.325 mL
PI	10 mg	0.07	143	3.71 mg	371 μL
Chol.	∞	0.21	∞	11.13 mg	1.113 mL

Fig. 1. A table of supplier information for components of the vesicle system described in this study.

2.2. Proteins

1. Rho GTPases must be expressed in eukaryotic cells in order to be post-translationally modified (more specifically, isoprenylated), as bacteria do not possess the geranylgeranyl transferase enzyme necessary for C-terminal lipid modification. Viruses that enable expressing modified Rho GTPases in cultured Sf21 cells are produced following the Bac-2-Bac Baculoviral expression system (Invitrogen) according to the manufacturer's instructions. Sf21 cells can be grown in 175 cm² screw cap flasks (Corning, Inc) and infected with the baculovirus encoding the Rho GTPase of interest at a multiplicity of infection (MOI) of ~3:1 (see Note 1). Once test infections demonstrate acceptable levels of protein expression, as viewed on Western blots with the polyhistidine tags, Sf21 cell numbers can be scaled up according to the needs of the experiment. Intact cells can be stored as pellets indefinitely at −80°C and worked up as needed. Large scale expression of Rho GTPases in insect cells can be performed in spinner culture where a typical yield for Cdc42 is approximately 5 mg protein/L of culture medium (see Note 2). We routinely contract out large scale expression of protein to our outside supplier, Kinnakeet Biotechnology (Midlothian, VA).

2. Bacterial expression of GST or polyhistidine fusion proteins of RhoGDI or Cdc42, respectively, is carried out by growing *E. coli* that harbor expression plasmids for each protein. Cells are grown at 37°C until an $OD_{600} \sim 0.6$ is achieved, at which time protein expression is initiated by adding 1 mM isopropyl 1-thio-β-D-galactopyranoside. Cells are grown for an additional 3 h prior to pelleting by centrifugation ($6,000 \times g$ for 10 min). Cell pellets are resuspended in TBSM and sonicated immediately for protein purification or frozen as pellets at −80°C for later use.

3. Methods

3.1. Protein Purification

Carry out all procedures at 4°C unless otherwise noted.

3.1.1. Purification of His-Tagged Proteins

The protocol described below is for IMAC of His_6-tagged proteins from insect cells. For IMAC of bacterially expressed His_6-tagged proteins proceed as in Subheading 3.1.2, steps 1–3 and continue with the supernatant from step 5 below.

1. The insect cell pellet harvested from a liter of culture is resuspended in 40 mL of hypotonic buffer and Dounce homogenized, using a 40-mL glass homogenizer (see Note 3).

2. Particulates are pelleted from this homogenate at $150,000 \times g$ for 20 min. The supernatant from this spin contains the non-prenylated form of the Rho GTPase (e.g. Cdc42), which can be recovered by affinity chromatography at this stage if desired or discarded. The pellet contains the geranylgeranylated form of Cdc42 and this is resuspended and repelleted twice in order to thoroughly clear the membranes of unmodified, aqueous soluble protein.

3. Resuspend the pellet a third time with a final 50 mL of TBSM supplemented with 1% Triton X-100 and rotate at 4°C for 30 min to solubilize the membranes completely.

4. Spin down the remaining particulates for 20 min at $9,000 \times g$ at 4°C and save the supernatant. This supernatant contains the isoprenylated, His_6-tagged Cdc42 and is used for further purification using standard Ni^{2+} charged Sepharose beads.

5. Carry out IMAC by incubating the supernatant from step 1 with 2 mL of Ni^{2+} bead slurry with gentle rocking at 4°C for 30 min.

6. Allow the beads to settle on a 10×1 cm bore fritted glass column and allow the liquid to drain from the column (save this for later protein analysis).

7. Wash the column with 400 mL of ice-cold high salt buffer.

8. Elute the His$_6$-Cdc42 with 10 mL of elution buffer. Divide eluate into two samples and concentrate one sample down to 1 mL (see Note 4) using Amicon Ultra 10 K concentrators. If you are unable to concentrate the protein to this low of a volume, due to precipitation, take note of the specific concentration at which this occurred and concentrate the remaining fraction down to a safe volume.

9. Analyze fractions collected at different stages of the purification by SDS-PAGE analysis using Colloidal Blue staining to visualize protein (see Note 5).

10. Divide the eluates into 100 μL aliquots, snap freeze the tubes and store at –20°C (see Note 6).

3.1.2. Purification of GST RhoGDI

1. Resuspend the bacterial pellet (see Subheading 2.2, item 2) in 25 mL of ice-cold TEDA buffer.

2. Sonicate with a Sonic Dismembrator set at level 5 for 5 min with 15 s bursts followed by 15 s off. Carry out sonication on ice.

3. Clarify the resuspended, sonicated pellet by centrifugation at 20,000×g for 30 min at 4°C.

4. Transfer the supernatant to a fresh tube with 3 mL of glutathione bead slurry pre-equilibrated with TEDA buffer.

5. Rock at 4°C for 30 min to allow binding of the GST-RhoGDI and then load on to a 10×1 cm bore fritted glass column and allow the liquid to drain from the column.

6. Wash with 20 column volumes of TEDA supplemented with 500 mM NaCl.

7. Elute the GST-RhoGDI with TBSM supplemented with 10 mM glutathione and analyze the fractions for purity by SDS-PAGE. Choose the most concentrated and purest fractions and concentrate these with an Amicon Ultra 10 K concentrator.

3.2. Preparation of Lipid Vesicles for Fluorescence Spectroscopy

The standard lipid composition as mole fractions for "control" vesicles is 0.35 phosphatidylethanolamine (PE), 0.25 phosphatidylserine (PS), 0.05 phosphatidylinositol, and 0.35 cholesterol (see Note 7). See Fig. 1 for tabulated information concerning lipid constituents used in the preparation of vesicles.

1. Just prior to the experiment, lightly vortex the lipid stock and transfer 100 μL of the lipid/chloroform mixture of desired composition to a clean glass vial. Take care to immediately argon seal the lipid stock and return to –20°C. Direct a stream of argon gas to the 100 μL of lipid mixture until the chloroform is evaporated leaving a translucent lipid film (see Note 8).

Add 1 mL of TBSM to the dried lipid film and incubate for 1 h at room temperature in order to rehydrate the lipids.

2. After incubation and gentle flicking of the tube, the film should detach and be ready for extrusion (see Note 9). Pass the hydrated lipid solution through the 1 μm pore polycarbonate membrane mounted on the mini-extruder 15 times (see Note 10), as an odd number of passes ensures that the liposomes are collected on the opposite side of the loading side of the filter.

3. Finally, transfer the liposomes to a clean 1.5-mL microcentrofuge tube. The final concentration of lipids is 1 mg/mL.

3.3. FRET Measurements for Monitoring the Interaction of Rho GTPases with Lipid Membranes and RhoGDI

1. Add 100 μL of liposomes (see Subheading 3.2) to a 0.5-mL eppendorf tube and add 0.4 μL of 5 mM HAF with a P2 pipetman. Immediately vortex the mixture at a maximum speed for 5 s (see Note 11).

2. Briefly vortex the HAF/liposome or liposome control mixtures at maximum speed and transfer 20 μL to a new 0.5-mL eppendorf tube. If liposomes are not to be included this early, then add 20 μL of TBSM instead. Add at least a twofold greater molar concentration of Mant-nucleotides over the GTPase. For our samples, 1 μL of 50 μM Cdc42 is added along with 2.5 μL of 50 μM Mant-nucleotide (see Note 12). Add the appropriate amount of 500 mM EDTA to achieve a final concentration of 8.5 mM. Pipet mixture and set a timer for 5 min (see Note 13).

3. During or before nucleotide exchange, clean the fluorescent cuvette to be used, by rinsing thoroughly with distilled water, cleaning the stir bar as well. This is best done with a new pair of powder-free gloves, as particulates that wash off into the cuvette can lead to undesirable light scattering (see Note 14).

4. Fluorescence monitoring of the association between Cdc42:Mant-GDP and HAF-labeled vesicles using a fluorimeter in counting mode. Use a magnetic stirrer with a 1-mL quartz cuvette with 1,000 μL of TBSM at 25°C; the photomultiplier tube is set to 750 V. The excitation wavelength is set at 365 nm (5 nm slit width) and emission is monitored continuously at 440 nm (20 nm slit width) (Fig. 2) (see Note 15).

5. With 10 s left on the timer, add 1 M $MgCl_2$ to the GTPase–Mant-nucleotide–liposome mixture to achieve a final concentration of 15 mM. Immediately start the fluorescence run and add 80% of the total sample volume to the cuvette at a precise time point (see Note 16).

6. Many configurations of the FRET assay can provide information regarding the kinetics of protein exchange between membrane vesicles. To observe the rate of Mant-Cdc42

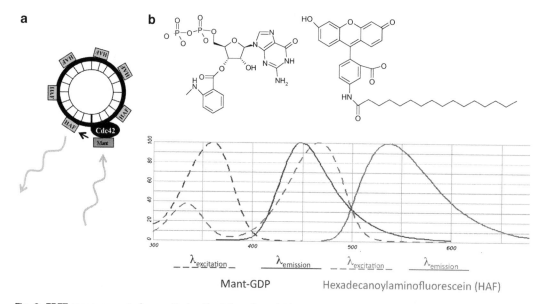

Fig. 2. FRET measurements for monitoring the interaction of Rho GTPases with lipid membranes. (a) The basis of the experiment is shown in this illustration where the co-localization of donor and acceptor probes on the surface of liposomes results in the quenching of donor fluorescence. (b) Elements of the synthetic lipid vesicle reconstitution and FRET pairing used to monitor the association–dissociation kinetics of Cdc42 and HAF-labeled vesicles in the presence and absence of RhoGDI. The chemical structures of the fluorophores used here and their excitation and emission spectra are shown.

incorporation into HAF-labeled vesicles, do not include any liposome during the in vitro nucleotide exchange reaction. Stir 1 µM Mant-Cdc42 in 1,000 µL of TBSM and monitor the Mant emission at 440 nm. At a later time point, carefully add the desired volume of HAF-labeled vesicles. The decrease in 440 emission reflects the degree of Mant-GDP:Cdc42 quenching and therefore the association and insertion of Cdc42 into the liposome.

3.3.1. Monitoring the Association of RhoGDI with Rho GTPases

To monitor the association of RhoGDI with MantGDP:Cdc42 bound to membranes and the subsequent translocation to the surrounding bulk medium, include 20 µL of the HAF/liposome mixture during the in vitro nucleotide exchange reaction of Cdc42 with Mant-nucleotide. As before, add this mixture to 1,000 µL of TBSM with stirring. After equilibration and signal stabilization (see Note 17), add a stoichiometric amount of RhoGDI. The subsequent increase in fluorescence provides a real-time read-out for the release of Cdc42 from the lipid surface as a complex of RhoGDI and Cdc42:Mant-GDP. A sample fluorescence emission trace following the above protocol is shown in Fig. 3.

3.3.2. Monitoring the Rate of Exchange of Rho GTPases Between Lipids

To monitor the rate of exchange of Cdc42 between lipid residues, prepare two populations of liposomes: one population that contains the HAF and a population of control liposomes that lack HAF. Pre-incubate the liposomes, devoid of HAF, with

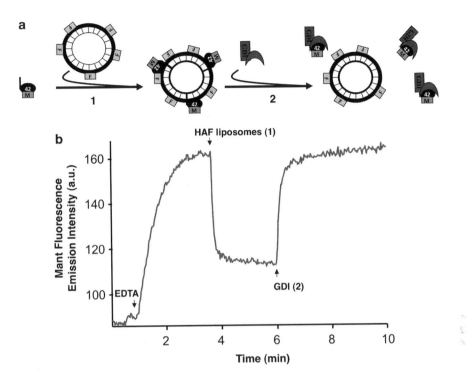

Fig. 3. Translocation experiment. (**a**) This schematic outlines the sequence of events in a typical translocation experiment. (**b**) Fluorescence emission trace showing the HAF-labeled vesicle quenching of Mant-GDP:Cdc42 upon HAF vesicle addition followed by the reversal of Mant emission with the addition of RhoGDI.

Cdc42:Mant-GDP as above and add to the cuvette with stirring. Upon addition of HAF-incorporated liposomes, the decrease in 440-nm fluorescence emission reflects the loss of Cdc42:Mant-GDP from one vesicle population to the other (see Fig. 4).

4. Notes

1. The MOI refers to the ratio of the number of virus particles to the number of Sf21 cells adherent to the flask or growing in spinner culture. A MOI of 3 is a good starting point for optimizing protein expression using baculovirus/Sf21 protein expression systems.

2. The 5 mg per liter of spinner culture is on the high end of what we expect typical yields to be for Rho GTPases in insect cell expression. Despite the similarity of the proteins, varying yields of the different Rho GTPases were consistently observed.

3. A thorough dissolution of the insect pellet requires both its immersion in the hypotonic buffer (described above) and

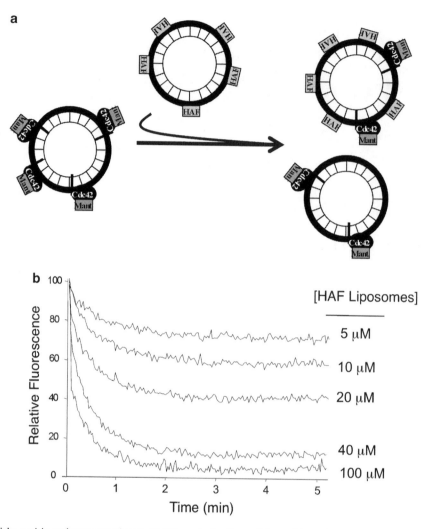

Fig. 4. Vesicle–vesicle exchange experiment. (a) Schematic for the vesicle–vesicle exchange observed in the absence of RhoGDI. (b) The kinetics of vesicle–vesicle exchange exhibits a well-behaved dose dependence on the amount of HAF vesicles added.

repeated disruption with a microspatula. To hasten this process, it is useful to break the pellet into multiple pieces and expose as much surface area to the solvent as possible. The homogenization process that follows is most efficient when the pellet has completely dissolved.

4. In purifying lipid-modified proteins, be wary of hydrophobic-induced aggregation. The protein concentration step, at the end of the purification, is most vulnerable to this. Once white fiber-like particulates are observed, the sample is no longer usable for functional assays. Therefore, approach this step with caution and concentrate only a fraction of the eluate in an Amicon 10 K MWKO membrane at 5 min intervals of 5,000 × g,

monitoring its status. If it precipitates from the solution, take note of the point at which this occurred and be careful when concentrating the remaining fractions.

5. Despite encompassing the majority of protein in this purification, the GTPase is not the only protein present in the eluate. Thus, the Bradford assay overestimates its concentration. A SDS-PAGE-assisted densitometric analysis, comparing the Coomassie blue-stained band from a particular volume of your concentrated sample with a known amount of BSA is an alternative way to determine the concentration of protein.

6. Mix the concentrate extensively prior to aliquoting, to avoid a heterogeneous configuration of sample concentrations. Producing many small volume aliquots helps to prevent the need to reuse a sample, as repeated freeze–thaw cycles reduce the quality of these modified proteins. 500-μL eppendorf tubes, with indented lids, are labeled by sample and date and placed on ice. After the concentrated protein is aliquoted and snap frozen, it can be stored indefinitely at −80°C.

7. This composition is aimed to mimic the inner leaflet of the plasma membrane, which is relevant for our purposes as the majority of Rho Family GTPases make important signaling contributions at this location.

8. Take care not to perturb the lipids during the solvent-evaporation process. We typically prepare at least two identical liposome samples per experiment, as liposomes will occasionally aggregate inappropriately, making them less ideal for extrusion.

9. This consists mostly of large multilamellar lipid bilayers. It should have a semi-translucent appearance, with no visible white particulates.

10. Always perform an odd number of passes, as it selects against any larger particulates that cannot pass through the membrane. Those particulates could be a source of scattering in the subsequent fluorescence experiments.

11. The liposome mixture should become yellow upon vortexing with HAF. This liposome preparation will be optimal for four trials. It is best not to let the HAF-liposome mixture sit for too long a period of time, as deviations from initial observations become evident over time. Preparing a fresh mixture of HAF and liposome every four trials allows for the most accurate and reproducible measurements.

12. Prenylated GTPases are best examined directly upon thawing, as they begin to behave differently when allowed to sit on ice for extended periods. Prior to an experiment, we always thaw one of our 100-μL aliquots of insect cell-expressed Cdc42 on

ice and divide it further into eighteen 5-μL aliquots (in 500-μL indented eppendorfs). These are snap-frozen and stored in the −80°C freezer. Once Cdc42 is needed, it can be directly added to the membranes without delay.

13. This step is GTPase specific and depends on both the protein and lipid composition. Certain Rho GTPases, in our experience, have been found to slowly associate with membranes, requiring a longer incubation period in order to bind to liposomes.

14. Cleaning the cuvette, after each experiment, involves rinsing with distilled water and drying the sides with Kimwipes. Avoid allowing tissue fibers from entering the cuvette, as they can cause erratic signals from light scattering.

15. Prior to use, perform a background analysis on the cuvette, filled with 1 mL TBSM. Set it as the blank and make sure the noise level is minimal. This ensures the experimental data will be as clean as possible. Once the noise level is acceptable, leave the cuvette to await the addition of your samples.

16. Depending on the particular assay, speed is often crucial at this point. Thus, it is useful to have all the micropipets, to be used, preset to their intended volumes.

17. This is a common variation among different protein and liposome preparations. For reasons that are not completely clear, Mant fluorescence can sometimes decline, failing to equilibrate to a steady value within a reasonable time frame. Occasionally, this can be remedied by preparing fresh liposomes. Otherwise, interpretable measurements can still be made, provided the signal reduction is slow relative to the measurement of interest.

References

1. Lemmon, M.A., Ferguson, K.M., O'Brien, R., Sigler, P.B., and Schlessinger, J. (1995) Specific and high-affinity binding of inositol phosphates to an isolated pleckstrin homology domain. *Proc Natl Acad Sci USA* 92, 10472–10476

2. Lemmon, M.A., and Ferguson, K.M. (2000) Signal-dependent membrane targeting by pleckstrin homology (PH) domains. *Biochem J* 350, 1–18

3. Heo, W.D., Inoue, T., Park, W.S., Kim, M.L., Park, B.O., Wandless, T.J., and Meyer, T. (2006) PI(3,4,5)P₃ and PI(4,5)P₂ lipids target proteins with polybasic clusters to the plasma membrane. *Science* 314, 1458–1461

4. Yeung, T., Terebiznik, M., Yu, L., Silvius, J., Abidi, W.M., Philips, M., Levine, T., Kapus, A., and Grinstein, S. (2006) Receptor activation alters inner surface potential during phagocytosis. *Science* 313, 347–351

5. Yeung, T., Gilbert, G.E., Shi, J., Silvius, J., Kapus, A., and Grinstein, S. (2008) Membrane phosphatidylserine regulates surface charge and protein localization. *Science* 319, 210–213

6. Young, B.P., Shin, J.J., Orij, R., Chao, J.T., Li, S.C., Guan, X.L., Khong, A., Jan, E., Wenk, M.R., Prinz, W.A., Smits, G.J., and Loewen, C.J. (2010) Phosphatidic acid is a pH biosensor that links membrane biogenesis to metabolism. *Science* 329, 1085–1089

7. Várnai, P., and Balla, T. (1998) Visualization of phosphoinositides that bind pleckstrin homology domains: calcium- and agonist-induced dynamic changes and relationship to Myo-[³H]

inositol-labeled phosphoinositide pools. *J Cell Biol* 143, 501–510

8. Stauffer, T.P., Ahn, S., and Meyer, T. (1998) Receptor-induced transient reduction in plasma membrane PtdIns (4,5) P2 concentration monitored in living cells. *Curr Biol* 8, 343–346

9. Botelho, R.J., Teruel, M., Dierckman, R., Anderson, R., Wells, A., York, J.D., Meyer, T., and Grinstein, S. (2000) Localized biphasic changes in phosphatidylinositol-4,5-bisphosphate at sites of phagocytosis. *J Cell Biol* 151, 1353–1368

10. Kozma, R., Ahmed, S., Best, A., and Lim, L. (1995) The Ras-related protein Cdc42Hs and bradykinin promote formation of peripheral actin microspikes and filopodia in Swiss 3 T3 fibroblasts. *Mol Cell Biol* 15, 1942–1952

11. Hall, A. (1998) Rho GTPases and the actin cytoskeleton. *Science* 279, 509–514

12. Cerione, R.A. (2004) Cdc42: new roads to travel. *Trends Cell Biol* 14,127–132

13. Michaelson, D., Silletti, J., Murphy, G., D'Eustachio, P., Rush, M., and Philips, M.R. (2001) Differential localization of Rho GTPases in live cells: regulation by hypervariable regions and RhoGDI binding. *J Cell Biol* 152, 111–126

14. Gosser, Y.Q., Nomanbhoy, T.K., Aghazadeh, B., Manor, D., Combs, C., Cerione, R.A., Rosen, M.K. (1997) C-terminal binding domain of Rho GDP-dissociation inhibitor directs N-terminal inhibitory peptide to GTPases. *Nature* 387, 814–819

15. Hoffman, G.R., Nassar, N., and Cerione, R.A. (2000) Structure of the Rho family GTP-binding protein Cdc42 in complex with the multifunctional regulator RhoGDI. *Cell* 100:345–356

16. Nomanbhoy, T., Erickson, J.W., and Cerione, R.A. (1999) Kinetics of Cdc42 Membrane Extraction by Rho-GDI monitored by real-time fluorescence resonance energy transfer. *Biochemisry* 38, 1744–1750

17. Johnson, J.L., Erickson, J.W., and Cerione, R.A. (2009) New insights into how the Rho guanine nucleotide dissociation inhibitor regulates the interaction of Cdc42 with membranes. *J Biol Chem* 284, 23860–23871

Part III

Functional Assays

Chapter 9

Rho GTPases and Cancer Cell Transendothelial Migration

Nicolas Reymond, Philippe Riou, and Anne J. Ridley

Abstract

Small Rho GTPases are major regulators of actin cytoskeleton dynamics and influence cell shape and migration. The expression of several Rho GTPases is often up-regulated in tumors and this frequently correlates with a poor prognosis for patients. Migration of cancer cells through endothelial cells that line the blood vessels, called transendothelial migration or extravasation, is a critical step during the metastasis process. The use of siRNA technology to target specifically each Rho family member coupled with imaging techniques allows the roles of individual Rho GTPases to be investigated. In this chapter we describe methods to assess how Rho GTPases affect the different steps of cancer cell transendothelial cell migration in vitro.

Key words: Endothelium, Transmigration, Adhesion, Cancer, ECM, HUVEC, Confocal, FACS

1. Introduction

To form new tumor foci known as metastases, cells that have shed from a primary tumor need to successfully complete a succession of sequential and critical steps. Cancer cells first invade and migrate through their surrounding tissues either individually or collectively as groups towards the circulation (1–3). They then enter the lymphatic vessels and/or the blood vessels (intravasation) and disseminate throughout the body while resisting the high shear stress in the bloodstream (4–6). Once they stop and adhere to the vascular endothelial cells (EC) that line blood vessel walls, they transmigrate across EC (transendothelial migration or extravasation) into the neighboring tissues (7–10). After invading the extracellular matrix (ECM), cells then need to survive and proliferate within this new environment to form metastases (11–13).

Francisco Rivero (ed.), *Rho GTPases: Methods and Protocols*, Methods in Molecular Biology, vol. 827,
DOI 10.1007/978-1-61779-442-1_9, © Springer Science+Business Media, LLC 2012

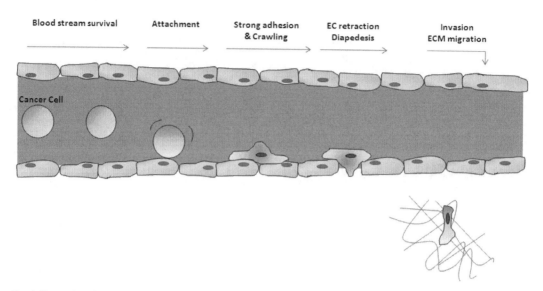

Fig. 1. Illustration of the different steps occurring during the cancer cell extravasation process. Cancer cells circulating in the blood stream adhere to EC that line blood vessels and then spread on them before inducing EC retraction and completing their diapedesis. Once extravasated, they invade the ECM and form new foci known as metastases.

Cancer invasion has been extensively studied both in vitro and in vivo, but cancer cell transendothelial cell migration (TEM) is still under-studied (14). Cancer cells were initially thought to extravasate at given sites mostly because they were trapped in nearby capillary beds due to size restriction (7, 8). However, in the past few years, many studies have demonstrated that a specific adhesion code also contributes to the organ specificity of metastasis, based on the adhesion receptor paradigm for leukocyte TEM (15). Indeed prostate cancer metastasizes preferentially to bone marrow, lung, and liver while breast cancer disseminates to bone, lungs, liver, and brain (12). At a molecular level this is explained by the expression of specific adhesion molecules on both cancer and vascular endothelial cells. For example, endothelial E-selectin interacts with its ligand PSGL-1 expressed on prostate cancer cells to control their adhesion and drive their tropism to form bone metastases (16). Chemokines and their receptors may also determine the metastatic destination of tumor cells, such as CXCL12/CXCR4 for breast cancer metastasis to regional lymph nodes and lung (17, 18).

Rho GTPases are involved in several cellular processes that contribute to cancer initiation and progression, including cytoskeletal dynamics and turnover of cell–cell and cell–ECM adhesions, cell cycle progression, transcriptional regulation, and cell survival (14). Their expression and/or their activity level are frequently deregulated in tumors and metastases (14). This often correlates with a poor clinical prognosis for patients (19).

In this chapter we describe different assays to monitor the involvement of Rho GTPases in multiple steps of cancer cell extravasation in vitro (Fig. 1): adhesion to EC, migration on and intercalation between EC, adhesion to and interaction with the ECM, and an overall TEM assay. We use PC3 prostate cancer cells as a model cancer cell line here, but any cancer cells that adhere to EC can be used. Since most of these assays are based on the coculture of cancer cells and EC, small-molecule reversible inhibitors of Rho GTPase pathways are not suitable. We therefore couple the assays with an siRNA-based strategy that allows down-regulation of each Rho GTPase family member specifically either in cancer cells or endothelial cells.

2. Materials

2.1. Cell Culture, Transfection, and Staining

1. Pooled primary human umbilical vein endothelial cells (HUVECs) (Lonza, CC-2519).

2. Endothelial cell basal medium-2 (EBM-2) (Clonetics®, Lonza, CC-3156) supplemented with growth factors (ascorbic acid, R3-IGF-1, heparin, rhFGF-B, hydrocortisone, GA-1000, rhEGF and VEGF, and 2% fetal bovine serum). Supplements are provided as part of a bullet kit, purchased with the basal medium (CC-41756). Medium plus supplements is referred to as EGM-2. Store at 4°C.

3. Phosphate-buffered saline (PBS) containing calcium and magnesium (Gibco, Invitrogen), referred as $PBS^{+/+}$.

4. Phosphate-buffered saline (PBS) without calcium and magnesium (Gibco, Invitrogen), referred as $PBS^{-/-}$.

5. 10× Phosphate-buffered saline (PBS) containing calcium and magnesium, referred as 10× $PBS^{+/+}$.

6. Human plasma fibronectin (Calbiochem) diluted in $PBS^{+/+}$ to 10 µg/mL. Store at −20°C (see Note 1).

7. Human bone marrow endothelial cells (HBMECs) kindly provided by Babette Weksler, Weill-Cornell School of Medicine, New York (20).

8. Dulbecco's modified Eagle's medium (DMEM) containing 4.5 g/L of glucose, L-glutamine, 25 mM HEPES, and supplemented with 10% fetal calf serum (FCS), 100 U/mL penicillin, 100 µg/mL streptomycin, and 1 mM sodium pyruvate and MEM acids mix (Gibco). Store at 4°C.

9. Bovine gelatin diluted in $PBS^{+/+}$ to 0.2% (w/v). Store at 4°C.

10. Bovine Collagen-1 (Purecol, Advanced Biomatrix) diluted in $PBS^{+/+}$ to 50 µg/mL. Store at 4°C.

11. Murine Matrigel (BD Biosciences) diluted in $PBS^{+/+}$ to 100 µg/mL. Store at 4°C.

12. PC3 prostate cancer cells (ATCC).

13. Roswell Park Memorial Institute 1640 medium (RPMI-1640) supplemented with 10% fetal calf serum (FCS), 100 U/mL penicillin and 100 µg/mL streptomycin (Gibco, Invitrogen, 21875–034). Store at 4°C.

14. Trypsin–EDTA. Store at 4°C.

15. 1× Non-enzymatic Cell Dissociation Solution (Sigma-Aldrich, C5914).

16. Hemocytometer or cell counter.

17. Carboxyfluorescein diacetate succinimidyl ester (CFSE) (Molecular Probes, C1157). Store at –20°C.

18. Cell tracker red (CMPTX) (Molecular Probes, C34552). Store at –20°C.

19. CXCL12/SDF-1α (Peprotech). Store at –20°C.

20. Human HGF (Peprotech). Store at –20°C.

21. Flasks, 6-, 24-, and 96-well dishes. Boyden chambers made of Transwell insert (8-µm pore size and 6.5-mm diameter) (BD biosciences). FALCON FACS tubes (BD Biosciences).

22. Optimem medium (Gibco, Invitrogen, 31985).

23. Oligofectamine (Invitrogen, 12252–011).

24. Rho GTPase-specific siRNAs (Dharmacon).

25. Plate reader (with fluorescence filters).

26. Time lapse Microscope with a climate chamber and an acquisition system. We use a TE2000 inverted microscope (Nikon) equipped with a motorized stage (Prior), 10×, 20×, or 40× objectives and Metamorph software (Molecular Devices).

27. Flow cytometer. We use a FACSCalibur 3.7 (BD Biosciences). Samples are processed using Cell Quest software and results are processed using Cell Quest or Flow Jo software.

2.2. Confocal Microscopy

1. Microscope cover slips (13-mm diameter).

2. Glass slides (Super Premium, 1- to 1.2-mm thickness).

3. PBS$^{+/+}$: See Subheading 2.1, item 3.

4. 4% (w/v) Paraformaldehyde: dissolve 4 g of paraformaldehyde in 75 mL of distilled deionized water and 1 mL of 1 M NaOH. Stir the mixture gently on a heating block (around 65°C) under a fume hood until paraformaldehyde is dissolved. Add 10 mL of 10× PBS$^{+/+}$ and wait until the solution comes back to RT. Adjust pH to 7.4 with 1 M HCl and add distilled deionized water to a final volume of 100 mL. Filter using 0.2-µm membrane filters, aliquot, and store at –20°C (see Note 2).

5. Permeabilization solution: 0.1% (v/v) Triton X-100 in $PBS^{+/+}$. Store at 4°C.

6. Antibody dilution buffer: 5% (v/v) FCS in $PBS^{+/+}$.

7. Mouse anti-human VE-Cadherin antibody (2 μg/mL working concentration) (BD Transduction Laboratories, 610251); rabbit anti-human β-catenin antibody (2 μg/mL working concentration) (Sigma, C2206).

8. Alexa Fluor-647-conjugated goat anti-mouse IgG (1 μg/mL working concentration) and Alexa Fluor-647-conjugated-goat anti-rabbit IgG (1 μg/mL working concentration) (Molecular Probes, Invitrogen).

9. 4′,6-Diamidino-2-phenylindole (DAPI) solution.

10. Alexa Fluor-546-conjugated phalloidin (5 μg/mL working concentration) (Molecular Probes, Invitrogen).

11. Mounting medium.

12. Confocal microscope chamber. We use a Zeiss 510 LSM confocal microscope with a 40× or a 63× objective.

13. Image analysis was initiated using the Zen software (Zeiss) and processed with Adobe Photoshop.

3. Methods

3.1. HUVEC Culture

Cultured HUVECs are flat, polygonal, and have cell–cell junctions similar to those of endothelial cells from large veins (21). HUVECs are primary cells whose optimal growth and endothelial phenotype depends on the composition of the extracellular matrix on which they are plated. We obtain the most successful results by culturing HUVECs on fibronectin-coated surfaces.

1. Coat culture flasks with 10 μg/mL human fibronectin in $PBS^{+/+}$ for 1 h at 37°C (see Notes 1 and 3).

2. Defrost cell aliquots in a water bath and add to 10 mL of EGM-2 medium. One cell aliquot of 500,000 cells is suitable for a 75-cm^2 flask, and should reach confluence in 2–3 days grown at 37°C and 5% CO_2 (see Notes 4 and 5).

3. Replace medium after 24 h to stimulate cell growth. HUVECs should be subcultured when they reach 70–80% of confluence, and used for experiments only between passages 1 and 4 (see Note 6).

4. Passage cells by removing the medium from the flask and washing cells once with 2.5 mL of $PBS^{-/-}$.

5. Trypsinize cells for 3 min (see Note 7) with 1 mL of trypsin–EDTA, and collect cells with 9 mL of EGM-2 medium to inactivate the trypsin.

6. Centrifuge the cells at $170 \times g$ for 4 min and aspirate the supernatant to remove residual trypsin.

7. Resuspend the cell pellet by adding fresh EGM-2 medium and seed in fresh culture flasks/dishes. Do not dilute cells more than 1:4 during passage, as this leads to unhealthy and clumped cell growth, and HUVECs could enter senescence earlier than normally, hence changing their phenotype.

If the cells do not require passaging, change the medium every 2 days to maintain a healthy culture. If growth is slow following defrosting, change the medium more frequently to stimulate HUVECs.

3.2. HBMEC Culture

Cultured HBMECs are flat, polygonal, and have mature cell–cell junctions (20). For optimal proliferation and endothelial phenotype, we obtain the most successful results by culturing HBMECs on gelatin-coated surfaces.

1. Coat plastic dishes and flasks with 0.2% human gelatin in PBS$^{+/+}$ for 1 h at 37°C.

2. Thaw cells in a water bath from frozen aliquots then add to 10 mL warm DMEM containing 10% FCS.

3. Centrifuge cells at $170 \times g$ for 4 min, then aspirate the supernatant to remove DMSO or other cryopreservatives.

4. Resuspend the cell pellet in 10 mL of fresh DMEM containing MEM acid mix and seed into a 75-cm^2 flask. Maintain cells at 37°C and 5% CO_2, and replace medium after 24 h.

5. When cells reach confluence (see Note 8), passage them using trypsin–EDTA as described in Subheading 3.1, step 5 and plate them into new culture flasks or use them for experiments. Split cells between 1:5 and 1:10 and change the medium every 2–3 days (see Note 9).

3.3. PC3 Cell Culture

PC3 cells are a prostate cancer cell line isolated from a bone marrow metastasis (22). They grow as single cells and are described to be highly invasive and tumorigenic (23).

1. Thaw cells in a water bath from frozen aliquots, then add to 10-mL warm RPMI containing 10% FCS.

2. Centrifuge cells at $170 \times g$ for 4 min, then aspirate the supernatant to remove DMSO or other cryopreservatives.

3. Resuspend the cell pellet in 10 mL of fresh RPMI and seed into a 75-cm^2 flask. Maintain cells at 37°C and 5% CO_2, and replace medium after 24 h.

4. When cells are 80% confluent, passage using trypsin–EDTA as described in Subheading 3.1, step 5, and seed into new culture flasks or use in experiments. Split cells between 1:5 and 1:10 and change the medium every 2–3 days.

To improve experimental consistency, discard cells after 1 month and defrost a fresh cell aliquot.

3.4. siRNA Transfection

The choice of reliable siRNAs to target each Rho GTPase is crucial (see Note 10). Several companies sell multiple gene-specific siR-NAs. We use siRNAs from Dharmacon, but other companies are likely to have equally effective siRNAs.

3.4.1. EC Transfection with siRNA

1. First coat the appropriate number of wells of a 6-well dish with 10 μg/mL fibronectin in PBS$^{+/+}$ (300 μL) for 1 h at 37°C. This ensures that EC spread homogeneously in the wells. Plate EC at a concentration of 1.25×10^5 cells per well. Incubate 16–18 h at 37°C.

2. Transfect cells using Oligofectamine according to the manufacturer's instructions. Change the medium after 6 h of incubation with the siRNAs.

3. After 24 h, transfect the EC again following the same procedure. This step is optional if the efficiency of the knock down is satisfactory after just one transfection.

4. Trypsinize EC 48 h after transfection and replate them for the assays described below, using the appropriate number of cells.

5. Use EC 72 h after transfection for the assays. Lyse a small aliquot to determine the level of protein depletion by western blotting or, if no antibodies are available, of mRNA depletion by quantitative PCR.

3.4.2. PC3 Cell Transfection with siRNA

1. Plate PC3 cells at a concentration of 1.25×10^5 cells per well in a 6-well dish. Incubate 16–18 h at 37°C.

2. Transfect cells using Oligofectamine according to the manufacturer's instructions. Change the medium after 4–6 h of incubation with the siRNAs.

3. After 48 h, trypsinize PC3 cells if they have reached confluence and replate at a lower density so they do not reach full confluence overnight. Incubate for 16–18 h at 37°C.

4. Use PC3 cells 72 h after transfection for the assays described below (see Note 11). Lyse a small aliquot to determine the level of protein depletion by western blotting or mRNA depletion by quantitative PCR.

3.5. PC3 Cell Labeling

Stain PC3 cells attached on flasks or plates using 10 μM CFSE diluted in PBS$^{+/+}$. Incubate for 20 min at 37°C and then wash twice with PBS$^{-/-}$ to remove the excess dye. Detach cells using non-enzymatic dissociation solution for 10 min at 37°C. Collect and resuspend cells using RPMI containing 0.1% FCS. It may be necessary to tap the bottom of the flask/plate to loosen cells. Calculate the concentration of cells per mL. Whilst quantifying, leave the cells in suspension at 37°C.

3.6. Adhesion Assay to EC and to ECM Components

PC3 cell adhesion to EC or ECM occurs within the first 15 min under static conditions. The two adhesion assays are both carried out in 96-well plates with either a confluent monolayer of EC or with wells pre-coated with ECM components.

3.6.1. PC3 Cell Adhesion Assay to EC

HUVECs and HBMECs are both suitable for cancer cell adhesion assays using this protocol.

1. Coat the appropriate number of wells of a 96-well plate with 50 μL of 10 μg/mL fibronectin (HUVECs) or 0.2% gelatin (HBMECs) in PBS$^{+/+}$ for 1 h at 37°C. This ensures that EC will form a uniform monolayer. The experiments should be planned using quadruplicates per condition of Rho GTPase depletion. An extra control condition (in quadruplicate) should be added to measure the auto-fluorescence of EC (fluorescence background) when measuring adhesion levels with a plate-reader. These wells are treated the same as all the other wells except that no cancer cells are added to the EC. Add EC so that they form a confluent monolayer within 2–3 days and have mature cell–cell junctions (see Note 12). It is important that EC are ready for 3 days after siRNA transfection.

2. Once a confluent EC monolayer has formed, remove medium from the wells, and replace with 100 μL of fresh RPMI containing 1% FCS. Place the cells back in the incubator at 37°C to recover after a medium change, which often induces transient cell contraction.

3. Stain PC3 cells with 10 μM CFSE, then collect and count as above (see Subheading 3.5).

4. For each Rho GTPase knockdown, add 2.5×10^4 CFSE-labeled PC3 cells in 100 μL of RPMI containing 0.1% FCS to the EC in each well except the quadruplicate wells reserved to determine EC auto-fluorescence. Incubate for 15 min at 37°C. Avoid opening the incubator door so that the assay is not disturbed by temperature changes and vibrations.

5. Remove the medium by aspiration and wash once carefully with PBS$^{+/+}$ (see Note 13). Cells need to be checked on a microscope at this stage. Depending on the strength of the adhesion, a second wash can be necessary. After the final wash, add PBS$^{+/+}$ (100 μL) in all the wells including the wells that will be used for EC auto-fluorescence.

6. Read fluorescence values using a plate reader able to detect CFSE fluorescence (see Note 14).

7. Analyze results and normalize cell number regarding fluorescence (see Subheading 3.6.3).

3.6.2. PC3 Cell Adhesion Assay to ECM Components

1. Coat the appropriate number of wells (quadruplicate for each condition) of a 96-well plate with 50 μL of the chosen extracellular matrix proteins for 1 h at 37°C. Use fibronectin at a

concentration of 10 μg/mL, Collagen-1 at a concentration of 50 μg/mL and Matrigel at a concentration of 100 μg/mL. An extra condition should be included to measure ECM protein auto-fluorescence when measuring adhesion levels with a plate reader. These wells are treated the same as all the others except that PC3 cells are not added.

2. After 1 h remove the liquid from the wells and replace with 100 μL of RPMI containing 1% FCS. Place the plate back in the incubator at 37°C.

3. Stain PC3 cells with 10 μM CFSE, then collect and count as above (see Subheading 3.5).

4. For each Rho GTPase depletion, add 2.5×10^4 CFSE-labeled PC3 cells in 100 μL of RPMI containing 0.1% FCS to each well except for the quadruplicates used for fluorescence background measurement. Incubate the cells for 15 min at 37°C. Avoid opening the incubator door during the assay.

5. Aspirate, wash, and measure fluorescence as described in Subheading 3.6.1, steps 5–7 (see Note 13).

3.6.3. Normalization of Cell Number and Analysis of Adhesion Results

The main issue with the cell adhesion assays is to determine whether the same number of PC3 cells has been added in all the wells, both between the four wells for each condition and between the different conditions, e.g., between control siRNA and Rho GTPase siRNA (see Note 15).

1. In parallel to the adhesion assays, add CFSE-labeled PC3 cells to an uncoated 96-well plate (normalization plate) to assess the total number of cells used for each condition. Load exactly the same volume of cells as was added to wells coated with EC or matrix components. Measure the fluorescence values without washing the cells. Use triplicates of each condition and determine the mean fluorescence value.

2. To analyze results, first determine the mean EC auto-fluorescence background level. Remove this background from all the raw data. Using the normalization plate, calculate the relative number of PC3 cells for each condition with respect to the control siRNA cells. The ratio of each condition to control normally varies between 0.5 and 1.5. Divide each fluorescence value in the adhesion plates by this corresponding relative number of cells from the normalization plate. The mean of the control cells on the adhesion plate is then set as 100%. The fluorescence value from each well is calculated as a percentage of this control. Calculate the mean of quadruplicates for each condition in %.

3. Repeat the experiment at least three times to do statistic analysis comparing each condition to control (two-tailed *t*-tests).

3.7. Analysis of PC3 Cell Intercalation and Spreading by Time-Lapse Microscopy

Cancer cell intercalation (see Subheading 3.7.3 for definition) between EC under static conditions occurs between 30 min and 6 h after cells have started to interact with EC. Cancer cell spreading on ECM components occurs within the first 5 min. The two similar assays are both carried out in 24-well plates with either a confluent monolayer of EC or ECM components pre-coated wells. HUVECs and HBMECs are both suitable for intercalation assays using the same protocol.

3.7.1. Time-Lapse Microscopy Analysis of PC3 Cell Intercalation Between EC

1. Coat the appropriate number of wells of a 24-well plate with 300 μL of 10 μg/mL fibronectin in PBS$^{+/+}$ (HUVECs) or 0.2% gelatin in PBS$^{+/+}$ (HBMECs) for 1 h at 37°C to ensure that EC will form a proper and uniform monolayer. Depending of the experiment and the number of conditions to be tested, single of duplicate wells can be used for each Rho GTPase depletion. An extra condition should be added to observe EC alone, which is treated the same as other wells but no cancer cells are added. Add the adequate number of EC so they form a confluent monolayer within 2–3 days and have mature and intact junctions (see Note 10). It is important that EC are ready for day 3 post-siRNA transfection. An appropriate number of movies need to be acquired in order to have enough cancer cell–endothelium interactions monitored (see Note 16).

2. Once a confluent EC monolayer has been formed, remove medium from the wells, and replace with 500 μL of fresh EGM-2 medium. Put the cells in the humidified chamber (at 37°C and 5% CO_2) of the time-lapse microscope so that they can recover after the medium change and so the plate is warmed up in the chamber avoiding any condensation on the lid, which adversely affects the images. Choose appropriate fields focusing on homogenous regions of the EC monolayer and save the coordinates of the different positions that will be recorded. Everything needs to be ready before PC3 cells are added, so that data can be acquired as fast as possible (see Note 17).

3. Stain PC3 cells with 10 μM CFSE, then collect and count as above (see Subheading 3.5).

4. For each Rho GTPase depletion, add 3.5×10^4 CFSE-labeled PC3 cells to confluent EC monolayers in the 24-well plate directly in the chamber of the time-lapse microscope. This allows recording of the events occurring when cancer cells interact with EC as early as possible. Cells are monitored by time-lapse microscopy for up to 6 h. Both phase contrast and fluorescent images are taken every 5 min or less. It is crucial to adjust the focus on EC cells since PC3 cells will very rapidly come in focus as they interact with EC (see Note 18).

5. Analyze images as described in Subheading 3.7.3.

3.7.2. Analysis of PC3 Cell Spreading on ECM Components by Time-Lapse Microscopy

1. Coat the required number of wells of a 24-well plate with 300 μL of the appropriate extracellular matrix proteins chosen for the assay for 1 h at 37°C (see Subheading 3.6.2, step 1). Depending on the experiment, single wells or duplicates can be used for each Rho GTPase depletion. An appropriate number of movies need to be acquired in order to have enough cancer cell–ECM interactions monitored (see Note 16).

2. After 1 h remove the different coatings from the wells and replace with 500 μL of fresh RPMI containing 1% FCS. Put the plate in the humidified chamber (at 37°C and 5% CO_2) of the time-lapse microscope so the plate is warmed up in the chamber avoiding any condensation that could affect the image.

3. Stain PC3 cells with 10 μM CFSE, then collect and count as above (see Subheading 3.5).

4. For each Rho GTPase depletion, add 3.5×10^4 CFSE-labeled PC3 cells in 100 μL of RPMI containing 0.1% FCS to each well directly in the chamber of the time-lapse microscope and monitor their behavior as describe above (see Subheading 3.7.2, step 4). For more detailed experiments of the early events of spreading, movies can be made of each condition individually (see Note 19).

5. Analyze movies as described in Subheading 3.7.3.

3.7.3. Analysis of Time-Lapse Movies

1. To quantify PC3 cell intercalation, a cell is considered to have intercalated when (1) it changes from phase bright to phase dark and (2) its shape is no longer round and it is clearly part of the EC monolayer and spread (Fig. 2). The CFSE fluorescence can be used to follow PC3 cells during the intercalation process. Cells are tracked manually using ImageJ software to measure the time to complete their intercalation, their migration speed, and the migration distance before they complete their intercalation (Fig. 3). T_{50}, when 50% of a given population has intercalated, is also determined.

2. To analyze cell spreading, cells plated on ECM components and imaged by time-lapse microscopy for up to 6 h can be fixed at different time points. Cell surface area of over 100 cells acquired in each of three independent experiments is measured using ImageJ software. Kymographs of cell spreading can also be made using Metamorph software.

3.8. Detailed Analysis of Events That Follow PC3 Cell Adhesion to EC

3.8.1. Immunofluorescence After PC3 Cell Adhesion Assays to EC

1. Place the appropriate number of coverslips (each in one well) in a 24-well plate and coat with 300 μL of 10 μg/mL fibronectin in PBS$^{+/+}$ (HUVECs) or 0.2% gelatin in PBS$^{+/+}$ (HBMECs) for at least 12 h at 37°C to ensure that EC will adhere, spread, and form a proper and uniform monolayer. The experiments should be planned using several coverslips per condition of Rho GTPase depletion as cocultures will be fixed with PFA at different time points after PC3 cells have been added to EC.

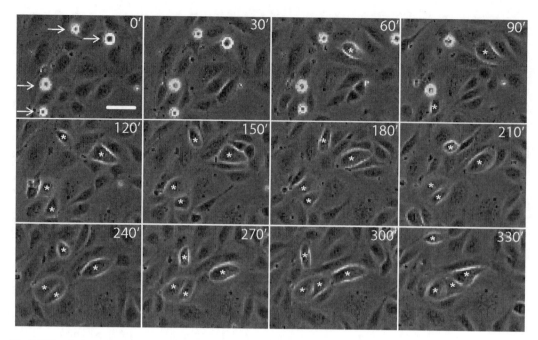

Fig. 2. PC3 cells intercalate between HUVECs. Images taken from time-lapse movies of CFSE-labeled PC3 cells (indicated by the *arrows*) adhering to and intercalating between EC by opening EC junctions. Cells were added to confluent HUVEC monolayers for 330 min and were filmed during their intercalation. Cells are indicated with an asterisk once they have intercalated. Scale bar = 50 μm.

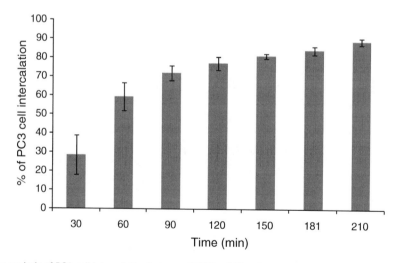

Fig. 3. Tracking analysis of PC3 cell intercalation between HUVECs. PC3 cells were added to confluent HUVEC monolayers for 210 min and were tracked individually over time. Data are expressed as %. Values are means ± SEM ($n > 3$). For each experiment, at least 100 different single cells were analyzed in at least three different fields recorded.

Extra coverslips should be included to observe EC alone. The latter will be treated the same all along except that EC will not receive cancer cells. Add the adequate number of EC so they form a confluent monolayer within 2 to 3 days and let EC junctions mature 2–3 days at confluency. Change the medium every day once cells are confluent (see Note 20). It is important that EC are ready for day 3 post-siRNA transfection.

2. Once a confluent EC monolayer has been formed, remove medium from the top chamber, and replace with 500 µL of fresh EGM-2 medium. Place back the cells in the incubator at 37°C as the cells need to recover after a medium change as it often induces cell contractility.

3. Stain PC3 cells with 10 µM CFSE, then collect and count as above (see Subheading 3.5).

4. For each Rho GTPase depletion, add 3.5×10^4 CFSE-labeled PC3 cells in 100 µL of EGM-2 to EC in each well except for the duplicate wells used to monitor EC monolayer integrity. Incubate the coculture for 30 or 60 min or longer at 37°C. Avoid opening the incubator door so that the assay is not disturbed by temperature changes and vibrations.

5. Cells are fixed by transferring individual coverslips into another 24-well dish in which each well contains 300 µL of 4% PFA in $PBS^{+/+}$. Fix cells at different time points (30 and 60 min) for 20 min (see Note 21).

6. After fixation, cells are permeabilized with 0.1% Triton X-100 for 5 min at 4°C and then blocked with 5% FCS in $PBS^{+/+}$ for 20 min at RT.

7. Incubate samples with β-catenin or VE-cadherin primary antibodies for 60 min at room temperature then wash three times in 5% FCS in $PBS^{+/+}$ for 5 min. Incubate samples with Alexa 543 goat anti rabbit and goat anti-mouse antibodies respectively, together with Alexa-647-Phalloidin and DAPI for 45 min at room temperature then wash again three times in $PBS^{+/+}$ for 5 min. VE-cadherin stains endothelial cell junctions specifically while β-catenin stains both endothelial cell junctions and cancer cells. Phalloidin stains F-actin in EC and cancer cells while DAPI stains EC and cancer cell nuclei.

8. Samples are mounted onto slides with mounting medium. Allow mounting medium to set to dry before imaging.

9. Acquire images using a confocal microscope. Images are processed using Adobe Photoshop software.

3.8.2. Analysis of the Results

Cells are analyzed based on three different criteria using confocal images as presented in Fig. 4.

1. By observing the CFSE staining it is possible to determine the precise localization of the PC3 cell adhesion sites: cells are classified to be on top of EC junctions, on top of the EC body, or fully intercalated.

136   N. Reymond et al.

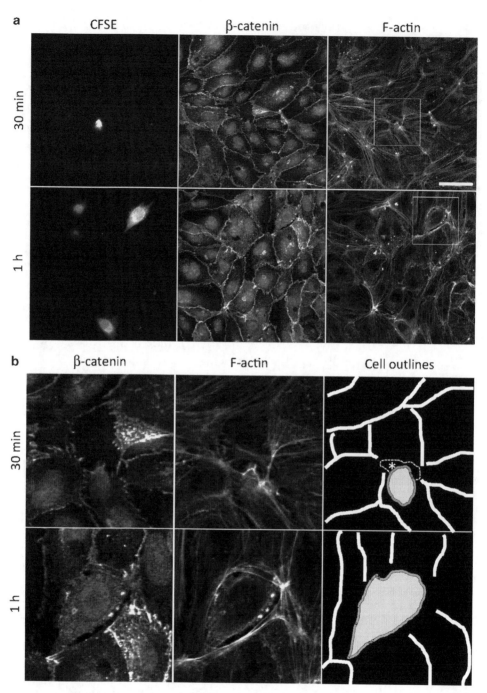

Fig. 4. PC3 cells induce the creation of local gaps between HUVECs before intercalating. (**a**) Confocal images of CFSE-labeled PC3 cells added to HUVECs for 30 or 60 min. The cocultures were fixed and stained for β-catenin and F-actin (Alexa-540-labeled phalloidin). Scale bar = 50 μm. (**b**) Zoom on single cells shown in (**a**). Cell outlines are indicated.

2. The status of EC junctions (open or closed) underneath cancer cells is determined using β-catenin or VE-cadherin staining, DAPI and Alexa-546-conjugated phalloidin.

3. The site of intercalation is determined by observing if PC3 cells complete their intercalation between two adjacent EC junctions (radial spreading) or at multicellular corner junctions (axial spreading).

4. Quantify at least 50 cells per condition. Normalize each dataset in percentage relative to control. Repeat the experiment three times to do statistic analysis comparing each condition to control (two-tailed t-tests).

3.9. PC3 Cell TEM Through Boyden Chambers

PC3 cell TEM occurs between 6 and 24 h. To avoid cancer cell division and to use minimum amount of cells, we use a modified Boyden chamber assay using Transwell inserts coupled to a flow cytometry detection approach to count very precisely the number of cells that have extravasated (Fig. 5a).

3.9.1. TEM Assay

1. Coat Transwell inserts with 10 μg/mL fibronectin for 1 h at 37°C. Plate HUVECs at 7.5×10^4 cells/well in EGM-2 medium (200 μL in the upper chamber and 600 μL in the lower chamber) (see Note 22). HUVECs should form a confluent monolayer within 2 days. It is important that EC are ready for 3 days after siRNA transfection. Plan the experiments using quadruplicates per Rho GTPase depletion, an appropriate siRNA control and including 4 control wells that will not receive any cancer cells.

2. On the day of the assay, change medium in both chambers and stimulate HUVECs in the upper chamber with 300 ng/mL of SDF1α in EGM-2 medium for 10 min (see Note 23). Keep normal EGM-2 in the lower chamber. After 10 min, remove medium and replace with fresh EGM-2 medium.

3. Stain PC3 cells with 10 μM CFSE, then collect and count as above (see Subheading 3.5). Resuspend cells in fresh EGM-2 medium.

4. For each Rho GTPase depletion, add 2.5×10^4 CFSE-labeled PC3 cells per Transwell insert in the top chamber in a total volume of 200 μL in EGM-2 medium (see Note 24).

5. Add HGF (40 ng/mL) as a chemo-attractant in the lower chamber in a total volume of 600 μL (Note 22). Allow cells to transmigrate for 6 h at 37°C.

6. Remove medium in the bottom chamber. PC3 cells remain attached on the filter and do not fall down into the bottom well. Recover PC3 cells from the bottom of the filter by inverting the filter and adding 50 μL of warm Trypsin-EDTA directly on the bottom part of the filter. Pipette up and down everywhere

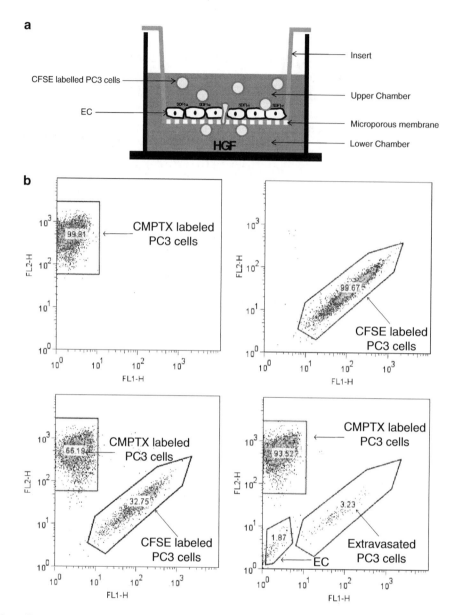

Fig. 5. Quantification of PC3 cell extravasation using FACS. (**a**) HUVECs are stimulated with SDF1α. HGF was used as a chemo-attractant in the lower chamber. CFSE-labeled PC3 cells are added to the upper compartment of a Boyden chamber and are allowed to transmigrate for 8 h at 37°C. Extravasated cells were then trypsinized from the bottom part of the filter and counted by FACS analysis. (**b**) FACS analysis of CMPTX-labeled cells alone (*top left panel*), CFSE-labeled cells alone (*top right panel*), CMPTX-labeled cells and CFSE-labeled cells mixed together (*lower left panel*) and an example of extravasated cells (*lower right panel*).

on the filter. Transfer the 50 µL to a FACS tube containing 300 µL of PBS$^{-/-}$ containing 5% FCS.

7. CMPTX-labeled PC3 cells are used to determine the exact number of CFSE-labeled PC3 cells that have extravasated using flow cytometry. During the 6-h Transwell incubation,

stain untransfected PC3 cells, attached on flasks or plates, with 10 µM CMPTX diluted in PBS$^{-/-}$. Incubate for 20 min at 37°C then collect and count as above (see Subheading 3.5). Resuspend the cells at 10^6 cells/mL in PBS$^{-/-}$. Add 100 µL of CMTPTX-labeled cell suspension to each FACS tube that contains extravasated cells.

3.9.2. Analysis of Results by Flow Cytometry

Cells that have extravasated are counted using flow cytometry.

1. Prepare three control tubes with CMPTX-labeled cells only, CFSE-labeled cells only and a mix of both. These controls will be used to set up compensation settings before acquiring data. The CFSE fluorescence can bleed through into the channel used to detect the CMPTX fluorescence. Compensation should be used in order to separate CMPTX- and CFSE-labeled cells well, so that extravasated CFSE-labeled cells can be counted (see Fig. 5b). Different gates are used for CMPTX-labeled cells, for HUVECs that may have grown on the bottom of the filter and for CFSE-labeled cells.

2. Set up the flow cytometer to record 5,000 or 10,000 events in gate G1 automatically. As this will be done for all the samples and as all of them have the same number of CMPTX-labeled cells then the total number of CFSE-labeled transmigrated cells will be determined and will be comparable between them.

3. Analyze results and normalize cell number for fluorescence as described in Subheading 3.6.3: divide each value by its corresponding relative number of cells. The mean of control cells is set as 100%. Normalize each data point in percentage relative to control, and then determine the mean of quadruplicates in %. Repeat the experiment three times to allow statistical analysis of differences (two-tailed t-tests).

4. Notes

1. Fibronectin is purchased already dissolved. Make aliquots and freeze as 50× stocks (500 µg/mL). Thaw aliquots as needed and dilute in PBS$^{+/+}$ to working concentration (10 µg/mL) and store at 4°C for a few weeks.

2. PFA is stable at 4°C for a week. Do not freeze/thaw aliquots as it is detrimental for the fixation.

3. Cover the bottom of culture flasks/dishes with a sufficient volume of fibronectin solution so that it is evenly coated (e.g., 2.5 mL of fibronectin solution is enough to coat a 75-cm^2 flask).

4. Do not centrifuge HUVECs to remove DMSO as this reduces viability.

5. HUVECs are extremely sensitive to any change of temperature and CO_2 concentration.

6. HUVECs grown beyond passage 4 gradually lose their endothelial phenotype, become larger and more spread and eventually enter senescence. HUVEC should not be grown to high confluence as this leads to irregular cell morphology.

7. Incubation of HUVECs with trypsin–EDTA for too long drastically reduces their viability.

8. The HBMEC cell line can be maintained at confluence for longer than HUVECs without adversely affecting their morphology.

9. HBMECs grow much faster than HUVECs hence cells can be split at higher dilutions.

10. At least two different siRNA oligos per Rho GTPase should be used for all assays, to ensure reproducibility and reduce the possibility of off-target effects. One irrelevant control siRNA should also be used in all experiments. If the single oligos have not been validated previously, it is possible to work with pools of oligos, together with a pool of control siRNAs.

11. We have observed that the efficiency of Rho GTPase depletion is optimal 72 h after transfection in terms of morphology and protein knockdown.

12. Our laboratory has found that HUVECs require at least 24 h to form mature adherens junctions, as determined by endothelial permeability and staining.

13. Aspiration and washes should be done quickly with an efficient aspiration. Multi-well aspirators are sold by many companies. Avoid direct contact between the cells and the tips and wash by adding PBS$^{+/+}$ on the walls of the wells.

14. When making fluorescence measurements with a plate-reader, it is important to first do a fast scan of the plate to localize the regions with the highest fluorescence. The apparatus settings can then be adjusted to ensure no well will be saturated for fluorescence detection as no differences between wells will be observed if the system is saturated.

15. To be sure that any change observed (increase or decrease) is a consequence of a change of adhesion and not of a difference in the number of cells added at the beginning ensure that (1) a multichannel pipette is used to add cancer cells to EC and/or to ECM components and a multichannel aspirator is used for the washes and (2) the fluorescence level of each sample is also measured using another plate. This will allow the fluorescence

value measured during the adhesion assay to be divided by the relative number of cells used as the input so the different conditions can be compared accurately.

16. When acquiring images using a time-lapse microscope, the best phase contrast images are obtained in the center of the wells.

17. In order to record the maximum number of events as early as possible, it is better to choose the fields that will be analyzed by time-lapse microscopy first. Focus on EC before adding the cancer cells. PC3 cells will initially be out of focus but will progressively become in focus when they adhere to EC. It is best to use the auto-focus function every hour, if available on the microscope.

18. Depending on the magnification, different number of movies can be acquired. If using a 10× objective, three movies are normally enough to observe at least 100 cells. If using a 20× objective, then five movies should be acquired. The 40× objective is used to observe fine details of cell morphology rather than for tracking.

19. When assessing cell spreading, add the cells to the different wells and wait until they make their first contact with the substrate at the bottom of the wells. As soon as one of the cells starts to spread, it will extend protrusions. Adjust the focus in each field with these first protrusions.

20. HUVECs can be kept at high confluence for 2–3 days if the medium is changed every day and their morphology carefully observed on a microscope to check they do not become irregular.

21. In order to keep the maximum number of PC3 cells on top of EC, transfer each coverslip to another well previously filled up with 4% PFA, without washing. PC3 cells that are weakly adhered will thereby remain attached to the EC and can be analyzed.

22. It is important to use the volumes indicated by the Transwell providers (normally 200 μL for the top chamber and 600 μL for the lower chamber respectively).

23. SDF1α stimulation of EC for 10 min is sufficient to ensure that it binds to the EC surface and thereby increases the binding of PC3 cells to EC.

24. The number of PC3 cells to add to EC grown on Transwell filters is crucial. Adding too many PC3 cells to EC induces detachment of EC, and PC3 cells can then migrate freely without interacting with EC.

References

1. Wolf, K., Wu, Y.I., Liu, Y., Geiger, J., Tam, E., Overall, C., Stack, M.S., and Friedl, P. (2007) Multi-step pericellular proteolysis controls the transition from individual to collective cancer cell invasion. *Nat Cell Biol* 9, 893–904.

2. Friedl, P., and Gilmour, D. (2009) Collective cell migration in morphogenesis, regeneration and cancer. *Nat Rev Mol Cell Biol* 10, 445–57.

3. Giampieri, S., Manning, C., Hooper, S. Jones, L., Hill, C.S., and Sahai, E. (2009) Localized and reversible TGFbeta signalling switches breast cancer cells from cohesive to single cell motility. *Nat Cell Biol* 11, 1287–96.

4. Li, C.Y., Shan, S., Huang, Q., and Dewhirst, M.W. (2000) Initial stages of tumor cell-induced angiogenesis: evaluation via skin window chambers in rodent models. *J Natl Cancer Inst* 92, 143–7.

5. Dadiani, M., Kalchenko, V., Yosepovich, A., Margalit, R., Hassid, Y., Degani, H., and Seger, D. (2006) Real-time imaging of lymphogenic metastasis in orthotopic human breast cancer. *Cancer Res* 66, 8037–41.

6. Wyckoff, J.B., Jones, J.G., Condeelis, J.S. and Segall, J.E. (2000) A critical step in metastasis: in vivo analysis of intravasation at the primary tumor. *Cancer Res* 60, 2504–11.

7. Ito, S., Nakanishi, H., Ikehara, Y. Kato, T., Kasai, Y., Ito, K., Akiyama, S., Nakao, A., and Tatematsu, M. (2001) Real-time observation of micrometastasis formation in the living mouse liver using a green fluorescent protein gene-tagged rat tongue carcinoma cell line. *Int J Cancer* 93, 212–7.

8. Naumov, G.N., Wilson, S.M., MacDonald, I.C. Schmidt, E.E., Morris, V.L., Groom, A.C., Hoffman, R.M., and Chambers, A.F. (1999) Cellular expression of green fluorescent protein, coupled with high-resolution in vivo videomicroscopy, to monitor steps in tumor metastasis. *J Cell Sci* 112, 1835–42.

9. Im, J.H., Fu, W., Wang, H., Bhatia, S.K., Hammer, D.A., Kowalska, M.A., and Muschel, R.J. (2004) Coagulation facilitates tumor cell spreading in the pulmonary vasculature during early metastatic colony formation. *Cancer Res* 64, 8613–9.

10. Tsuji, K., Yamauchi, K., Yang, M., Jiang, P., Bouvet, M., Endo, H., Kanai, Y., Yamashita, K., Moossa, A.R., and Hoffman, R.M. (2006) Dual-color imaging of nuclear-cytoplasmic dynamics, viability, and proliferation of cancer cells in the portal vein area. *Cancer Res* 66, 303–6.

11. Chambers, A.F., Groom, A.C., and MacDonald, I.C. (2002) Dissemination and growth of cancer cells in metastatic sites. *Nat Rev Cancer* 2, 563–72.

12. Nguyen, D.X., Bos, P.D., and Massague, J. (2009) Metastasis: from dissemination to organ-specific colonization. *Nat Rev Cancer* 9, 274–84.

13. Joyce, J.A., and Pollard, J.W. (2009) Microenvironmental regulation of metastasis. *Nat Rev Cancer* 9, 239–52.

14. Vega, F.M., and Ridley, A.J. (2008) Rho GTPases in cancer cell biology. *FEBS Lett* 582, 2093–101.

15. Madsen, C.D., and Sahai, E. (2010) Cancer dissemination--lessons from leukocytes. *Dev Cell* 19, 13–26.

16. Dimitroff, C.J., Descheny, L., Trujillo, N., Kim, R., Nguyen, V., Huang, W., Pienta, K.J., Kutok, J.L., and Rubin, M.A. (2005) Identification of leukocyte E-selectin ligands, P-selectin glycoprotein ligand-1 and E-selectin ligand-1, on human metastatic prostate tumor cells. *Cancer Res* 65, 5750–60.

17. Muller, A., Homey, B., Soto, H., Ge, N., Catron, D., Buchanan, M.E., McClanahan, T., Murphy, E., Yuan, W., Wagner, S.N., Barrera, J.L., Mohar, A., Verástegui, E., and Zlotnik, A. (2001) Involvement of chemokine receptors in breast cancer metastasis. *Nature* 410, 50–6.

18. Cardones, A.R., Murakami, T., and Hwang, S.T. (2003) CXCR4 enhances adhesion of B16 tumor cells to endothelial cells in vitro and in vivo via beta(1) integrin. *Cancer Res* 63, 6751–7.

19. Kusama, T., Mukai, M., Tatsuta, M., Nakamura, H., and Inoue, M. (2006) Inhibition of transendothelial migration and invasion of human breast cancer cells by preventing geranylgeranylation of Rho. *Int J Oncol* 29, 217–23.

20. Schweitzer, K.M., Vicart, P., Delouis, C. Paulin, D., Dräger, A.M., Langenhuijsen, M.M., and Weksler, B.B. (1997) Characterization of a newly established human bone marrow endothelial cell line: distinct adhesive properties for hematopoietic progenitors compared with human umbilical vein endothelial cells. *Lab Invest* 76, 25–36.

21. Aird, W.C. (2003) Endothelial cell heterogeneity. *Crit Care Med* 31, S221–30.

22. Kaighn, M.E., Narayan, K.S., Ohnuki, Y., Lechner, J.F., and Jones, L.W. (1979) Establishment and characterization of a human prostatic carcinoma cell line (PC-3). *Invest Urol* 17, 16–23.

23. Webber, M.M., Bello, D., and Quader, S. (1997) Immortalized and tumorigenic adult human prostatic epithelial cell lines: characteristics and applications Part 2. Tumorigenic cell lines. *Prostate* 30, 58–64.

Chapter 10

Quantitative and Robust Assay to Measure Cell–Cell Contact Assembly and Maintenance

Sébastien Nola, Jennifer C. Erasmus, and Vania M.M. Braga

Abstract

Epithelial junction formation and maintenance are multistep processes that rely on the clustering of macromolecular complexes. These events are highly regulated by signalling pathways that involve Rho small GTPases. Usually, when analysing the contribution of different components of Rho-dependent pathways to cell–cell adhesion, the localisation of adhesion receptors at junctions is evaluated by immuno-fluorescence. However, we find that this method has limitations on the quantification (dynamic range), ability to detect partial phenotypes and to differentiate between the participation of a given regulatory protein in assembly and/or maintenance of cell–cell contacts. In this chapter, we describe a suitable method, the aggregation assay, in which we adapted a quantitative strategy to allow objective and reproducible detection of partial phenotypes. Importantly, this methodology estimates the ability of cells to form junctions and their resistance to mechanical shearing forces (stabilisation).

Key words: Cadherin adhesion, Epithelia, Aggregation assay, Quantification methods, Junctions

1. Introduction

Cadherin-mediated cell–cell contacts are regulated by different small GTPase family members such as Rho, Rac, Cdc42, Ras, Arf6 and Rap1 (1, 2). However, the specific effectors involved downstream of each GTPase are not well understood. Mechanistically, it is important to differentiate between effector proteins that regulate the localisation of cadherins to junctions (i.e., transport to cell surface and clustering) to those that contribute to the stabilisation of receptors already present at cell–cell contacts (association with the cytoskeleton, strength of cell–cell contacts).

Francisco Rivero (ed.), *Rho GTPases: Methods and Protocols*, Methods in Molecular Biology, vol. 827,
DOI 10.1007/978-1-61779-442-1_10, © Springer Science+Business Media, LLC 2012

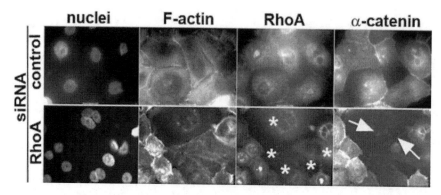

Fig.1. Depletion of RhoA prevents assembly of cell–cell adhesion. Keratinocytes grown in low calcium medium were treated with control or RhoA siRNA oligos and induced to assemble cell–cell contacts for 30 min. Cells were fixed and stained with DAPI (nuclei), phalloidin (F-actin) anti-RhoA and anti-α-catenin to label cell–cell contacts. Arrows show the absence of α-catenin re-localisation to junctions; asterisks mark cells with reduced levels of RhoA in the monolayer.

Perturbation of GTPase function by different means gives a clear phenotype on disruption of junctions when assessed by localisation of cadherin complexes by immunofluorescence (Fig. 1). Yet, multiple GTPase effectors may cooperate to promote stable cell–cell adhesion (3). Thus, interfering with a single effector protein would give a partial phenotype that may be beyond the detection limit of visual inspection. The above issue raises the need to develop more sensitive, quantitative methods that allow determination of partial phenotypes reproducibly.

Aggregation assays have been used for many years in the adhesion field and were essential in the original characterisation of the properties of different cadherin receptors (4). In the original method, cells aggregate in suspension using a rotator shaker and the number of cells in each aggregate is quantified manually, which is laborious and time consuming (4). An alternative method is the "hanging drop" method, in which cell suspension droplets are hanged by surface tension from a Petri dish lid and the cells fall to the bottom of the droplet by gravity to form aggregates (hence the name "hanging drop method") (Fig. 2).

The hanging drop and rotator shaker methods give comparable results (5–7). However, during gyratory aggregation assays, cells are subjected to constant mechanical stress, which has been described to perturb GTPase activity by modifying cell shape and cytoskeleton (see for example ref. 8). Hence, the gravity-based hanging drop assay is preferable to study the role of GTPase in forming cell–cell junctions to avoid non-specific effects.

Although the hanging drop method is technically easier to perform, issues still remain with the analysis of the aggregates, which can be subjective, laborious and limit the number of samples to be analysed. Qualitative analysis of the hanging drop method

a Lateral View

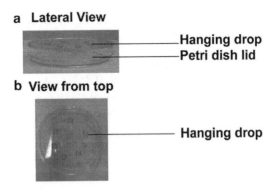

Hanging drop
Petri dish lid

b View from top

Hanging drop

Fig. 2. Image of hanging drops from the lid of a Petri dish. Lateral view (**a**) and top view (**b**) of the Petri dish containing six hanging drops on the lid.

has been done via photographs of aggregates (9–13). Alternatively, the ability of cells to aggregate can be quantified by two different methods. First, the number of cells in each aggregate is scored and estimated visually (14, 15). Second, the size of aggregates is evaluated with image processing tools by manually selecting the regions of interest (16–21). However, measuring the area of each aggregate by manual selection is time consuming and can lead to variability issues across experiments. Thus, a more simple and precise method for quantification of aggregate size is welcomed to allow fast, objective and reproducible data analysis.

We have adapted the hanging drop method and used it successfully to determine the participation of different Rho GTPase effectors in E-cadherin dependent signalling. Using free software (ImageJ), we have designed an automated method for quantification of aggregates that allows precise measurements of partial phenotypes in an objective manner. As such, this method quantifies the ability of cells to adhere to each another (assembly) and the stability of junctions following mechanical stress (maintenance).

Our method has several advantages:

- Aggregation assays are very good for subtle phenotypes upon RNAi, even when there is no obvious change using conventional immunofluorescence assays.

- Phenotypes in which cell–cell adhesion is enhanced (larger aggregates) or perturbed (smaller aggregates) can be observed following RNAi, thereby identifying positive or negative regulators of junction stability.

- To determine the specificity of the RNAi phenotype, rescue experiments can also be performed with RNAi-insensitive cDNAs, even if transfection efficiency is poor (i.e., in primary cells).

- By coupling aggregation and trituration of pre-formed aggregates, assessment of assembly and maintenance of cell–cell contacts can be inferred.

Our method has, however, the disadvantages that:

- Because adhesion is stimulated by calcium ions, other calcium-dependent receptors such a desmosomes also contribute to the size of aggregates. Yet, if E-cadherin adhesion is perturbed at any degree, it has a knock-on effect on other adhesive systems. This caveat can, however, be avoided and the specificity of the phenotype to a particular receptor determined by addition of inhibitory antibodies (22, 23) or siRNA.

- Handling of droplets is critical: special care and attention should be taken during the optimisation process to insure good reproducibility across experiments.

- This method is not suitable for large-scale studies as the handling of droplets can be limiting.

2. Materials

1. Cells: primary keratinocytes isolated from human foreskins (24).

2. Tissue culture dishes with lids: 2-cm^2 and 19.6-cm^2 dishes (Fisher).

3. Standard calcium medium for keratinocytes: standard medium (DMEM:F12; Sigma) supplemented with 1.8 mM $CaCl_2$, 100 U/mL penicillin, 100 μg/mL streptomycin, 10% foetal calf serum, 5 mM glutamine, 5 μg/mL insulin, 10 ng/mL epidermal growth factor (EGF), 0.5 μg/mL hydrocortisone and 0.1 nM cholera toxin (24).

4. Low calcium medium: same composition as the standard medium, but with no calcium added. The serum used for this medium must be pretreated with Chelex-100 resin (BioRad, catalogue # 143-2832) to chelate the calcium ions (final concentration 0.1 mM) (25).

5. Specific siRNA oligos (Dharmacon; custom made) and control scrambled oligo (Dharmacon: D-001206-13) and siRNA transfection reagents (INTEFERin; Peqlab). RNAi is conducted following the manufacturer's instructions or according to procedures previously established in the laboratory.

6. Phosphate buffered saline (PBS): NaCl 8 g/L; KCl 0.2 g/L; Na_2HPO_4 1.44 g/L; KH_2PO_4 0.24 g/L, pH 7.4.

7. Trypsin: NaCl 8 g/L; KCl 0.2 g/L; Na_2HPO_4 1.15 g/L; KH_2PO_4 0.2 g/L; trypsin 2.5 g/L; EDTA 0.2 g/L; phenol red 0.0015 g/L.

8. Versene: 0.53 mM EDTA in PBS.

9. Trypsinisation buffer: 60% versene (v/v), 0.01% trypsin, 1 mM $CaCl_2$.

10. Haemocytometer (Fisher).

11. Lysis buffer: 10 mM Tris–HCl pH 7.5, 1% NP-40, 150 mM NaCl, 1 mM $MgCl_2$, 50 mM NaF, 1 mM DTT, 5 µM Leupeptin, 1 mM PMSF.

12. Laemmli buffer: 60 mM Tris–HCl pH 6.8, 2% SDS, 10% glycerol, 5% β-mercaptoethanol, 0.01% bromophenol blue.

13. Wide-field microscope with 5× or 10× phase contrast objective and digital camera attached.

14. Imaging software such as ImageJ (available free at http://www.rsbweb.nih.gov/ij/).

15. Microsoft Excel.

3. Methods

A key aspect of the aggregation method is to avoid cleavage of cadherins by trypsin. This is achieved by providing low levels of calcium ions to stabilise cadherin extracellular domains against cleavage, but not enough to allow homophilic adhesion (requires over 250 µM). If the above is not performed, a considerable lag in aggregate formation is seen, as intact cadherin receptors have to be synthesised and transported to the cell surface to support cell–cell adhesion. This lag is not desirable, as many types of epithelial cells become apoptotic in suspension, with obvious implications for the interpretation of the phenotype.

The use of a hanging drop to aggregate minimises the number of cells required and avoids the problem of available shakers at 37°C. We describe here a simplified and adapted version of a protocol published for lymphocyte aggregation (26). One important aspect is the provision of an automated macro to identify aggregates in phase contrast images, exclude single cell/doublets and quantify the area of thresholded aggregates (Fig. 3). Furthermore, the area of aggregates before and after trituration in each sample are quantified and normalised to allow comparison between experiments and control for variations in cell densities. When coupled with methods to interfere with the function of endogenous proteins (overexpression, RNAi), the quantification shown herein is particularly suitable for detection of partial phenotypes and to differentiate between a function in assembly or maintenance of cell–cell contacts (Fig. 4).

This procedure has the advantages that:

- Aggregate area is directly proportional to the number of cells present in the aggregate.

- Provides an automated and reproducible selection of the aggregates.

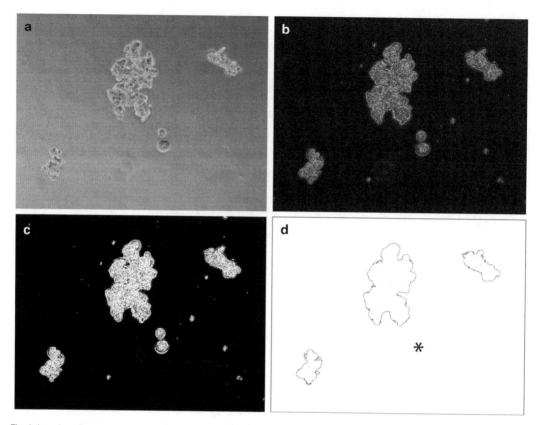

Fig. 3. Imaging of aggregates. (**a**) Image of aggregates obtained from phase contrast microscopy and converted to greyscale. (**b**) Image after "Find edges" process. Note that the background is more uniform and the contrast between aggregates and background is increased. (**c**) Image of aggregates after thresholding (binary mask). (**d**) Outlines of selected aggregates after particles analysis. Aggregates smaller than 3 cells are excluded (*asterisk*).

- Minimises inaccuracy or poor reliability of manual selection of each aggregate.

- Reduces time of analysis.

- Does not depend on an arbitrary thresholding or visual cell counting.

- Can be done using free image processing software such as ImageJ.

It should be nevertheless kept in mind that the quantification of aggregate area from phase contrast images:

- Does not provide the precise number of cells in each aggregate.

- Assumes that cell size is constant within each cluster and does not consider cell shape variability.

- Does not consider the level of compaction of the aggregates, which can reduce the area of a given aggregate. However, in

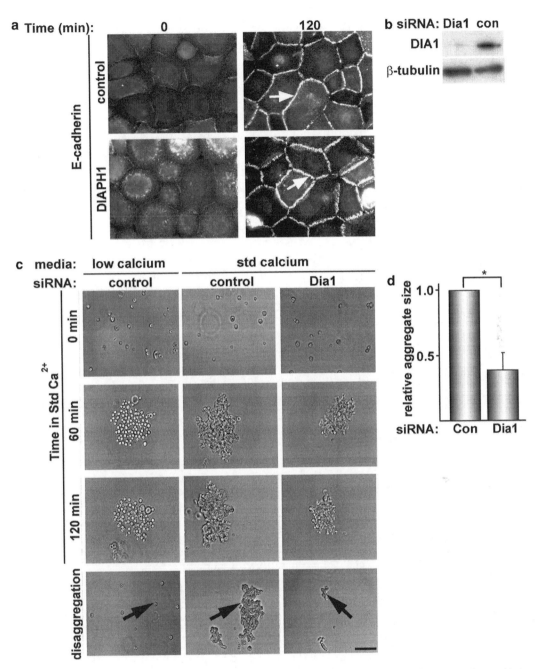

Fig. 4. Aggregation assay. Confluent keratinocytes grown in low calcium medium were treated with control scrambled or Dia1 oligo. (a) Cells were fixed in the absence (0 min) or presence (120 min) of cell–cell contacts and stained for E-cadherin. *Arrows* shows junctions that do not appear to be perturbed by Dia1 depletion. (b) Western blots of lysates prepared from parallel samples were probed with anti-Dia1 antibodies and β-tubulin as loading control. (c) Keratinocytes were trypsinised and a single-cell suspension made. Cells were maintained in low calcium medium or allowed to aggregate for 120 min after which they were pipetted up and down to disrupt aggregates. Images were taken before and after trituration. (d) Quantification of aggregation assays. Area of each aggregate was quantified as described and values are expressed relative to controls (arbitrarily set as 100%). A difference between the aggregates remaining after trituration is observed between the scrambled control oligo and the Dia1 oligo treated cells. No difference was observed in the aggregates formed after 60 min or 120 min (not shown). Scale bar = 200 μm.

our experience, the latter does not pose problems with the reproducibility of the experiments, but rather underestimates the extent of the phenotype.

- Might detect a few false positive particles.

3.1. Aggregation Assay

1. Seed keratinocytes into 4-well dishes (2-cm^2) in standard calcium medium to perform the aggregation assays and a set of parallel samples to check depletion by Western blots. Allow cells to grow until colonies reach about 20 cells.

2. Switch cells to low calcium medium and grow until about 90% confluence.

3. Treat with desired siRNA or inhibitor for optimised amount of time. RNAi is optimised previously for each oligonucleotide (incubation time and concentration). After 4 h, the transfection solution is removed and fresh medium placed on the cells. The cells are then incubated for the optimised time for protein depletion by siRNA. Wash cells once with warm versene.

4. Process a set of samples for Western blot to confirm depletion using standard procedures (protein extraction in lysis buffer, clarification of lysates at $8,800 \times g$ for 10 min at 4°C, addition of Laemmli buffer and SDS-PAGE).

5. On the remaining samples, trypsinise cells in each well in 500 µL of trypsinisation buffer (see Note 1).

6. Allow cells to detach – this can take up to 20 min (see Note 2).

7. Pipette the suspension gently to disrupt any clusters. Keratinocytes can attach to each other very strongly and one will need to press the tip of the pipette down onto the bottom of the dish to obtain a single-cell suspension. Do not add any medium to the cells at this stage.

8. Take a small aliquot of the suspension (10 µL) for counting cells using a haemocytometer.

9. Centrifuge cells at $36 \times g$ for 5 min. Remove the supernatant very carefully using a 1,000-µL pipette. Take great care not to disrupt the cell pellet (see Note 3).

10. Re-suspend cells to a density of 5×10^4 cells/mL in standard calcium medium or in low calcium medium (negative control) (see Note 4).

11. Fill the bottom of a Petri dish (19.6-cm^2) with sterile water to create a humid chamber. One Petri dish will be needed per sample. Invert the lid upside down.

12. Pipette six drops of 20 µl of the cell suspension of each sample (5×10^4 cells/mL) onto the internal side of the lid of the dish (see Fig. 2).

13. Place the lid back on the bottom part of the Petri dish and label each droplet for future quantification and imaging. Return the dishes to an incubator at 37°C, 5% CO_2.

14. To take pictures, place the lid of the Petri dish onto the stage of the microscope in an inverted position (upside down; see Note 5).

15. Take photographs of the control and aggregates at 0, 60 and 120 min (Fig. 2, the lid must be inverted) (see Note 6).

16. After 120 min incubation each droplet suspension should be gently pipetted 10 times with a 200-µL micropipette to triturate the aggregates. Be careful not to press down the tip of the pipette onto the bottom of the lid as this will disrupt the cluster completely (see Note 7).

17. Take images of all cell aggregates remaining after disruption in a droplet (see Notes 8 and 9).

18. Quantification of the size of each aggregate is performed as described in Subheading 3.2.

3.2. Quantification of Aggregation Assays

3.2.1. Analysis Using ImageJ

Analysis in ImageJ can be performed automated and recorded as a macro to speed up the analysis (steps 1–12). In case of very poor acquisition pictures (lots of bubbles, acquisition using another microscope, etc.) proceed to manual analysis (steps 14–17):

1. Open the image of interest in ImageJ.

2. To speed up the analysis, images of the same sample can be pooled in a stack and processed at the same time. Open all images of interest in ImageJ and press "Stacks", "Image to Stack" (see Note 10).

3. Create a montage from the resulting stack of images ("Image", "Stacks", "Make montage", make sure "Border width" and "Scale factor" are set to 0 and 1, respectively).

4. Convert to greyscale by pressing "Image", "Type", "8-bit" (see Note 11).

5. Automatically detect the edges of the cells by selecting "Process", "Find Edges".

6. Adjust the threshold ("Image", "Adjust", "Threshold") so that the borders of the cells are shown in red (selection) and the background in black. Press Enter or "Apply" in the threshold window to create the binary mask (Fig. 3).

7. In "Analyze", "Set measurements", make sure "Area" and "Limit to threshold" are ticked.

8. In "Analyze", "Analyze Particles", set the "Size" to 3,000–150,000 square-pixels. For keratinocytes and using a 10× objective, this range allows selecting aggregates containing

more than three cells but need to be adjusted depending on the cell type and objective used (see Note 12).

9. Set the "Circularity" to 0.00–1.00.

10. Select "Outlines" if you want to visualise the selected area after analysis.

11. Make sure "Exclude on edges", "Includes holes" and "Display results" are ticked.

12. Press OK and the area of each aggregate within the range of size is reported in a "Results" window. Each aggregate is automatically numbered and outlined in the "Drawing" window (see Note 13).

13. Copy and paste these values in a Microsoft Excel spreadsheet and process the data as required (see Subheading 3.2.2).

14. For the manual analysis, open the image of interest in Image J and convert it to greyscale (press "Image", "Type", "8-bit") (see Note 11).

15. Zoom in/out (by pressing + or -) and manually outline one aggregate using the freehand selection tool and press "T" to add to the ROI manager.

16. Process all aggregates this way and then press "Measure" in the ROI manager window to report all areas in the "Result" window.

17. Copy and paste the values in a Microsoft Excel spreadsheet and process the data as required (see Subheading 3.2.2).

3.2.2. Analysis in Microsoft Excel

The area of each aggregate before and after trituration is imported from ImageJ to a Microsoft Excel spreadsheet. Two parameters are calculated (see Note 14):

1. The aggregation index (i.e., ability of the cells to assemble junctions and form aggregates). For each sample, areas of the initial aggregates within the six suspension drops are pooled and averaged. The aggregation index of each sample is expressed relative to the control, arbitrarily set to 100%.

2. The disaggregation index (i.e., ability of cell–cell junctions to be maintained when submitted to mechanical stress). For each sample, the areas of all aggregates after trituration within the six suspension drops are pooled and averaged. The same calculation is carried out on the images of the corresponding aggregates before dissociation. For each sample, a ratio is determined by dividing the mean area of triturated aggregates by the averaged area of the original aggregates in the six droplets before dissociation. To obtain the disaggregation index, the disaggregation ratio of each sample is expressed relative to the control, which is set to 100%. Note that this calculation corrects for variability on cell density and allows comparison between experiments, as the sizes of the aggregates are normalised within each sample.

4. Notes

1. Addition of calcium to the trypsinisation buffer is essential as if this is omitted, cadherins are cleaved by trypsin and aggregates cannot form properly (4).

2. Cells may take a while to detach; however, this time should be kept to a minimum. To help cells detach quicker from the substratum, gently tap the bottom of the dish on a hard surface during trypsinisation.

3. The pellet obtained is very fragile. Do not spin cells at a higher speed to have a denser pellet, as this may damage the cells. By removing the liquid with a pipette, disruption of the pellet is minimised.

4. EGTA can be added to standard calcium medium to chelate calcium ions. This is performed by adding EGTA to a final concentration of 4 mM to media.

5. Aggregates before trituration are formed at the meniscus of the drop and exhibit a flat or slightly concave shape that facilitates imaging. However, if the aggregate is wry or vertical during imaging, gently tap the lid on a hard surface to make it horizontal.

6. Time lapses between imaging can be adjusted depending on the cell type and kinetics of junction assembly and has to be determined during optimisation of the technique in distinct cell types.

7. The disaggregation step should be performed very carefully in terms of pressure and speed to ensure consistent results across the experiments.

8. Care should be taken when imaging disaggregates. Each aggregate should be photographed in full (i.e., not cut in half on the edges of the image) and not imaged twice to avoid quantification mistakes.

9. Depending on the objective used and the dispersion of aggregates, between 5 and 20 images are usually required to cover the entire drop.

10. This method is only recommended if special care has been taken during acquisition to avoid detection of false positive particles (see Note 8). Note that the number of images to open simultaneously in ImageJ can be limited by the specifications of the computer used for the analysis.

11. Direct image acquisition in 8-bit format can avoid the conversion step.

12. When proceeding to aggregate detection (see Subheading 3.2.1, step 6), the size range of analysed particles might need to be

adjusted. The lower value of this threshold can be empirically adjusted by carrying out preliminary trials on several images showing aggregates containing less than three cells. Note that the higher value of this threshold has to be increased to measure larger particles such as aggregates before dissociation. Once calibrated, it is highly recommended to use the same thresholding values, microscope and objective across the set of experiments.

13. If some artefacts (shadows or bubbles) still appear at the end of the process, zoom in the resulting "Drawing" window (see Subheading 3.2.1, step 12), and note the number of this false positive aggregate. Then, delete this value in the Excel file.

14. Note that the aggregation and disaggregation indexes can be displayed as dot plots (i.e., each dot represents the size of an aggregate, expressed in square pixels) by plotting the raw area of each aggregate per condition.

Acknowledgements

We would like to thank the Wellcome Trust (VIP award and GR081357MA) and Cancer Research UK (C1282/A11980) for the generous support to the lab. We also thank Dr. A. Wheeler for images shown in Fig. 1.

References

1. Fukata, M., and Kaibuchi, K. (2001) Rho-family GTPases in cadherin-mediated cell-cell adhesion. *Nat Rev Mol Cell Biol* 2, 887–897

2. Braga, V.M. (2002) Cell-cell adhesion and signalling. *Curr Opin Cell Biol* 14, 546–56

3. Braga, V.M., and Yap, A.S. (2005) The challenges of abundance: epithelial junctions and small GTPase signalling. *Curr Opin Cell Biol* 17, 466–74

4. Takeichi, M. (1977) Functional correlation between cell adhesive properties and some cell surface proteins. *J Cell Biol* 75, 464–474

5. Shoshani, L, Contreras, R.G., Roldán, M.L., Moreno, J., Lázaro, A., Balda, M.S., Matter, K., and Cereijido, M. (2005) The polarized expression of Na+,K+–ATPase in epithelia depends on the association between β-subunits located in neighboring cells. *Mol Biol Cell* 16, 1071–1081

6. Katsamba, P, Carroll, K., Ahlsen, G., Bahna, F., Vendome, J., Posy, S., Rajebhosale, M., Price, S., Jessell, T.M., Ben-Shaul, A., Shapiro, L.,and Honig, B.H. (2009) Linking molecular affinity and cellular specificity in cadherin-mediated adhesion. *Proc Natl Acad Sci USA* 106, 11594–11599

7. Wendeler, MW, Praus, M., Jung, R., Hecking, M., Metzig, C., and Gessner, R. (2004) Ksp-cadherin is a functional cell-cell adhesion molecule related to LI-cadherin. *Exp Cell Res* 294, 345–355

8. Tzima, E. (2006) Role of small GTPases in endothelial cytoskeletal dynamics and the shear stress response. *Circ Res* 98, 176–185

9. Kim, J-B, Islam, S., Kim, Y.J., Prudoff, R.S., Sass, K.M., Wheelock, M.J., and Johnson, K.R. (2000) N-Cadherin extracellular repeat 4 mediates epithelial to mesenchymal transition and increased motility. *J Cell Biol* 151, 1193–1206

10. Bardella, C, Costa, B., Maggiora, P., Patane', S., Olivero, M., Ranzani, G.N., De Bortoli, M., Comoglio, P.M., and Di Renzo, M.F. (2004) Truncated RON tyrosine kinase drives tumor cell progression and abrogates cell-cell adhesion through E-cadherin transcriptional repression. *Cancer Res* 64, 5154–5161

11. Carrozzino, F, Soulié, P., Huber, D., Mensi, N., Orci, L., Cano, A., Féraille, E., and Montesano, R. (2005) Inducible expression of Snail selectively increases paracellular ion permeability and differentially modulates tight junction proteins. *Am J Physiol Cell Physiol* 289, C1002–C1014

12. Yanagisawa, M., and Anastasiadis, P.Z. (2006) p120 catenin is essential for mesenchymal cadherin–mediated regulation of cell motility and invasiveness. *J Cell Biol* 174, 1087–1096

13. Fang, WB, Ireton, R.C., Zhuang, G., Takahashi, T., Reynolds, A., and Chen, J. (2008) Overexpression of EPHA2 receptor destabilizes adherens junctions via a RhoA-dependent mechanism. *J Cell Sci* 121, 358–368

14. Ehrlich, J.S., Hansen, M.D.H., and Nelson, W.J. (2002) Spatio-temporal regulation of Rac1 localization and lamellipodia dynamics during epithelial cell-cell adhesion. *Dev Cell* 3: 259–270

15. Katuri, V, Tang, Y., Marshall, B., Rashid, A., Jogunoori, W., Volpe, E.A., Sidawy, A.N., Evans, S., Blay, J., Gallicano, G.I., Premkumar Reddy, E., Mishra, L., and Mishra, B. (2005) Inactivation of ELF//TGFβ signaling in human gastrointestinal cancer. *Oncogene* 24, 8012–8024

16. Chakraborty, S, Mitra, S., Falk,M.M., Caplan, S.H., Wheelock, M.J., Johnson, K.R., and Mehta, P.P.. (2010) E-cadherin differentially regulates the assembly of connexin43 and connexin32 into gap junctions in human squamous carcinoma cells. *J Biol Chem* 285, 10761–10776

17. Crooke, C.E, Pozzi, A., and Carpenter, G.F. (2009) PLC-γ1 regulates fibronectin assembly and cell aggregation. *Exp Cell Res* 315, 2207–2214

18. Govindarajan, R, Chakraborty, S., Johnson, K.E., Falk, M.M., Wheelock, M.J., Johnson, K.R., and Mehta, P.P. (2010) Assembly of Connexin43 into gap junctions is regulated differentially by E-cadherin and N-cadherin in rat liver epithelial cells. *Mol Biol Cell* 21, 4089–4107

19. Qin, Y, Capaldo, C., Gumbiner,B.M., and Macara, I.G. (2005) The mammalian Scribble polarity protein regulates epithelial cell adhesion and migration through E-cadherin. *J Cell Biol* 171, 1061–1071

20. Pappas, D.J. and Rimm, D.L. (2006) Direct interaction of the C-terminal domain of α-catenin and F-actin is necessary for stabilized cell-cell adhesion. *Cell Comm Adh* 13, 151–170

21. Sarrio, D, Palacios, J., Hergueta-Redondo, M., Gómez-López, G., Cano, A., and Moreno-Bueno, G. (2009) Functional characterization of E- and P-cadherin in invasive breast cancer cells. *BMC Cancer* 9, 74

22. De Corte, V, Bruyneel, E., Boucherie, C., Mareel, M., Vandekerckhove, J., and Gettemans, J. (2002) Gelsolin-induced epithelial cell invasion is dependent on Ras-Rac signaling. *EMBO J* 21, 6781–6790

23. Hatta, K., Okada, T.S., and Takeichi, M. (1985) A monoclonal antibody disrupting calcium-dependent cell-cell adhesion of brain tissues: possible role of its target antigen in animal pattern formation. *Proc Natl Acad Sci USA* 82, 2789–2793

24. Watt, F.M. (1994) Cultivation of human epidermal keratinocytes with a 3 T3 feeder layer. In: Celis, J.E. (ed) Cell biology: a laboratory handbook. Academic Press, London

25. Hodivala, K.J. and Watt, F.M. (1994) Evidence that cadherins play a role in the downregulation of integrin expression that occurs during keratinocyte terminal differentiation. *J Cell Biol* 124, 589–600

26. Neelamegham, S., and Zygourakis, K. (1997) A Quantitative assay for intercellular aggregation. *Ann Biomed Engineer* 25, 180–189

Chapter 11

Rho GTPase Knockout Induction in Primary Keratinocytes from Adult Mice

Esben Pedersen, Astrid Basse, Tine Lefever, Karine Peyrollier, and Cord Brakebusch

Abstract

Primary keratinocytes are an important tool to investigate the molecular mechanism underlying the skin phenotype of mice with null mutations in Rho GTPase genes. If the RhoA gene deletion is conditional, the knockout can be induced in vitro by transfection with cre-IRES-GFP and sorting for GFP positive cells by flow cytometry. Such in vitro knockout will allow determining the cell autonomous functions of the Rho GTPase, independent of any in vivo interactions. Using the same method, also other expression vectors or knockdown constructs can be introduced into primary mouse keratinocytes.

Key words: Primary keratinocytes, Cre recombinase, Knockout

1. Introduction

Keratinocyte-restricted knockouts of Cdc42, Rac1, and RhoA in skin revealed specific functions of these Rho GTPases in skin development and maintenance. Loss of Cdc42 in keratinocyte resulted in an impairment of hair follicle-specific differentiation, corresponding to an impaired Wnt signaling (1). Cdc42 was found to bind to a complex of Par6, Par3, and aPKCζ, leading to the activation of aPKCζ. aPKCζ then phosphorylates and inactivates GSK3β, which is part of a β-catenin degradation complex. This complex, consisting in its core of GSK3β, axin, and APC, binds free cytoplasmic β-catenin, phosphorylates it and thus targets it for proteasomal degradation. β-Catenin has two roles in the cells, one as an integral part of the adherens junctions and a second one as a transcription factor, driving in keratinocyte progenitor cells differentiation into the hair follicle lineage. Reduction of cytoplasmic β-catenin,

Francisco Rivero (ed.), *Rho GTPases: Methods and Protocols*, Methods in Molecular Biology, vol. 827,
DOI 10.1007/978-1-61779-442-1_11, © Springer Science+Business Media, LLC 2012

as in the Cdc42 knockout, will therefore quickly lead to a decrease of nuclear β-catenin and a stop of hair follicle differentiation, and in the long run also to a depletion of junctional β-catenin, resulting in impaired cell–cell contacts and intraepidermal blisters (1).

Also Rac1 was shown to interact with Par6 in keratinocytes (2). Loss of Rac1 in keratinocytes, however, does not replicate the Cdc42-deficient phenotype, suggesting that in vivo Cdc42 is the dominant regulator of aPKCζ activity via Par6. Mice lacking Rac1 in keratinocytes show a loss of hair follicles, corresponding with a loss of hair follicle progenitors and morphological alterations of the hair follicles (3). Depending on the model used to induce the skin-specific knockout of Rac1, slightly different phenotypes have been observed. Mice in which the deletion of the Rac1 gene was mediated by tamoxifen-induced activation of a cre-ER fusion protein expressed under the control of the strong keratin 14 promoter showed a complete loss of skin, which was explained by loss of all keratinocyte stem cells (4). Stemness of keratinocytes was suggested to be controlled via Rac1, Pak2, and c-myc. Keratinocyte-restricted deletion of the Rac1 gene driven by cre expressed under the control of the keratin 5 promoter, however, resulted only in the loss of the lower parts of the hair follicle, as two independent groups reported (3, 5). Differentiation and maintenance of the interfollicular epidermis was not altered in these mice and no alteration in c-myc levels was detected (3). Malformed Rac1-deficient hair follicles were shown to be removed by a transient infiltration of macrophages into the skin. Interestingly, crossing of the first conditional Rac1 mutant mouse with mice expressing the cre recombinase under the control of the keratin 5 promoter resulted in a much milder phenotype, where mice showed either no interfollicular epidermis phenotype or a patchy one (6). These results suggest that the onset of cre expression can have a big impact on the phenotype. In addition, tamoxifen-induced defects (3) and potentially differences in genetic background and off-target effects of cre might contribute to the different phenotypes observed (7).

Loss of RhoA in keratinocytes did not result in any major alterations in skin differentiation and maintenance, although RhoA-ROCK dependent signaling pathways were clearly reduced (8). However, these mice show an increased sensitivity toward skin inflammation, which is currently analyzed.

Although mice with keratinocyte restricted deletions of Rho GTPase genes are excellent models to understand the in vivo function of Rho GTPases in skin, analysis limited to the use of tissue for immunofluorescence staining and Western blots often remains at a descriptive level, correlating alterations in signal transduction with phenotypic impairments. In order to prove these correlations, it is therefore mandatory to investigate primary keratinocytes in vitro, where cells can be manipulated by inhibitors and activators, and transfected with mutant proteins activating or inactivating certain signaling pathways. Primary keratinocytes can be isolated from newborn and adult mice for direct analysis by Western blot or microarrays, grafting, FACS sorting, or plating for cellular assays such as directed migration or cell–cell contact formation.

This is not without problems, since primary keratinocytes grown under standard, low calcium conditions in vitro, show differences in gene expression and activation of signaling pathways when compared to basal keratinocytes in vitro. For example, keratinocytes lacking Cdc42 showed normal levels of β-catenin and GSK3β phoshporylation, due to an increased activation of Akt under in vitro conditions (1). After inhibition of Akt activation, Cdc42-null keratinocytes were clearly different from wildtype control cells and showed decreased β-catenin and GSK3β phosphorylation, as observed in vivo. Also the opposite phenomenon can be observed, i.e., that keratinocytes show in vitro a more severe defect than in vivo. This is the case for RhoA null keratinocytes, which in vitro have a twofold increased cell area on glass and show severe migration defects, although in vivo cell area and wound healing are normal (8). This latter difference correlates with differential crosstalk of RhoA with Cdc42 and Rac1 in vivo and in vitro.

Another possibility is that the in vitro keratinocyte phenotype is not cell autonomous, but critically dependent on previous interactions in vivo with nonkeratinocytes, such as dermal fibroblasts or immune cells (9). In conditional knockouts carrying a floxed gene, inducing the gene deletion in vitro by transient transfection with cre recombinase can test this.

Despite these principal caveats, primary keratinocytes and their manipulation in vitro is an invaluable tool complementing the in vivo analysis of mutant mice. Yet, longer in vitro culture of mouse keratinocytes, for example, selection of stably transfected cells by antibiotics, should be avoided, since it will result in loss of keratinocyte characteristic differentiation. In addition, primary keratinocytes grow badly at low cell densities. To circumvent this problem, we use co-expression of fluorescent proteins as a marker for transfected cells and select them by preparative FACS sorting, which is described in detail below.

Here we present the method used in our laboratory for the induction of Rho GTPase knockouts in primary keratinocytes from adult mice. The same method can of course be used to introduce any foreign gene into primary keratinocytes. For an excellent and more extensive description of the isolation of primary mouse keratinocytes, we would like to refer to recent articles (10, 11).

2. Materials

2.1. Coating of Plates

1. Coating medium: mix 25 mL of minimum essential medium Eagle (Sigma #M8167), 2.5 mL of BSA Fraction V (1 mg/mL), 500 µL of 1 M Hepes, pH 7.3, 250 µL of Vitrogen 100 collagen (Cohesion FXP-019), 250 µL of fibronectin (1 mg/mL) (Invitrogen #33016-015), and 290 µL of 100 mM $CaCl_2$. All components should be sterile.

2. Plates: 10-cm tissue culture dishes and 6-well tissue culture plates.

2.2. Isolation of Keratinocytes

1. Instruments: curved forceps, scissor, scalpel (sterilize with 70% ethanol).

2. Solutions: 70% ethanol, 1 mM iodine solution (Sigma #35089), PBS (137 mM NaCl, 2.7 mM KCl, 8.1 mM Na_2HPO_4, 1.76 mM KH_2PO_4, pH 7.4).

3. Bacteriological dishes.

4. Ab/PBS: per mouse mix 25 mL of PBS with 0.5 mL of Pen/Strep (penicilin 10,000 U/mL streptomycin 10,000 µg/mL) (Invitrogen #15140), 0.5 mL of Nystatin (10,000 U/mL) (Sigma #N1638), and 0.5 mL of Fungizone (250 µg/mL) (Invitrogen #15290). Pass through a 0.2-µm filter to sterilize it. Prepare fresh each time.

5. 0.8% Trypsin: dissolve 0.2 g of trypsin 1:250 (Invitrogen #27250-018) in 25 mL of PBS per mouse. Pass through a 0.2-µm filter to sterilize it. Prepare fresh each time.

6. 100 µm Cell strainer (Becton Dickinson #352360).

7. Chelated fetal calf serum (FCS): add 40 g of Chelex (Biorad #142-2832) to 1 L of dH_2O. Adjust pH to 7.4 using 4 M HCl (see Note 1). Filter the solution through a folded filter (Schleicher&Schuell #314856). Add the chelex resin to 100 mL of FCS. Stir overnight at 4°C. To remove the Chelex, filter two times through a folded filter. To sterilize the FCS, pass it through a 0.2-µm filter. Aliquot and store at −20°C.

8. Minimal keratinocyte medium: mix 500 mL of minimum essential medium Eagle, 5 mL of GlutaMAX (100× Invitrogen #35050), 5 mL of Pen/Strep (100× Invitrogen #15140), 40 mL of chelated FCS. Sterilize the medium by passing it through a 0.2-µm 500-mL filter unit.

9. Full keratinocyte medium: Mix 500 mL minimum essential medium Eagle, 500 µL of insulin (5 mg/mL in 4 mM HCl) (Sigma #I5500), 25 µL of EGF (200 µg/mL in PBS) (Sigma #E9644), 1,000 µL of transferrin (5 mg/mL in PBS) (Sigma #T8158), 500 µL of phosphoethanolamine (10 mM in PBS) (Sigma #P0503), 500 µL of ethanolamine (10 mM in PBS)

(Sigma #E0135), 36 μL of hydrocortisone (5 mg/mL in ethanol) (Calbiochem #386698), 5 mL of GlutaMAX (100× Invitrogen # 35050), 5 mL of Pen/Strep (100× Invitrogen # 15140), 40 mL of chelated FCS. Sterilize the medium by passing it through a 0.2-μm 500-mL filter unit.

10. Hemocytometer (Hausser BRIGHT-LINE).

11. 0.4% Trypan blue solution.

2.3. Transfection

1. Transfection reagent (Mirus #MIR 2800).

2. Serum-free keratinocyte medium: the same keratinocyte medium as in Subheading 2.2, item 9 without serum.

3. CRE-GFP expressing plasmid (see Note 2).

2.4. Sorting

1. 0.1% Trypsin–EDTA (Invitrogen #15400-054).

2. Sorting buffer. 1% BSA in PBS.

3. 50-μm Cup filter (Becton Dickinson #340630).

4. FACS Aria with BD FACSDiva software and a 12-color 70-μm configuration.

3. Methods

3.1. Coating of Plates

Dispense 10 mL of coating medium on a dish and spread the liquid over the whole surface, collect the medium, leaving around 1.0 mL per 10-cm tissue culture dish. There is enough solution to coat around twenty 10-cm dishes. Incubate the dishes 2–4 h at 37°C. Wrap the plates in parafilm and store at 4°C (see Note 3). Wash the plates two times with PBS before use. 6-well, 24-well, and other plate types are coated in the same manner.

3.2. Isolation of Keratinocytes

1. Kill mouse by cervical dislocation (see Note 4).

2. Shave the mouse or cut off hair with scissors.

3. Dip the mouse in iodine solution for 1 min. Rub over the fur to make sure that the skin becomes wet. Transfer mouse to a sterile laminar flow bench.

4. Dip the mouse in 70% ethanol for 1 min.

5. Peel off the skin from the trunk. Start by making an incision across the lower abdomen. Insert the blade of a pair of scissors under the skin and open and close the scissors to loosen the skin from the underlying tissue. Then cut the abdominal skin from the tail to the head and start peeling the skin off. Cut around limbs and tail (see Note 5).

6. Place the skin in a sterile bacteriological dish with dermis up. Use a scalpel to completely remove all fat (see Note 6).

7. Put the skin with dermis side up in a bacteriological dish containing 25 mL of Ab/PBS solution and incubate for 5 min at room temperature.

8. Transfer the skin to a dish with 25 mL of 0.8% trypsin/PBS, epidermis side up.

9. Incubate the skin for 50 min at 37°C.

10. Transfer the skin to a new bacteriological dish and separate epidermis from dermis. Carefully scrape off hair and keratinocytes from dermis using small curved forceps. The epidermis should come off very easily.

11. Place epidermis scrape in a 50-mL tube containing 25 mL of minimal keratinocyte medium.

12. Pipette carefully three times up and down with a 5-mL pipette.

13. Filter the suspension through a 100-μm cell strainer.

14. Centrifuge the suspension for 5 min at $300 \times g$, aspirate supernatant.

15. Resuspend the cells in 5 mL of minimal keratinocyte medium.

16. Take a sample of 10 μL from the cell suspension and mix with 90 μL of trypan blue solution. Apply the mix to a hemocytometer and count the white living cells (see Note 7).

17. Plate the cells on coated dishes. For each mice, plate cells in one well of a 6-well plate and on two to three 10-cm dishes (see Note 8). Plate approximately 1×10^6 cells in a 6-well and 5×10^6 cells on a 10-cm dish. Use around 2 mL of minimal keratinocyte medium in a 6-well and 10 mL in a 10-cm dish.

18. Incubate plates at 34°C, 5% CO_2 95% humidity.

19. Change medium the next day to full keratinocyte medium. Wash with PBS to remove dead cells. Only the nondifferentiated cells are able to attached and spread, this is approximately 10% of the cells (see Fig. 1). Afterwards medium should be changed every 2–3 days.

3.3. Transfection

1. Transfect cells when they are 50–70% confluent (see Note 9).

2. Prepare plasmid mix. For transfection of one 10-cm plate: add 40 μL of transfection reagent to 1 mL of serum-free keratinocyte medium and mix thoroughly by vortexing. Incubate at room temperature for 5 min. Add 10 μg of a CRE-GFP expressing plasmid and mix by gentle pipetting up and down. Incubate at room temperature for 20 min.

3. Remove medium from cells and replace with 6 mL of fresh keratinocyte medium.

4. Add 1 mL of the plasmid mix dropwise to cells, and gently rock the dish.

5. Add 5–6 mL of keratinocyte medium 4–6 h later.

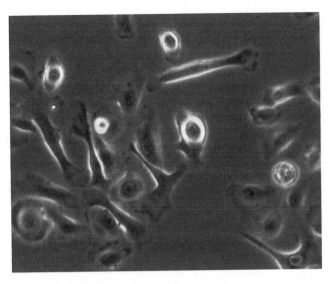

Fig. 1. Primary mouse keratinocytes after 3 days culture in vitro.

3.4. Sorting

1. Sort cells 2–3 days after transfection (see Note 10).

2. Wash cells two times with PBS and incubate them for 10–15 min in 1 mL of 0.1% trypsin–EDTA at 37°C.

3. Collect detached cells in keratinocyte medium. Pool three 10-cm plates and spin cells down for 5 min at $300 \times g$.

4. Resuspend control cells in 250 µL of sorting buffer, put them through a 50-µm Cup filter and rinse the filter with 250 µL of sorting buffer. Do the same for pooled cells of each specimen, but use 500 µL of sorting buffer in each step instead, to obtain a total sample volume of 1 mL per specimen. Keep cells on ice (see Note 11).

5. For the FACS, use a density filter of 2.0 (gray) and a sheath pressure of 70 psi. The frequency should be approximately 90, and the flow rate should be set to 3. The voltage parameters are 270 for FSC, 325 for SSC, and 325 for GFP. Cool down the sorted cells (see Note 12).

6. Gate for living cells on the two-dimensional dot plots of forward scatter (FSC) and side scatter (SSC) first. Then make a gate for GFP positive cells, use untransfected cells as negative controls (see Fig. 2).

7. Sort the 1 mL samples and collect the GFP positive cells in a 15-mL tube containing 5–6 mL of cold medium (see Note 13).

8. Spin sorted cells down for 5 min at $300 \times g$ and resuspend in fresh keratinocyte medium.

9. Replate cells (see Note 14). Verify absence of the targeted Rho GTPase using western blot analysis (see Note 15).

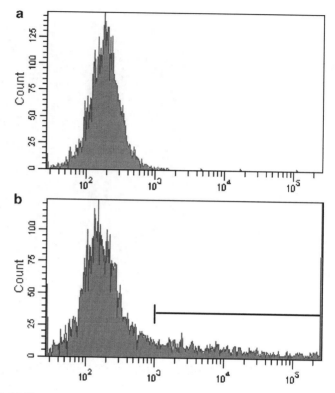

Fig. 2. (**a**) Histogram showing autofluorescence of not transfected keratinocytes. (**b**) Histogram showing fluorescence of keratinocytes 2 days after transfection with cre-EGFP. Black bar marks the cells sorted as GFP positive.

4. Notes

1. The pH of the chelated FCS solution should be between 7.35 and 7.45. Add HCl slowly. pH takes a long time to stabilize, especially when preparing a large batch. If the pH gets too low, add NaOH to adjust.

2. We use a CRE-IRES-EGFP plasmid with a pRRLSIN.cPPT. PGK/GFP.WPRE (Addgene) backbone. Adenoviral transduction is more efficient, but can induce chemokines and cytokines, which could influence further experiments.

3. The plates can be stored for long periods at 4°C, but should not be used if the coating medium has dried out.

4. Do not use mice at the age of 4–6 weeks, as they are in anagen phase (growth phase) of the hair cycle. Keratinocytes isolated from mice in anagen phase do not grow well in culture. After the age of 8 weeks, the hair cycle is no longer synchronized and only smaller patches of the skin will be in this phase. In mice with pigmented hair, anagen areas can be recognized after shaving by a darkened skin.

5. If needed, the skin can be now transported to another workspace in a 50-ml tube with PBS on ice.

6. Scrape hard, skin from adult mice is quite resistant, but try not to make holes. When you remove the fat, the dermis changes color from glossy light yellow to matt white. If the fat is not removed properly, the efficiency of trypsin will be very low and the epidermis cannot be easily scraped off in step 10.

7. Normally from one adult mouse, one can obtain around 20×10^6 cells.

8. The cells in the 6-well plate will be used as untransfected negative controls for the FACS sort.

9. The transfection day will be 3–5 days after isolation, depending on how fast the cells grow, the efficiency of transfection is highest when the cells are 50% confluent.

10. Keratinocytes do not like to be in suspension. But if they are cooled down they are quite resistant, therefore the FACS Aria, medium, and sorting buffer should be cooled down to +4°C.

11. If only less than three 10-cm plates from each mouse are available, the amount of sorting buffer can be reduced to two times 300 μL.

12. The settings are what we find to be optimal and can be different for different FACS machines.

13. Depending on the transfection efficiency, 20–40% of the cells will be GFP positive. In the reanalysis, approximately 90% GFP positive cells should be expected.

14. If around 400,000 cells are plated in a 6-well or 60,000 cells in a 24-well plate, they will be confluent the following day.

15. The time it takes for the cells to become functionally knockout depends on the half life of the protein of interest. For the small Rho GTPase Rac1, the protein is gone 2 days after sorting meaning 4 days after transfection (see Fig. 3).

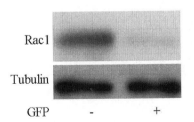

Fig. 3. Western blot analysis for Rac1 of the GFP negative and GFP positive fraction of Rac fl/fl keratinocytes transfected with cre-GFP, cultured for 2 days after sorting. Western blot for tubulin was carried out as a loading control.

Acknowledgments

This work was supported by the Novo Nordisk Foundation, the Lundbeck Foundation, and the Danish Research Council.

References

1. Wu, X., Quondamatteo, F., Lefever, T., Czuchra, A., Meyer, H., Chrostek, A., Paus, R., Langbein, L., and Brakebusch, C. (2006) Cdc42 controls progenitor cell differentiation and beta-catenin turnover in skin. *Genes Dev* 20, 571–585

2. Mertens, A. E., Rygiel, T. P., Olivo, C., van der Kammen, R., Collard, J. G. (2005) The Rac activator Tiam1 controls tight junction biogenesis in keratinocytes through binding to and activation of the Par polarity complex. *J Cell Biol* 170, 1029–1037.

3. Chrostek, A., Wu, X., Quondamatteo, F., Hu, R., Sanecka, A., Niemann, C., Langbein, L., Haase, I., and Brakebusch, C. (2006) Rac1 is crucial for hair follicle integrity but is not essential for maintenance of the epidermis. *Mol. Cell. Biol* 26, 6957–6970.

4. Benitah, S. A., Frye, M., Glogauer, M., and Watt, F. M. (2005) Stem cell depletion through epidermal deletion of Rac1. *Science* 309, 933–955.

5. Castilho, R. M., Squarize, C. H., Patel, V., Millar, S. E., Zheng, Y., Molinolo, A., and Gutkind, J. S. (2007) Requirement of Rac1 distinguishes follicular from interfollicular epithelial stem cells. *Oncogene* 26, 5078–5085.

6. Benitah, S. A., Watt, F. M. (2007) Epidermal deletion of Rac1 causes stem cell depletion, irrespective of whether deletion occurs during embryogenesis or adulthood. *J Invest Dermatol* 127, 1555–1557.

7. Schmidt-Supprian, M., and Rajewsky, K. (2007) Vagaries of conditional gene targeting. *Nat Immunol* 8, 665–668.

8. Jackson, B., Peyrollier, K., Pedersen, E., Basse, A., Karlsson, R., Wang, Z., Lefever, T., Ochsenbein, A., Schmidt, G.., Aktories, K., Stanley, A., Quondamatteo, F., Ladwein, M., Rottner, K., van Hengel, J., and Brakebusch, C. (2011) RhoA is dispensable for skin development, but crucial for contraction and directed migration of keratinocytes. Mol Biol Cell in press.

9. Lefever, T., Pedersen, E., Basse, A., Paus, R., Quondamatteo, F., Stanley, A. C., Langbein, L., Wu, X., Wehland, J., Lommel, S., and Brakebusch, C. (2010) N-WASP is a novel regulator of hair-follicle cycling that controls antiproliferative TGFβ pathways. *J Cell Sci* 123, 128–140.

10. Lichti, U., Anders, J, and Yuspa, S. H. (2008) Isolation and short –term culture of primary keratinocytes, hair follicle populations and dermal cells from newborn mice and keratinocytes from adult mice for in vitro analysis and for grafting to immunodeficient mice. *Nat Protocols* 3, 799–810.

11. Lorz, C., Segrelles, C., Garin, M., and Paramio, J. M. (2010) Isolation of adult mouse stem keratinocytes using magnetic cell sorting. In Turksen, K. (ed) Methods in Molecular Biology, Humana Press, Totowa, NJ.

Chapter 12

Rho GTPase Techniques in Osteoclastogenesis

Roland Leung and Michael Glogauer

Abstract

Historically, in vitro culturing of primary osteoclasts involved co-culturing of mononuclear monocytes with bone marrow stromal cells, thereby providing the cytokines required for osteoclast formation and multinucleation. Since the identification and cloning of receptor activator of nuclear factor kappa B ligand (RANKL), culturing primary osteoclasts in vitro has become much simplified. It has become apparent that the actin cytoskeleton is extremely important for the osteoclast, not only in terms of structural support, but also for adhesion, polarization, and migration. Rho family GTPases are key regulators of the actin cytoskeleton. In this chapter, we describe simple techniques in culturing primary osteoclasts from murine bone marrow cells, evaluating the activation states of Rho GTPases in osteoclasts, measuring the migratory abilities of monocytes, and introducing proteins of interest into osteoclasts using the TAT construct.

Key words: Osteoclast, Monocyte, Rho GTPases, TAT fusion protein, Migration

1. Introduction

Osteoclasts are multinucleated, bone-resorbing cells that play important roles in bone homeostasis, a parameter that is normally maintained via the balanced activities of osteoclasts and bone-producing osteoblasts (1). Osteoclast activity in excess of that of osteoblasts results in a net loss of bone mass, whereas the opposite results in osteopetrosis. Pathologically, osteolytic diseases such as periodontitis, osteoporosis, and rheumatoid arthritis exhibit excessive osteoclast activities. A unique feature of osteoclast formation, or osteoclastogenesis, involves multiple fusions of mononuclear monocytes/preosteoclasts differentiated from granulocytic progenitors in the bone marrow (2). Fusion is preceded by cell migration which brings cells in close proximity (3). In vivo osteoclastogenesis is supported by cell-to-cell contact between the receptor activator of nuclear factor kappa B ligand (RANKL) on the surface of

Francisco Rivero (ed.), *Rho GTPases: Methods and Protocols*, Methods in Molecular Biology, vol. 827,
DOI 10.1007/978-1-61779-442-1_12, © Springer Science+Business Media, LLC 2012

osteoblasts and/or bone marrow stromal cells, and its receptor RANK on the surface of osteoclast precursor cells (2). Two critical ligands, RANKL and macrophage-colony stimulating factor (M-CSF), are essential for osteoclastogenesis: mice with ablated genes for RANKL or its receptor, RANK, are completely devoid of osteoclasts (4). Further, the addition of RANKL and M-CSF to cells of the mononuclear phagocyte lineage in vitro is sufficient for the generation of mature functional osteoclasts (5).

It is well known that the actin cytoskeleton plays an important role in mature osteoclast biology as osteoclasts are dynamic, adherent cells that undergo migration, polarization during resorption, and transmigration through cell layers (6). Actin is especially crucial for osteoclast's primary adhesive structures known as podosomes, which make up the most prominent component of the actin cytoskeleton in monocyte-derived cells (7). Actin polymerization and depolymerization within podosomes mediate osteoclast adhesion to the substratum (8). Cellular migration and membrane fusion during osteoclastogenesis also require dynamic actin cytoskeleton reorganization, and it has been firmly established that the Rho family small GTPases Rac1, Rac2, Cdc42, and RhoA regulate the actin cytoskeleton (9). These GTPases control actin structures and actin-based cell processes both during osteoclast formation and in mature osteoclast function, including actin filament elongation during chemotaxis, formation of actin ring and sealing zone, cell spreading, and mature osteoclast polarization to form the resorption lacunae (10–14). Following stimulation, GTPases release GDP and bind to GTP, a reaction mediated by guanine nucleotide exchange factors (GEFs). In their active GTP-bound state, Rho GTPases interact with effector proteins to promote cellular responses. The active state of GTPases is transient because of their intrinsic GTPase activity, which is stimulated further by GTPase activating proteins (GAPs).

Many studies have shown the importance of Rho GTPases for osteoclast formation and/or function. The role of Rho GTPases in podosome regulation is supported by the presence of p190rhoGAP at these sites (6). The signaling role of Rho GTPases in osteoclast migration toward M-CSF is illustrated by the activation of RhoA and Rac downstream of the β_3 integrin (12), a main osteoclast adhesive and signaling molecule. We showed that osteoclast progenitors lacking Rac1 suffered significant defects in osteoclast formation with concomitant defects in migration, defective actin growth, and reactive oxygen species generation (10). Others have shown the importance of Rac1 in osteoclast function and survival (15, 16). In osteoclast cell spreading and cytoskeletal remodeling studies, Sakai et al. (14) showed that Rac1 induced lamellipodia formation and Cdc42 induced moderate filopodia/lamellipodia formation, whereas RhoA had the opposite effect of inducing cell retraction. A very recent paper described role of Cdc42 in osteoclast resorption activity, activation and differentiation signaling induced by M-CSF and RANKL, and polarization (17).

In this chapter, we describe a method to differentiate monocytes/preosteoclasts isolated from the murine bone marrow into mature, functional osteoclasts. To study the activation of Rac, Cdc42, and RhoA, we describe a pulldown assay that assesses the active, GTP-bound forms of the GTPases, using M-CSF as the agonist that has been shown to activate these GTPases (12, 14, 15). Second, since it can be difficult to introduce new transgenes into primary cells, including osteoclasts, we also describe a simple method for protein transduction into primary preosteoclasts/osteoclasts using a TAT construct coupled to a protein of interest. This method could be used to introduce Rho GTPases into preosteoclasts/osteoclasts.

2. Materials

2.1. Mouse Dissection and Preosteoclast Isolation

1. Dissection tools: 1 pair of surgical scissors and tissue forceps (sterilized in an autoclave).
2. 70% Ethanol.
3. 10-mL Syringes, 15-mL conical tubes, sterile 60-mm petri dishes and sterile surgical gloves (optional) and sterile needles (21-gauge and 30-gauge).
4. Alpha minimal essential medium (α-MEM, Life Technologies).
5. Ficoll-Paque PLUS (Amersham Biosciences).
6. Cell counter: Z1 Coulter counter (Coulter Electronics) or a hemocytometer.

2.2. In Vitro Osteoclast Culture

1. Phosphate-buffered saline (PBS) (Sigma).
2. α-MEM growth medium, containing 10% fetal bovine serum (FBS) and antibiotics (final concentrations: 164 IU/mL penicillin G, 50 μg/mL gentamicin, and 0.25 μg/mL fungizone).
3. M-CSF (10 μg, Sigma). Dissolve in 100 μL of sterile water and 900 μL of PBS to yield a stock concentration of 10 ng/μL, and store in single-use aliquots at $-20°C$.
4. RANKL (10 μg, Peprotech). Dissolve in 500 μL of PBS with 0.05% bovine serum albumin (BSA).
5. 6-Well tissue culture plates.
6. 4% Paraformaldehyde (PFA) fixative solution.
7. Naphthol AS-BI phosphate and fast red TR salt (Sigma) in 0.2 M acetate buffer (pH 5.2) containing 100 mM sodium tartrate.

2.3. Rho GTPase Activation Assay

1. PBS.

2. RIPA buffer (Sigma, ready-to-use solution containing 150 mM NaCl, 1.0% IGEPAL® CA-630, 0.5% sodium deoxycholate, 0.1% SDS, 50 mM Tris–HCl, pH 8.0).

3. PAK-GST glutathione sepharose beads (Cytoskeleton, Inc.) reconstituted in 500 μL of water to yield a stock concentration of 1 mg/mL, and stored in single-use aliquots at –80°C.

4. Rhotekin-RBD GST glutathione sepharose beads (Cytoskeleton, Inc.) reconstituted in 600 μL of water to yield a stock concentration of 3.3 mg/mL, and stored in single-use aliquots at –80°C.

5. 50× protease inhibitor cocktail (BD Pharmingen).

6. 1 mM Phenylmethylsulfonyl fluoride (PMSF) dissolved in ethanol.

7. Tube rotator.

8. PBD/RBD wash buffer (20 mM Hepes, pH 7.4, 142.5 mM NaCl, 4 mM EGTA, 4 mM EDTA, 1% NP-40, 10% glycerol; store at 4°C).

9. Investigator's choice of SDS-PAGE and immunoblot equipment.

10. Precast 12% polyacrylamide gels commercially available from various suppliers.

11. Laemmli buffer. 2× buffer: 0.125 M Tris–HCl, 20% glycerol, 4% SDS, 10% 2-mercaptoethanol, 0.004% bromophenol blue, pH 6.8; 5× buffer: 0.3 M Tris–HCl, 50% glycerol, 10% SDS, 25% 2-mercaptoethanol, 0.01% bromophenol blue, pH 6.8.

12. Nitrocellulose membrane (Amersham Biosciences).

13. Tris-buffered saline (137 mM NaCl, 20 mM Tris–HCl, pH 7.6) with 0.05% (v/v) Tween-20 (TBS-T).

14. Chemiluminescence reagent (Amersham ECL Plus, GE Healthcare).

15. X-ray films.

16. Primary antibodies for immunoblot: mouse monoclonal anti-Rac1 (23A8, Upstate Biotechnologies; 1:2,000 in TBS-T/5% milk powder); rabbit polyclonal anti-Rac2 (Upstate Biotechnologies; 1:5,000 in TBS-T/5% milk powder); mouse monoclonal anti-RhoA (26C4, Santa Cruz Biotechnology, Inc.; 1:200 in TBS-T/5% BSA); mouse monoclonal anti-Cdc42 (B-8, Santa Cruz Biotechnology, Inc; 1:100 in TBS-T/5% BSA).

17. Secondary antibodies for immunoblot (all diluted in TBS-T/5% milk powder): horse raddish peroxidase (HRP)-conjugated sheep anti-mouse IgG (610-603-002, Rockland Inc; 1:4,000 for Rac1, 1:1,000 for Cdc42, 1:2,000 for RhoA); HRP-conjugated donkey anti-rabbit IgG (NA934V, GE Healthcare; 1:2,000).

2.4. TAT Protein Constructs and Migration Rescue

1. 6His-pTAT-HA expression vector, generous gift from the late Dr. Gary M Bokoch, Scripps Research Institute, La Jolla, CA.

2. BL21(DE3)pLysS competent *Escherichia coli* strain (Promega).

3. Luria broth (LB): 10 g of tryptone, 5 g of yeast extract, 10 g of NaCl per liter of double distilled water. Autoclave and store at 4°C.

4. LB plates with 100 μg/mL ampicillin. Autoclave LB with 1.5% agar, allow to cool to approximately 50–60°C, then add ampicillin and pour about 10 mL into 10-cm sterile petri dishes. Allow to solidify and store at 4°C.

5. Isopropyl β-D-1-thiogalactopyranoside (IPTG, Sigma).

6. QIAexpress Ni-NTA Fast Start Kit (#30600, Qiagen). For the beginner, this kit offers all of the reagents necessary to easily and quickly purify 6His-tag recombinant proteins.

7. Slide-A-Lyzer cassettes (Thermo Scientific) for dialysis.

8. PBS.

9. 5-mL Sterile syringe and 21-gauge needle, cotton swabs, scalpel, and forceps.

10. BCA protein assay kit (#23225, Pierce).

11. Anti-HA polyclonal antibody, HRP conjugated (Thermo Scientific).

12. Transwell permeable supports with 5 μm membrane pores in 24-well plates (Corning Life Sciences).

13. 0.165 μM 4′6,-Diamidino-2-phenylindole (DAPI) (Sigma).

14. Microscope slides.

15. Fluorescence mounting medium (Dako).

16. Fluorescence microscope equipped with a filter set for DAPI (Nikon Eclipse E400 or equivalent).

3. Methods

3.1. Mouse Dissection, Preosteoclast Isolation, and In Vitro Osteoclast Culture

1. Euthanize animals in accordance with ethical guidelines employed at the researcher's institution. Two commonly used methods for euthanasia are cervical dislocation or carbon dioxide asphyxiation.

2. Euthanized mice should be immersed in, or sprayed well with 70% ethanol to decrease the chance of contamination of cell cultures. All procedures involving mice dissection and tissue culture should be done aseptically in a laminar airflow hood.

3. Using a pair of sterile forceps, lift the skin on the ventral surface of the mouse at the midpoint between the front and

hind legs. Ensure that you do not lift the deeper peritoneal lining with the skin.

4. With sterile scissors, make a small 5-mm cut in the skin where it has been lifted.

5. Using sterile surgical gloves or gloves that have been well-sprayed with 70% ethanol, grab the skin on either side of the cut with your thumb and index finger and pull firmly in opposite directions, until the hind legs pull through the skin. This exposes the muscle of the hindlegs.

6. While holding the hindfoot with your nondominant hand, de-flesh the femur and tibia using scissors as much as possible. The fibula can be discarded at this stage (see Note 1).

7. Separate the hind legs from the body at the femur-ilium joint. By holding the leg with forceps, cut off the hindfoot.

8. Separate the tibia and femora at the joint using scissors. Continue to de-flesh ensuring that the bones are clean and free of soft tissue. Place the cleaned bones in cold α-MEM in a petri dish and proceed to isolate the bones from the other leg (see Note 2).

9. Prepare a 10-mL syringe with α-MEM and attach a 30-gauge needle.

10. Cut both ends of the tibia/femora to expose the bone marrow. Hold the bone securely at mid-shaft with forceps, insert the tip of the needle into one end of the bone, and flush the bone marrow over a 60-mm sterile petri dish. Repeat this process by flushing the bone marrow from the opposite cut-end. The bone should appear chalky-white once the bone marrow has been flushed out. Repeat for the other bones.

11. Once all the bone marrow has been flushed into the petri dish, attach a 21-gauge needle to the same syringe and aspirate the cell suspension several times to break up cell aggregates.

12. Pipette the cell suspension into a 15-mL conical tube avoiding tissue debris and bone spicules.

13. Pellet bone marrow cells by spinning at $1,500 \times g$ for 5 min at room temperature.

14. Aspirate the supernatant from the tube, being careful not to disturb the cell pellet. Wash the pellet once with fresh α-MEM, and resuspend the pellet in 10 mL of α-MEM with 10% FBS and antibiotics. Culture cells overnight in a 100-mm petri dish in a humidified incubator at 37°C with 5% CO_2. This removes stromal cells, while most osteoclast precursors will remain in suspension.

15. Collect the supernatant into a 15 mL conical tube. Pellet the cells at $1,500 \times g$ for 5 min.

16. Resuspend the pellet containing osteoclast precursors with 4 mL of α-MEM. Using an automatic pipette on its slowest setting, or using sterile Pasteur pipettes attached to a bulb, gently layer the cell suspension over 4 mL of Ficoll-Paque PLUS in a 15-mL tube, being careful not to disturb the solution interface. The α-MEM should form a distinct layer above the Ficoll (see Note 3).

17. Centrifuge at $350 \times g$ for 30 min at 4°C. The cell layer at the solution interface is enriched in osteoclast precursor cells. Using a sterile Pasteur pipette attached to a bulb, transfer the cells at the interface into a 15-mL conical tube.

18. Wash the cells three times with α-MEM and resuspend with 5 mL of α-MEM growth medium. Count the cells either using an electronic cell counter or a hemocytometer. Adjust the cell concentration to 0.25×10^6 cells/mL.

19. Add 2 mL of cells per well to a 6-well plate, resulting in 0.5×10^6 cells/well. Supplement with 20 ng/mL M-CSF and 100 ng/mL RANKL. Culture for 6 days, with a change of medium and cytokine supplementation every other day. Increase initial plating density if larger or more osteoclasts are desired (see Note 4).

20. Under light microscopy, search for large, multinucleated osteoclasts. Alternatively for easier identification, osteoclasts can be stained for tartrate-resistant acid phosphatase (TRACP). Wash culture wells once with prewarmed PBS and fix cells with 4% PFA for 15 min at room temperature. Wash fixed cells with PBS, stain with TRACP staining solution for 10–15 min at 37°C, then wash twice with water. TRACP-positive osteoclasts will appear pink-red.

3.2. Rho-GTPase Activation Assay

1. Plate osteoclast precursors isolated from Ficoll centrifugation (see Subheading 3.1, step 18) in two 100-mm petri dishes and culture for 2 days in α-MEM growth medium without cytokine supplementation. On average, bone marrow from four mice should be isolated to yield adequate cells for each experiment (totaling about 2×10^7 cells plated per dish) (see Note 5).

2. After 2 days, aspirate nonadherent cells, and gently wash adherent cells three times with PBS (see Note 6).

3. Immediately add α-MEM growth medium to one petri dish (control cells) and to the other α-MEM growth medium supplemented with M-CSF (20 ng/mL) and incubate for 4 h at 37°C. M-CSF concentration and length of activation should be optimized for each investigator (see Note 7).

4. Aspirate culture medium, immediately lyse cells with 200 μL of ice-cold RIPA buffer supplemented with 1 mM PMSF and protease inhibitor, and collect by cell scraping using a rubber policeman. Transfer control and activated cell lysates to prechilled 1.5 mL microfuge tubes.

5. Centrifuge lysates for 1 min at $13,000 \times g$ at 4°C to pellet cellular debris.

6. To assess total Rho GTPase content in control and activated cells, add 20 μL of cell lysates to 5 μL of 5× Laemmli sample buffer, boil the mixture for 5 min and reserve for subsequent SDS-PAGE and immunoblotting.

7. To assess activated (GTP-bound) Rho GTPase content, add the remainder of the cell lysates to either 20 μg of PAK-GST glutathione sepharose beads (for Rac1, Rac2, and Cdc42) or 60 μg Rhotekin-RBD GST glutathione sepharose beads (for RhoA). Secure tubes to a rotator and incubate for 1 h at 4°C and 40 rpm to allow for binding.

8. Pellet the beads for 1 min at $13,000 \times g$ and at 4°C. Wash beads three times with ice-cold wash buffer and aspirate without disturbing the bead pellet.

9. Add 20 μL of 2× Laemmli sample buffer and boil for 5 min to release bound proteins. These proteins are GTP-bound forms of the Rho GTPases. Spin down and pellet the beads.

10. Fractionate the total GTPase and activated GTPase samples by SDS-PAGE on a 12% polyacrylamide gel. Load the total protein and "pulled-down" protein samples in neighboring wells, being careful to exclude the beads.

11. Transfer the fractionated proteins onto a nitrocellulose membrane using your desired apparatus.

12. Perform immunoblotting for the desired Rho GTPase: Rac1, Rac2, or cdc42 if PAK-GST beads are used, and Rho A if Rhotekin-RBD beads are used. Antibodies and their recommended starting dilutions are listed in Subheading 2.3.

13. Immunoreactive proteins are detected by chemiluminescence and exposure to film.

14. Develop the film and quantify the band intensities by densitometry.

15. For both the control and activated lysates, normalize the band intensity of the pulled-down, activated protein against the corresponding total protein level. Express the degree of Rho GTPase activation by dividing the normalized value of the activated sample by the normalized value of the control sample (see Fig. 1).

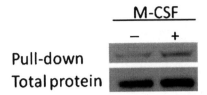

Fig. 1. Pulldown of active GTP-bound Rac1 in resting and M-CSF-stimulated monocytes. After activation by M-CSF (20 ng/mL for 4 h), significantly more Rac1-GTP is pulled down by the RBD beads.

3.3. TAT Protein Constructs and Migration Rescue

This method can be used to introduce Rho GTPases at all stages of osteoclastogenesis. The value of protein transduction is the ability to modulate the global activation states of these GTPases by introducing constitutively active and dominant negative mutants into osteoclasts, and monitoring the associated changes in the cells.

1. Clone the cDNA of the protein of interest and its mutants into the multiple cloning site of the 6His-pTAT-HA expression vector. The 6His tag allows for affinity purification of the recombinant protein.

2. Transform the ligated TAT plasmids into BL21(DE3)pLysS competent cells.

3. Streak a LB agar plate containing the appropriate antibiotic with the ligated plasmid and incubate at 37°C overnight.

4. Pick a colony and inoculate 10 mL of LB with antibiotic. Grow overnight at 37°C in a shaking incubator.

5. Add 1 mL of overnight culture to 100 mL of LB with antibiotic. Incubate at 37°C with shaking until OD_{600} reaches 0.6.

6. Add IPTG (1 mM final concentration) to induce protein expression and grow for 3–4 h.

7. Pellet bacteria at $4,000 \times g$ for 15 min at 4°C.

8. Purify TAT fusion proteins using Ni-NTA columns (QIAexpress Ni-NTA fast Start Kit) under denaturing conditions.

9. Dialyze eluted proteins in Slide-A-Lyzer dialysis cassettes in a large volume of PBS overnight at 4°C. Repeat dialysis after a change of PBS for several more hours (see Note 8).

10. Precipitates are commonly observed after dialysis; remove the protein sample with a syringe and needle and spin down the precipitate.

11. Measure protein concentration in the supernatant using the BCA protein assay kit.

3.3.1. Optimizing TAT Fusion Protein Entry into Cells

1. Prepare several 1.5-mL microfuge tubes each with 0.5×10^6 monocytes in 1 mL of α-MEM growth medium.

2. Add increasing concentrations of purified TAT proteins into each tube, from 50 nM to 1 μM, and incubate at 37°C for 5–30 min.

3. Pellet cells by centrifugation at maximal speed, and wash three times with prewarmed PBS.

4. Add 2× Laemmli sample buffer and boil for 5 min.

5. Perform immunoblotting on the HA tag to identify the optimal concentration and incubation time in order for maximal TAT protein entry. We find that entry reaches a maximum level after 10 min, and the optimal protein concentration to elicit function is 500 nM, which is in a similar range of concentrations used by others (18–20).

3.3.2. Studying the Effects of TAT Fusion Proteins on Migration

1. Incubate Transwell inserts in α-MEM growth medium for 30 min at 37°C.

2. Add 0.5×10^6 cells (in 200 μL) to Transwell inserts and incubate for 2 h at 37°C to allow for cell attachment to the membrane.

3. Remove nonadherent cells by turning over the inserts.

4. Place inserts in 600 μL of growth medium containing 100 ng/mL M-CSF and further incubate for 2 h at 37°C to allow for migration.

5. Fix the cells on the membrane with 4% PFA.

6. Stain cell nuclei with DAPI solution for 10 min in the dark, then wash thoroughly with water.

7. With gentle wiping using a cotton swab and frequent rinsing with PBS, remove cells that have adhered to the top of the membrane but have not migrated through the membrane.

8. Cut out the membrane from the insert and mount onto a glass slide using fluorescence mounting medium.

9. Using a fluorescent microscope at 200× magnification, count the number of cell nuclei in ten random fields of view in order to quantify the mean number of cells that have migrated through the membrane.

10. To study the effects of the TAT fusion proteins on migration, preincubate monocytes with 500 nM TAT proteins for 10 min at 37° prior to being added to Transwell supports. Evaluate the level of migration in control cells and in the presence of TAT proteins (see Fig. 2).

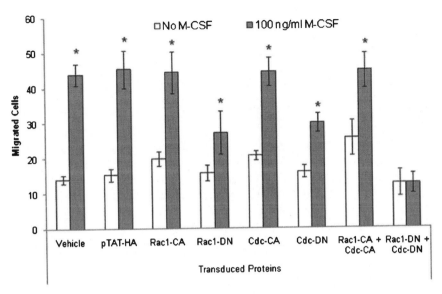

Fig. 2. The ability of selective Rho GTPases and their mutants to affect the migration of monocytes toward a 100 ng/mL M-CSF chemoattractant gradient is examined by TAT protein transduction. The TAT fusion proteins are pTAT-HA (empty vector), and constitutively active (CA) and dominant negative (DN) forms of Rac1 and Cdc42. When there is no chemoat-tractant, Rac1-CA and Cdc42-CA increases cell migration slightly when added individually and in combination. In the presence of M-CSF, neither Rac1-CA nor Cdc42-CA further increases migration even when added in combination. Rac1-DN and Cdc42-DN significantly reduces migration toward M-CSF when added individually, and completely inhibit all migration when added in combination (*p < 0.05 comparing M-CSF vs. no M-CSF for each treatment group).

4. Notes

1. Removing soft tissue while the legs are still attached to the body makes it much easier, since one can pull the leg straight while the weight of the body holds the mouse in place.

2. Separate the femur and tibia by feeling for a depression in the joint with scissors. To remove the remaining soft tissue, place the cutting edge of the scissors at mid-shaft and scrape toward the ends.

3. To ensure that the solution interface is not disturbed, place the tip of the pipette against the side of the tube when layering cells.

4. Osteoclastogenesis performed using this method routinely results in close to 100% of cells staining for TRACP by day 6.

5. Rho GTPase activity can also be assessed during later-stage osteoclast differentiation. Culture osteoclasts with M-CSF and RANKL supplementation for the desired number of days. Wash cells with PBS, and incubate cells with α-MEM growth

medium alone to return Rho GTPases to their inactive state. The investigator should optimize the length of time that cells are starved of cytokines, as prolonged periods of time without cytokine supplementation will result in cell death. Proceed with step 3.

6. Nonadherent cells are lymphocytes and do not adhere to tissue culture plastic.

7. The amount of GTPase activation can be variable depending on experimental conditions. Investigators should plan control experiments with increasing concentrations of M-CSF and varying activation times in order to identify optimal activation conditions.

8. Different pore sizes are available; choose a pore size that is smaller than the predicted size of the TAT protein to prevent sample loss due to dialysis. Because the protein sample is injected into the cassette through a syringe port using a small needle and syringe, one must be careful not to pierce the membrane during injection to prevent accidental loss of the sample.

References

1. Sims, N.A., and Gooi, J.H. 2008. Bone remodeling: Multiple cellular interactions required for coupling of bone formation and resorption. *Semin Cell Dev Biol* 19, 444–451.

2. Boyle, W.J., Simonet, W.S., and Lacey, D.L. 2003. Osteoclast differentiation and activation. *Nature* 423, 337–342.

3. Chen, E.H., Grote, E., Mohler, W., and Vignery, A. 2007. Cell-cell fusion. *FEBS Lett* 581, 2181–2193.

4. Kong, Y.Y., Yoshida, H., Sarosi, I., Tan, H.L., Timms, E., Capparelli, C., Morony, S., Oliveira-dos-Santos, A.J., Van, G., Itie, A., Khoo, W., Wakeham, A., Dunstan, C.R., Lacey, D.L., Mak, T.W., Boyle, W.J., and Penninger, J.M. 1999. OPGL is a key regulator of osteoclastogenesis, lymphocyte development and lymph-node organogenesis. *Nature* 397, 315–323.

5. Miyamoto, T., and Suda, T. 2003. Differentiation and function of osteoclasts. *Keio J Med* 52, 1–7.

6. Saltel, F., Chabadel, A., Bonnelye, E., and Jurdic, P. 2008. Actin cytoskeletal organisation in osteoclasts: a model to decipher transmigration and matrix degradation. *Eur J Cell Biol* 87, 459–468.

7. Akisaka, T., Yoshida, H., Suzuki, R., and Takama, K. 2008. Adhesion structures and their cytoskeleton-membrane interactions at podosomes of osteoclasts in culture. *Cell Tissue Res* 331, 625–641.

8. Jurdic, P., Saltel, F., Chabadel, A., and Destaing, O. 2006. Podosome and sealing zone: specificity of the osteoclast model. *Eur J Cell Biol* 85, 195–202.

9. Jaffe, A.B., and Hall, A. 2005. Rho GTPases: biochemistry and biology. *Annu Rev Cell Dev Biol* 21, 247–269.

10. Wang, Y., Lebowitz, D., Sun, C., Thang, H., Grynpas, M.D., and Glogauer, M. 2008. Identifying the relative contributions of Rac1 and Rac2 to osteoclastogenesis. *J Bone Miner Res* 23, 260–270.

11. Saltel, F., Destaing, O., Bard, F., Eichert, D., and Jurdic, P. 2004. Apatite-mediated actin dynamics in resorbing osteoclasts. *Mol Biol Cell* 15, 5231–5241.

12. Faccio, R., Novack, D.V., Zallone, A., Ross, F.P., and Teitelbaum, S.L. 2003. Dynamic changes in the osteoclast cytoskeleton in response to growth factors and cell attachment are controlled by beta3 integrin. *J Cell Biol* 162, 499–509.

13. Ory, S., Brazier, H., Pawlak, G., and Blangy, A. 2008. Rho GTPases in osteoclasts: orchestrators of podosome arrangement. *Eur J Cell Biol* 87, 469–477.

14. Sakai, H., Chen, Y., Itokawa, T., Yu, K.P., Zhu, M.L., and Insogna, K. 2006. Activated c-Fms recruits Vav and Rac during CSF-1-induced cytoskeletal remodeling and spreading in osteoclasts. *Bone* 39, 1290–1301.

15. Fukuda, A., Hikita, A., Wakeyama, H., Akiyama, T., Oda, H., Nakamura, K., and Tanaka, S. 2005. Regulation of osteoclast apoptosis and motility by small GTPase binding protein Rac1. *J Bone Miner Res* 20, 2245–2253.

16. Yan, J., Chen, S., Zhang, Y., Li, X., Li, Y., Wu, X., Yuan, J., Robling, A.G., Kapur, R., Chan, R.J., and Yang, F.C. 2008. Rac1 mediates the osteoclast gains-in-function induced by haploinsufficiency of Nf1. *Hum Mol Genet* 17, 936–948.

17. Ito, Y., Teitelbaum, S.L., Zou, W., Zheng, Y., Johnson, J.F., Chappel, J., Ross, F.P., and Zhao, H. Cdc42 regulates bone modeling and remodeling in mice by modulating RANKL/ M-CSF signaling and osteoclast polarization. *J Clin Invest* 120, 1981–1993.

18. Zhang, H., Sun, C., Glogauer, M., and Bokoch, G.M. 2009. Human neutrophils coordinate chemotaxis by differential activation of Rac1 and Rac2. *J Immunol* 183, 2718–2728.

19. Dolgilevich, S., Zaidi, N., Song, J., Abe, E., Moonga, B.S., and Sun, L. 2002. Transduction of TAT fusion proteins into osteoclasts and osteoblasts. *Biochem Biophys Res Commun* 299, 505–509.

20. Becker-Hapak, M., McAllister, S.S., and Dowdy, S.F. 2001. TAT-mediated protein transduction into mammalian cells. *Methods* 24, 247–256.

Chapter 13

Assessment of Rho GTPase Signaling During Neurite Outgrowth

Daniel Feltrin and Olivier Pertz

Abstract

Rho GTPases are key regulators of the cytoskeleton during the process of neurite outgrowth. Based on overexpression of dominant-positive and negative Rho GTPase constructs, the classic view is that Rac1 and Cdc42 are important for neurite elongation whereas RhoA regulates neurite retraction in response to collapsing agents. However, recent work has suggested a much finer control of spatiotemporal Rho GTPase signaling in this process. Understanding this complexity level necessitates a panel of more sensitive tools than previously used. Here, we discuss a novel assay that enables the biochemical fractionation of the neurite from the soma of differentiating N1E-115 neuronal-like cells. This allows for spatiotemporal characterization of a large number of protein components, interactions, and post-translational modifications using classic biochemical and also proteomics approaches. We also provide protocols for siRNA-mediated knockdown of genes and sensitive assays that allow quantitative analysis of the neurite outgrowth process.

Key words: Neurite outgrowth, Effector pulldown assays, Immunofluorescence, Time-lapse imaging

1. Introduction

Proper functioning of the nervous system requires connections between neurons and their targets. This requires undifferentiated cells to extend cylindrical extensions in a process called neurite outgrowth (1). This process is a three-step event. First, the round shape of the cell is broken down and a filopodium-like extension is generated. Second, the extension elongates and it is transformed into a proper neurite. Finally, the neurite differentiates into an axon or a dendrite. Recently, genetic and biochemical approaches have been applied to identify and characterize molecular components involved in this process. A wealth of evidence has suggested that cytoskeletal (actin and microtubules), adhesion, and trafficking dynamics play central roles in neurite outgrowth initiation and

Francisco Rivero (ed.), *Rho GTPases: Methods and Protocols*, Methods in Molecular Biology, vol. 827,
DOI 10.1007/978-1-61779-442-1_13, © Springer Science+Business Media, LLC 2012

elongation. Thus, integrated approaches are necessary for an adequate understanding of neurite outgrowth and there is a need for tools that allow large-scale identification and characterization of components important for this morphogenetic event. Here, we present a method that allows the large-scale purification of neurites from the soma of differentiating neuronal cells and thus enables simple spatiotemporal measurements of the subcellular localization (neurite/soma) of protein components, protein–protein interactions, signaling activities, and post-translational modifications using classic biochemical techniques (2). This assay takes advantage of N1E-115 neuroblastoma cells but can in principle also be extended to other cell types. An example that will be discussed in this chapter are Rho GTPase activation effector pulldown assays of the neurite and soma fractions, which reveal that Rac1 and Cdc42 activation is virtually confined to the extending neurite. We also have combined this assay with large-scale quantitative proteomics to decipher the neurite and soma proteomes of N1E-115 cells. Using a bioinformatics approach, we have identified a potential neurite-localized Rho GTPase interactome that comprises several guanine nucleotide exchange factors (GEFs), GTPase activating proteins (GAPs) and downstream effectors (2). This suggests that Rac1 and Cdc42 activation in the neurite is the result of a complex signaling network. To understand the molecular significance of this complexity, we took a RNA interference approach to knockdown individually a subset of Rac1 and Cdc42 specific GEFs and GAPs that are enriched in the neurite. Surprisingly, whereas expression of dominant negative alleles of Rac1 and Cdc42 lead to potent phenotypes, such as complete loss of neurite outgrowth, siRNA-mediated knockdown of the GEFs and GAPs only affect neurite outgrowth in a very mild manner (e.g., loss of neurite outgrowth is never observed). In fact, we observe a large variety of more subtle phenotypes, affecting diverse functions such as filopodium stability and morphology, initiation of neurite outgrowth, neurite elongation, and neurite pathfinding. These complex phenotypes can only be understood with more sensitive assays such as high resolution imaging of the cytoskeleton and time-lapse experiments that evaluate the fine morphodynamic behavior of neurite outgrowth. These results point to a much more complex regulation of Rho GTPase signaling during neurite outgrowth than previously anticipated, and suggest a signaling modularity in which different GEFs and GAPs regulate distinct biochemical pools of Rac 1 and Cdc42 with different functions. These results are compatible with insights obtained using fluorescent Rho GTPase activation biosensors, which have revealed multiple subcellular pools of active GTPases with distinct functions operating simultaneously within one cell (3–5). In this article, we present some of the experimental approaches that were necessary to understand the complexity of spatiotemporal Rho GTPase signaling during neurite outgrowth. These includes protocols for siRNA transfection into N1E-115 cells, their neuronal differentiation, immunofluorescence techniques to visualize their actin and microtubule cytoskeleton, automated image analysis of neurite length and protocols for phase time-lapse microscopy to harness the fine morphodynamics of the neurite outgrowth process.

2. Materials

Prepare all the solutions using ultrapure water and analytical grade reagents. Use sterile media and reagents for the cell culture procedures.

2.1. Cell Culture

1. N1E-115 neuroblastoma cells (American Tissue Culture Collection).
2. Culture medium: Dulbecco's Modified Eagle Medium (DMEM), supplemented with 1% penicillin–streptomycin, 2% L-glutamine, and 10% fetal bovine serum (all from Sigma Aldrich).
3. Differentiation medium: Neurobasal medium (GIBCO, Invitrogen), supplemented with 1% penicillin–streptomycin, and 2% L-glutamine (see Note 1).
4. Phosphate buffered saline (PBS): 13.7 mM NaCl, 0.27 mM KCl, 10 mM Na_2HPO_4, 0.2 mM KH_2PO_4. Autoclave.
5. PUCK'S saline solution: 0.17 mM Na_2HPO_4, 0.22 mM KH_2PO_4, 20 mM Hepes, 138 mM NaCl, 5.4 mM Ca_2Cl, 5.5 mM glucose and 58.4 mM sucrose. Autoclave.

2.2. Microporous Transwell Filter System

1. Transwell Permeable Supports. Use polycarbonate membranes with 3.0 μm pore size (Costar, Corning Inc.). For imaging, use 6.5 mm diameter (12 inserts/24-well plate). For biochemical neurite purification, use 24 mm diameter (six inserts/ six-well plate) (see Note 2).
2. Laminin from Engelbreth–Holm–Swarm murine sarcoma basement membrane (Sigma Aldrich).
3. HPLC grade Methanol.
4. Cotton swabs.
5. Syringe needles (Microlance3, Becton Dickinson).

2.3. Lysis Buffers and Rho GTPase Pulldown

1. Native lysis buffer: 150 mM NaCl, 1.0% NP-40 or Triton X-100, 0.5% sodium deoxycholate, 0.1% sodium dodecyl sulfate (SDS), 50 mM Tris–HCl, pH 8.0.
2. Denaturing lysis buffer: 1% SDS, 2 mM sodium orthovanadate, 1 mM phenylmethanesulfonyl fluoride (PMSF), one tablet of proteases inhibitor cocktail (Roche) per 10 mL of buffer, pH 7.
3. Denaturing lysis buffer for proteomics analysis: 4 mM Tris–HCl pH 7.5, 8 M urea, 0.1% RapiGest (Waters Corporation), 1 mM

sodium orthovanadate, one tablet of PhosSTOP phosphatase inhibitor cocktail (Roche) per 10 ml of buffer (see Note 3).

4. Spin columns with paper filters for immunoprecipitation (Pierce, Thermo Fisher Scientific).

5. GST-PAK or GST-Rhotekin pulldown kits (Millipore Chemicon).

2.4. SiRNA Reagents

1. Pure DMEM medium (Sigma).

2. ON-TARGETplus SMARTpool siRNAs (Dharmacon) (see Note 4).

3. DharmaFECT 2 Transfection Reagent (Dharmacon) (see Note 5).

2.5. Immuno-fluorescence

1. BRB80 buffer: 80 mM PIPES pH 6.8, 1 mM $MgCl_2$, 1 mM EGTA.

2. Fixing solution: BRB80, 0.25% glutaraldehyde (Sigma Aldrich).

3. Permeabilization solution: BRB80, 0.25% glutaraldehyde, 0.1% Triton X-100.

4. Antibody buffer: PBS, 2% bovine serum albumin (BSA) (Sigma Aldrich), 0.1% Triton X-100.

5. Glutaraldehyde autofluorescence limiting buffer: PBS, 0.2% sodium borohydride (Sigma Aldrich) (see Note 6).

6. ProLong Gold antifade reagent (Invitrogen).

7. Antibodies and dyes. Tubulin antibody (monoclonal anti-α-tubulin clone DM 1A, developed in mouse) (Sigma Aldrich), Alexa Fluor 488 phalloidin (Invitrogen) 2 mg/mL stock, 4′,6-diamidino-2-phenylindole (DAPI) 1 mg/mL stock. Secondary antibody: Alexa Fluor 546 goat anti-mouse IgG /H + L (Invitrogen).

2.6. Neurite Outgrowth Analysis and Time-lapse Experiments

1. Coverslips (Menzel-Gläser Gmbh).

2. Glass-bottom plates (Mattech Co).

3. Neurobasal medium supplemented with 25 mM HEPES.

4. Epifluorescence microscope equipped with a temperature-controlled chamber.

5. Metamorph Imaging Software (Molecular Devices).

3. Methods

3.1. N1E-115 Cell Culture and Differentiation

1. Grow N1E-115 cells in DMEM, in 10-cm dishes at a density that never exceeds 70% of confluence, at 37°C and 5% CO_2.

2. For passaging, wash the cells once with warm PBS, aspirate PBS, incubate cells in 2 mL of warm PUCKS's saline solution

DMEM culture medium
(with serum)

Neurobasal differentiation
medium (serum-starved)

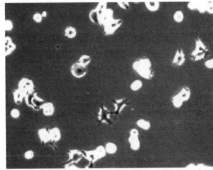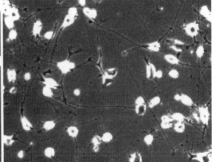

Fig. 1. Phase contrast pictures of nondifferentiated and differentiated N1E-115 cells.

at 37°C for 3–5 min, add 8 mL of DMEM and mechanically detach the cells by gently rinsing the plate and transfer the cell suspension to a 15-mL tube. Centrifuge at $800 \times g$ for 5 min and remove supernatant. Resuspend cell pellet in DMEM. Typically passage 1/5 three times a week.

3. For neuronal differentiation, serum-starve a 50% confluent 10-cm dish of N1E-115 cells by aspirating the DMEM culture medium, rinsing the dish once with 10 mL of sterile PBS, aspirating again and adding to the cells 10 mL of Neurobasal differentiation medium (see Note 7). Typical morphologies of undifferentiated and differentiated cells are shown in Fig. 1.

3.2. Neurite and Soma Purification on Microporous Transwell Filters

1. Grow approximately 15×10^6 N1E-115 cells and differentiate them by serum starvation in Neurobasal differentiation medium overnight (see Note 8).

2. Dilute the laminin in sterile PBS to a final concentration of 10 μg/mL.

3. Place the filters upside down on the lid of the six-well plate where the filters are stored and pipette 500 μL of the laminin solution on the bottom (outer) side of the filters. We typically use 12 filters (e.g., two 6-well plates) in one experiment. Two filters will be used for soma purification and ten filters for neurite purification (see Note 9).

4. Incubate at 37°C for 2 h or over night at 4°C. This should be performed in an environment as sterile as possible.

5. Aspirate the solution from the bottom of the filters and place them back in the six-well plate where 2 mL of Neurobasal medium have been previously added in the bottom chamber (i.e. the well).

6. Detach cells using PUCK's saline solution as explained in Subheading 3.1, step 2. Count the cells and resuspend the cells at a concentration of 10^6 cells per mL in Neurobasal differentiation medium.

7. Dispense 1.2 mL of this cell solution (1.2×10^6 cells) on the top of the filter.

8. Incubate at 37°C and 5% CO_2 for 24 h.

9. Cool down PBS in a beaker and pour 2 mL of methanol/well in a six-well plate on ice before starting the experiment.

10. Remove the six-well plate from the incubator and place it on ice for 10 min.

11. Holding a filter with a forceps, wash it by carefully dipping it in the beaker containing PBS and transfer the filter to a well of the six-well plate containing methanol.

12. Add 2 mL of methanol to the top chamber of the filters for cell fixation (see Note 10).

13. Incubate at 4°C for 20 min.

14. Remove the filters from the plate and dry them upside down.

15. Make one cotton swab slightly wet with some PBS. For soma purification, scrape the bottom of the filters with the swab. For neurite purification, scrape the top of the filter with the swab. Cut the filters at the edges using a syringe needle. Pool the respective neurite and soma filters in separate spin columns where 200–300 μL of lysis buffer have been pipetted previously. A schematic of the procedure is shown in Fig. 2a.

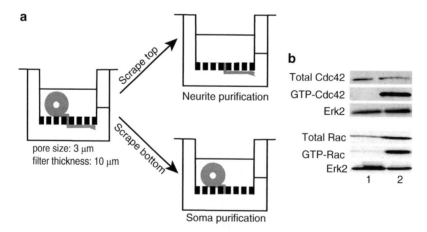

Fig. 2. Neurite purification procedure. (**a**) Schematics of the procedure. Note that the bottoms of the transwell filters are coated with laminin. (**b**) Typical western blot result of an effector pulldown assays of Rac1 and Cdc42 activity in neurite (1) and soma (2) fraction. While Rac1 and Cdc42 are more or less equally distributed in neurite and soma fractions, active pools of Rac1 and Cdc42 are solely restricted to the neurite. Erk2 is equally distributed in neurite and soma fractions and serves as loading control.

16. For lysis in denaturing buffer, boil the spin column at 100°C for 5 min. Centrifuge the spin column and recover lysate in the bottom tube. Measure protein concentration using a standard procedure. The lysate can be used for Western blot analysis to compare abundance of the protein of interest in equivalent amounts of neurite and soma lysate (see Note 11). In this case supplement the lysate with Laemmli buffer and carry out Western blot analysis using standard methods.

17. For lysis in denaturing buffer for proteomics analysis, do not boil but incubate filters with lysis buffer for 5 min at room temperature. Centrifuge spin columns and recover lysate in the bottom tube. Measure protein concentration using a standard procedure and use in the proteomics experiment of interest.

18. For lysis in native buffer, omit the methanol fixation (steps 9–14). Proceed as in step 15 and work at at 4°C. Centrifuge the spin columns and recover the soluble fraction of the cell lysate in the bottom tube. For a further clarification step, performed to get rid of the particulate fraction, the lysate that has been collected in the bottom of the tube can be reloaded on a fresh column and can be centrifuged once more. Measure protein concentration using a standard procedure.

19. For GST-PAK pulldown of active Rac1 or Cdc42, use 100 μg of neurite and soma lysates prepared in native buffer (step 18). Perform the pulldown according to the manufacturer's instructions (see Note 12). Typical results of Rac1 and Cdc42 effector pulldown assays are shown in Fig. 2b.

3.3. Transfection of siRNAs in N1E-115 Cells

The following transfection instructions have to be followed for any of the tested siRNAs as well as for the non-targeting control.

1. *Day 1: cell plating.* Plate 2×10^5 cells per well in a six-well plate in DMEM (see Note 13).

2. *Day 2: transfection.* In a 1.5-mL tube (tube A) pipette 100 μL of pure DMEM. Add to the medium 100 pmol of siRNA (5 μL of a 20-μM concentrated stock solution) (see Note 4).

3. In another tube (tube B) pipette 100 μL of pure DMEM. Add to the medium 6 μL of DharmaFECT 2 transfection Reagent.

4. Pipette the solution of tube A into tube B and mix gently (do not vortex).

5. Incubate for 20 min at room temperature.

6. Aspirate the medium from the cells in the six-well plate and rinse once with 1 mL of sterile PBS.

7. Aspirate the PBS and pipette 1 mL of DMEM (see Note 14).

8. Add the transfection mix dropwise to the corresponding well.

9. *Day 3: medium change.* Aspirate the medium from the cells. Rinse once with 1 mL of sterile PBS. Aspirate the PBS and pipette 2 mL of DMEM culture medium.

10. *Day 4: cell starving and differentiation.* Aspirate the complete DMEM from the cells. Rinse once with 1 mL of sterile PBS and aspirate it. Pipette 2 mL of Neurobasal differentiation medium to the cells.

11. *Day 5: cell plating for assays.* Coat 16-mm coverslips or 12-well glass bottom plates with 10 μg/mL laminin for 2 h at 37°C or at 4°C over night.

12. Aspirate the Neurobasal differentiation medium from the six-well plate that contains the serum-starved cells. Rinse the cells once with 1 mL of sterile PBS and aspirate it. Detach the cells using 1 ml of PUCK's saline solution by incubating at 37°C for 3–5 min. Resuspend in a given volume of neurobasal differentiation medium and count the cells. Centrifuge the cells at $800 \times g$ for 5 min, aspirate the supernatant and resuspend the cells in an appropriate volume.

13. Plate 4×10^4 cells per well (or per coverslip). The remainder of the cells can be used for other assays.

14. Place cells in the incubator for 3 h for the time-lapse assay (see Subheading 3.4.3) or for 16–24 h for immunostaining and the neurite outgrowth analysis (see Subheadings 3.4.1 and 3.4.2).

3.4. Evaluation of Neurite Outgrowth Responses

This protocol starts with cells that were allowed to extend neurites for 16–24 h (see Subheading 3.3) (see Note 15).

3.4.1. F-Actin/Tubulin Immunostaining

1. Fill a 12-well plate with 1 mL of PBS per well.

2. Carefully transfer the coverslips to the plate with PBS to wash them.

3. Fix the cells by transferring the coverslips to another 12-well plate containing the fixing solution. Incubate for 45 s.

4. Transfer the coverslips to another 12-well plate containing the permeabilization solution and incubate for 10 min.

5. Move the coverslips to a 12-well plate containing PBS. Incubate for 10 min. Repeat twice this step.

6. Transfer the coverslips a 12-well plate containing the 0.2% sodium borohydride solution. Incubate for 20 min.

7. Wash twice for 10 min with PBS.

8. Incubate for 10 min with antibody buffer.

9. Prepare the dilution of the primary antibody against tubulin (1:500) in antibody buffer.

10. Transfer the coverslips to a parafilm-covered surface and pipette 100 μL of the primary antibody dilution on each coverslip.

11. Incubate at room temperature for 30 min.

12. Transfer coverslips back to the 12-well plate and wash three times for 10 min with PBS.

13. Prepare the secondary antibody, diluting it 1:1000 in antibody buffer. AddDAPI (1:1,000 dilution of a 1-mg/mL stock solution) to the antibody dilution for staining of nuclei. Add eventually also a fluorescently conjugated phalloidin for the staining of F-actin (1:1,000 dilution of a 2-mg/mL stock solution).

14. Transfer the coverslips to the parafilm-covered surface and pipette 100 μL of the antibody buffer with secondary antibody and DAPI (and eventually phalloidin) on each coverslip.

15. Incubate at room temperature and in the dark for 20 min.

16. Wash three times with PBS (always protect from the light).

17. Prepare the slides to be mounted by pipetting one drop of antifade reagent per coverslips (typically one slide can host two coverslips).

18. Gently mount the slides by placing the coverslips upside down on the drop of antifade reagent.

19. Leave the mounted slides dry over night in the dark.

3.4.2. Neurite Outgrowth Analysis and High Resolution Imaging

1. Using a 10× objective acquire multiple tubulin and DAPI pictures (multiple fields of view) (see Note 16). We typically use the "Scan slide" module of the Metamorph software that allows stitching multiple fields of view in one image (see Fig. 3a).

2. Perform neurite outgrowth analysis using the Metamorph neurite outgrowth plugin. The module will use the DAPI image for the recognition of the cell bodies and the tubulin image for segmentation of the neurite. The plugin will ask for a panel of parameters that can be measured manually in metamorph (cell body width, fluorescence intensity, and area; neurite minimum and maximum width and fluorescence intensity). Once these parameters are determined, we test them and compare the original and the segmented images. The parameters are then fine-tuned for optimal results and all images can then be analyzed.

3. The software will provide binary images of the nuclei and of the neurites (Fig. 3a) as well as numerical results of including total neurite outgrowth, area of the cell bodies, number of neurites per cell body, neurite branching, etc. These numerical results can then be exported to Excel or other software for statistical analysis and graphical representations. Typical results of this procedure are shown in Fig. 3b.

4. For high resolution analysis, we use high numerical aperture oil immersion lenses. Typical results are shown in Fig. 4.

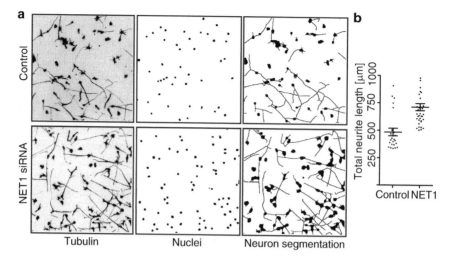

Fig. 3. Neurite outgrowth measurements. (**a**) Fluorescent micrographs of control and Net1 siRNA transfected cells that were replated on a laminin-coated coverslip and stained for tubulin (left panel). Net1 is a GEF for RhoA. The image is rendered with an inverted contrast. Middle and right panels display segmented binary images of the DAPI (nuclei) and tubulin (neurites) channels as analyzed using the neurite outgrowth algorithm. (**b**) Graph of total neurite length on per-cell basis measurements from segmented images in (**a**).

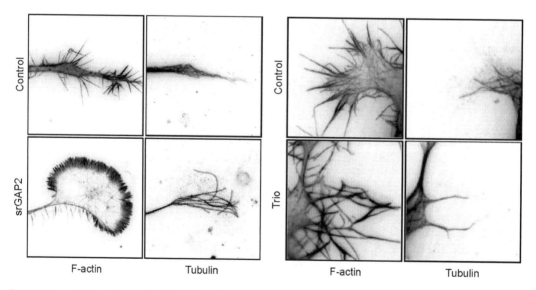

Fig. 4. High resolution micrographs of F-actin and tubulin-stained control or knockdown cells. Images were acquired with a 63× high NA objective and are rendered with inverted contrast. Note large filopodia arrays leading to spread growth cone in srGAP2 knockdown cells (GAP for Rac1). Note aberrant, intertwined filopodia found at the periphery of Trio knockdown cells (GEF for Rac1 and RhoA).

3.4.3. Time-lapse Analysis   After 3 h of incubation in a cell incubator, which allows for spreading and initial neurite outgrowth (see Subheading 3.3), the dynamics of the neurite outgrowth process can be analyzed using phase contrast time-lapse experiments (see Note 17).

1. Gently aspirate the medium from the wells using a pipette. Do not aspirate using a vacuum line (see Note 18).

2. Pipette very gently fresh neurobasal medium supplemented with 20 mM HEPES, filling up the well completely (see Note 19).

3. Quickly replace the lid avoiding trapping air bubbles.

4. Run a 16-h time-lapse experiment at the microscope using a multidimensional acquisition program (for example use the multi-dimensional acquisition module of Metamorph, that allows to save several positions, using several wavelengths). We typically perform multistage experiments that allow time-lapse capture of multiple positions across multiple wells (e.g., multiple siRNA experiments). Use a microscope equipped with a temperature-controlling box that keeps the temperature fixed at 37°C. It is also possible to perfuse the box with 5% CO_2 (in this case HEPES is dispensable) (see Note 20).

5. Time-lapse movies are visually inspected using Metamorph software for evaluation of neurite outgrowth morphodynamic phenotypes.

4. Notes

1. After a certain amount of time, the Neurobasal medium deterio-rates leading to cell death. Once we open a new Neurobasal medium bottle, we aliquot it and store it in 50-mL tubes at 4°C.

2. Some of the Costar Transwell filters are leaky. Be careful to test different batches of transwell filters before purchasing them. For that purpose, dispense 2 mL of water in the lower chamber (for 24-mm filters in six-well plates) and check if it leaks into the upper chamber. The 3-μm pore filter is appropriate for N1E-115 neuroblastoma cells which have a soma diameter of roughly 35 μm. Smaller pore sizes might apply for other cell lines or primary neurons that are typically smaller.

3. Detergents such as SDS are typically not tolerated by mass spectrometry machines. This is the best denaturing buffer we have tested so far for this application. Furthermore it is com-patible with subsequent trypsin digestion when diluted.

4. We use Dharmacon smart pool plus siRNAs. This typically allows for 70 % of knockdown efficiency at the mRNA level and for 90–95% knockdown efficiency at the protein level. We use a non-targeting siRNA smart pool as control.

5. A wide variety of siRNA reagents from different manufacturers function in N1E-115 cells. Dharmafect 2 is our favorite transfection reagent because of low toxicity and potency (100% transfection efficiency and siRNA being detected in the cells for the whole duration of the experiment).

6. This buffer has to be freshly prepared for each experiment. Bubbles should be apparent in the solution. Be careful, this is toxic!

7. Additional supplements such a B12 as commonly used for culture of primary neurons are not needed. Serum-free DMEM can also be used for differentiation but is less potent than Neurobasal medium.

8. Rather than starving cells on the filter, we find that a cycle of neuronal differentiation through serum starvation in Neurobasal differentiation medium and subsequent replating on the filter allows for very robust neurite outgrowth necessary for efficient neurite purification. Regarding cell number, a 50 % confluent dish typically yields 10^6 cells. In one experiment, 15–20 10-cm dishes are typically used. 15-cm dishes can also be used to reduce the amount of plates to handle.

9. For neurite purification, one filter typically yields 20–30 μg of cell lysate depending on the cell lysis buffer. For soma purification, 300 μg of cell lysate are typically obtained. This allows generating 200–300 μg of neurite and 600 μg of soma lysate in one experiment.

10. Methanol fixation allows to freeze the signaling state of the cell, and therefore to avoid the upregulation of stress signaling pathways during the neurite or soma scraping procedure. This is at the expense of the loss of membrane components during the fixation. This procedure is compatible with evaluation of phosphorylation events using phosphor-specific antibodies and western blot analysis. However this is not compatible with activation pulldown assays or co-immunoprecipitation experiments, which require native conditions.

11. Once neurite and soma lysates have been separated, equal lysate amounts can be loaded on a gel and probed using western blot. With the assumption that protein density (amount of protein per cell volume) is constant within a cell, this allows to measure relative protein density in each respective subcellular domain (neurite and soma). Classic quality controls for correct loading are done with Erk2 and phospho-Erk antibodies. Total Erk is equally abundant in neurite and soma lysates, whereas phospho-Erk is highly enriched in the neurite. It is important to mention that many proteins are found in the detergent insoluble fraction in the neurite and might show different degrees of enrichment depending if a denaturing or native lysis

buffer is used. Thus phospho-ERK signal in the neurite is typically lost when the neurite has been lysed in a native buffer.

12. Using this procedure, we were successful in detecting robust pools of active Rac1 and Cdc42 in the neurite fraction (Fig. 2b). However, this might still be an under-estimate since large pools of Rac1 and Cdc42 are observed in the particulate fraction. We were not able to detect any RhoA activation in the neurite because all RhoA was found in the particulate fraction and therefore is resilient to solubilization by native lysis buffers.

13. 2×10^5 cells/well allows to perform a series of experiments such as time-lapse analysis, immunofluorescence and quantitative RT-PCR to test siRNA mediated knockdown efficacy. If knockdown efficacy has to be tested using western blot, we advise to transfect an additional well.

14. Transfection is performed in DMEM culture medium. The presence of serum allows the cells to remain in the undifferentiated state and increases cell survival. Transfection in serum free differentiation medium leads to poor cell survival.

15. At this stage of the procedure, cells should be inspected and neurites should be observed if differentiation has occurred.

16. The immunofluorescence procedure presented here allows for excellent preservation of the actin and tubulin cytoskeleton. This procedure can be extended to other proteins. However, in this case, one has to keep in mind that glutaraldehyde can lead to loss of antigenicity of certain epitopes. Furthermore, a commonly observed artifact is a dim autofluorescence signal in the nucleus.

17. Because there is no need for high spatial resolution to get a global picture of neurite outgrowth, we typically work by binning the images 3×3. This yields better signal to noise ratios (which is not limiting here), but also allows drastically reducing the image size. This then allows performing automated image analysis much faster with simple computers.

18. We find that cells are highly sensitive to stress such as light in the initial spreading and neurite outgrowth phase. Once cells are adherent, they become light resistant and can be time-lapsed.

19. N1E-115 cells on a laminin-coated coverslip are extremely loosely adherent. They adhere to some extent through their soma but mostly through their growth cones. Any strong shear stress will therefore immediately lead to their detachment.

20. While glass bottom multiwell plates are preferable, phase contrast time-lapse imaging can also be performed using classic plastic dishes with excellent picture quality. We typically use 10× or 20× long working distance, phase contrast objectives to capture ten or three cells per field of view.

References

1. da Silva, J.S., and Dotti, C.G. (2002) Breaking the neuronal sphere: regulation of the actin cytoskeleton in neuritogenesis. *Nat Rev Neurosci* 3, 694–704.

2. Pertz, O.C., Wang, Y., Yang, F., Wang, W., Gay, L.J., Gristenko, M.A., Clauss, T.R., Anderson, D.J., Liu, T., Auberry, K.J., Camp, D.G. 2nd, Smith, R.D., and Klemke, R.L. (2008) Spatial mapping of the neurite and soma proteomes reveals a functional Cdc42/Rac regulatory network. *Proc Natl Acad Sci USA* 105, 1931–1936.

3. Nalbant, P., Hodgson, L., Kraynov, V., Toutchkine, A., Hahn, K.M. (2004) Activation of endogenous Cdc42 visualized in living cells. *Science* 305, 1615–1619.

4. Pertz, O. (2010) Spatio-temporal Rho GTPase signaling - where are we now? *J Cell Sci* 123, 1841–1850.

5. Pertz, O., Hodgson, L., Klemke, R.L., and Hahn K.M. (2006) Spatiotemporal dynamics of RhoA activity in migrating cells. *Nature* 440, 1069–1072.

Chapter 14

Assessment of the Role for Rho Family GTPases in NADPH Oxidase Activation

Kei Miyano and Hideki Sumimoto

Abstract

Rac, a member of the Rho family small GTPases, plays a crucial role in activation of Nox family NADPH oxidases in animals, enzymes dedicated to production of reactive oxygen species such as superoxide. The phagocyte oxidase Nox2, crucial for microbicidal activity during phagocytosis, is activated in a manner completely dependent on Rac. Rac in the GTP-bound form directly binds to the oxidase activator p67[phox], which in turn interacts with Nox2, leading to superoxide production. Rac also participates in activation of the nonphagocytic oxidase Nox1; in this case, GTP-bound Rac functions by interacting with Noxa1, a p67[phox]-related protein that is required for Nox1 activation. On the other hand, in the presence of either p67[phox] or Noxa1, Rac facilitates superoxide production by Nox3, which is responsible in the inner ear for formation of otoconia, tiny mineralized structures that are required for sensing balance and gravity. All the three mammalian homologs of Rac (Rac1, Rac2, and Rac3), but not Cdc42 or RhoA, are capable of serving as an activator of Nox1–3. Here, we describe methods for the assay of Rac binding to p67[phox] and Noxa1 and for the reconstitution of Rac-dependent Nox activity in cell-free and whole-cell systems.

Key words: NADPH oxidase, Nox1, Nox2, Nox3, Superoxide, Rac, p67[phox], Noxa1

1. Introduction

The membrane-integrated enzyme NADPH oxidase (Nox) is a flavocytochrome that deliberately produces reactive oxygen species (ROS) such as superoxide and hydrogen peroxide from molecular oxygen in conjunction with oxidation of NADPH (1). The Nox family NADPH oxidases participate in various biological processes including innate immunity, biosynthetic processes, and signal transduction. Members of this family are present in a wide variety of eukaryotes but absent in prokaryotes (1). The human genome contains seven distinct genes encoding NADPH oxidases: five homologous enzymes designated as Nox (Nox1 through Nox5)

Francisco Rivero (ed.), *Rho GTPases: Methods and Protocols*, Methods in Molecular Biology, vol. 827,
DOI 10.1007/978-1-61779-442-1_14, © Springer Science+Business Media, LLC 2012

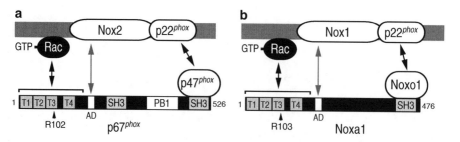

Fig. 1. Role of Rac in activation of Nox2 and Nox1. (**a**) Upon cell stimulation, Rac in the GTP-bound form directly binds to the N-terminal domain of p67^{phox} comprising four TPR motifs (T1–T4). The binding renders p67^{phox} able to interact with Nox2 via the activation domain (AD), leading to superoxide production. (**b**) GTP-bound Rac is also capable of binding to Noxa1, the AD of which probably interacts with Nox1, resulting in further production of superoxide by Nox1.

and two distantly related ones named dual oxidases (Duox1 and Duox2).

The founding and best-characterized member of the family is mammalian Nox2 (also known as gp91^{phox}), which is most highly expressed in phagocytes such as neutrophils albeit it is present in a variety of cells. Nox2 forms a heterodimer with the membrane-spanning protein p22^{phox}, which provides a docking site for soluble activation proteins. Nox2 is dormant in resting cells but activated during phagocytosis to produce superoxide, a precursor of powerful microbicidal ROS. Activation of Nox2 requires the small GTPase Rac in addition to the specialized proteins p67^{phox} and p47^{phox} (2–4); the three proteins translocate upon cell stimulation from the cytosol to the membrane to interact with the Nox2–p22^{phox} complex (the complex is known as flavocytochrome b_{558}), leading to superoxide production (see Fig. 1). The phagocyte oxidase can be activated in vitro by anionic amphiphiles such as sodium dodecyl sulfate (SDS) and arachidonic acid in a cell-free system reconstituted with flavocytochrome b_{558}, p67^{phox}, p47^{phox}, and Rac in the GTP-bound form (5).

The significance of the Nox2-based oxidase in host defense is evident because a genetic defect in any of the four proteins gp91^{phox}/ Nox2, p22^{phox}, p67^{phox}, and p47^{phox} causes chronic granulomatous disease, patients with which suffer from recurrent and life-threatening infections. A patient with decreased oxidase activity and other neutrophil function defects is known to have an inhibitory (dominant-negative) mutation in Rac2 (6, 7). Neutrophils from Rac-deficient mice are also defective in superoxide-producing activity (8, 9). In addition to these supportive proteins (p22^{phox}, p67^{phox}, p47^{phox}, and Rac), Nox2 activation requires cell stimulation with particulates (phagocytosis) or soluble stimulants such as phorbol 12-myristate 13-acetate (PMA), a potent activator of protein kinase C, which results in phosphorylation of p47^{phox} and conversion of Rac to the GTP-bound form (10, 11). In stimulated cells, the cytosolic proteins p67^{phox} and p47^{phox} are recruited together to the membrane in a manner dependent on phosphorylation-induced binding of p47^{phox} to p22^{phox} but independent of Rac. Rac dimerizes with RhoGDI in the cytosol of resting cells, but the GTPase dissociates upon cell stimulation from the partner and translocates to the membrane. At the membrane, Rac in the GTP-bound form directly binds to p67^{phox}, which results in superoxide production by the membrane-integrated enzyme Nox2 (12, 13).

Nox1, the first mammalian nonphagocytic oxidase to be cloned, is expressed most abundantly in epithelial cells of the colon, but also at lower levels in a variety of cells such as vascular smooth muscle cells (1). Nox1 in the vasculature is considered to participate in angiotensin II-elicited regulation of vascular tone, whereas its function in the colon remains to be elucidated. Although Nox1 is also complexed with p22^{phox}, its activation requires both of the soluble proteins Nox activator 1 (Noxa1) and Nox organizer 1 (Noxo1) instead of their respective homologs p67^{phox} and p47^{phox} (Fig. 1). GTP-bound Rac is capable of directly interacting with Noxa1 (14) thus facilitating superoxide production by Nox1 (15–17). In the presence of these supportive proteins, Nox1 produces a large amount of superoxide without cell stimulants. The production is, however, considerably increased when cells are stimulated with PMA.

Nox3 exists in the inner ear and the fetal kidney and, like Nox2, forms a stable heterodimer with p22^{phox}. In mice, this oxidase is crucial for the genesis of otoconia, tiny mineralized structures that are required for sensing balance and gravity. In contrast to Nox1 and Nox2, Nox3 constitutively produces a small but considerable amount of superoxide even in the absence of supportive proteins, although Nox3 activity is reinforced by p47^{phox}, Noxo1, or p67^{phox} (18). In the absence of both p47^{phox} and Noxo1, Rac enhances superoxide production by Nox3 in a manner dependent on the presence of either p67^{phox} or Noxa1; the enhancement requires Rac binding to these activator proteins (15, 19). On the other hand, Rac is incapable of regulating ROS-producing activity of the other mammalian oxidases (Nox4, Nox5, Duox1, and Duox2), although Rac homologs participate in activation of fungal and plant NADPH oxidases (1).

In mammals, there exist three closely related Rac proteins, namely, Rac1, Rac2, and Rac3. Rac1 is ubiquitously expressed, whereas Rac2 expression is mostly restricted to cells of hematopoietic lineage; Rac3 is expressed in various cells and is most abundant in the brain. All the three paralogs of Rac (20), but not Cdc42 or RhoA (21, 22), are capable of functioning as an activator of Nox1–3. These Rac proteins share high amino acid identity with an identical switch I region, which is different from those of Cdc42 and RhoA; the Rac GTPases differs primarily in the C-terminal ten residues, encompassing a polybasic sequence (amino acids 183–188) and a CAAX motif for geranylgeranylation (amino acids 189–192).

The Rac switch I region is essential for binding to p67^{phox} and Noxa1 and activation of Nox (22–24), both of which does not require the switch II region or the insert helix of Rac (20); the insert helix (a surface-exposed, α helix), also known as the insert region, is unique to the Rho family among the Ras-related super-family of small GTPases. The direct target of the Rac switch I region is the p67^{phox} N-terminal domain comprising four tetratrico-peptide repeat (TPR) motifs (12, 13). The evolutionarily conserved arginine residue in the third motif (corresponding to Arg-102 in p67^{phox} and Arg-103 in Noxa1) plays a crucial role in binding to Rac (12, 14) and also in activation of the phagocyte oxidase Nox2 as well as Nox1 and Nox3 (12, 15, 17, 19). Rac binding is consid-ered to induce a conformational change of p67^{phox}, which allows the activation domain, N-terminal to the Rac-binding TPR domain, to interact with Nox2, leading to superoxide production (25–27).

Here, we describe a relatively simple and highly reliable assay for estimation of binding of Rac to its effector proteins p67^{phox} and Noxa1, using a glutathione S-transferase (GST) pull-down method combined with immunoblot analysis. The binding is crucial for activation of Nox1–3, especially Nox2 (Rac binding to p67^{phox} is absolutely required for superoxide production by Nox2), but is considerably weak. It is thus very important to detect a bound protein with high sensitivity under the conditions where nonspe-cific binding does not occur. We also present protocols for the reconstitution of Rac-dependent Nox activity in both cell-free and whole-cell systems. Cell-free activation systems are available only for Nox2. Although a large amount of Nox2 can be prepared from neutrophil membranes, much smaller amounts of Nox oxidases are endogenously expressed in other types of cell, and it is presently difficult to express Nox in transfected cells at a level comparable to Nox2 in neutrophils. On the other hand, whole-cell reconstitution systems are used not only for Nox2 but also for Nox1 and Nox3.

2. Materials

2.1. Buffers

1. HEPES-bufferd saline (HBS): 120 mM NaCl, 5 mM KCl, 5 mM glucose, 1 mM MgCl$_2$, 0.5 mM CaCl$_2$, and 17 mM HEPES, pH 7.4.

2. Binding buffer (Rac): 100 mM KCl, 0.005% Triton X-100, and 100 mM potassium phospate, pH 7.0.

3. Washing buffer (Rac): 100 mM KCl, 0.1% Triton X-100, and 100 mM potassium phospate, pH 7.0.

4. Elution buffer (Rac): 20 mM glutathione, 200 mM NaCl, 0.1% Triton X-100, and 200 mM Tris–HCl, pH 8.0.

5. Sonication buffer: 75 mM NaCl, 170 mM sucrose, 1 mM MgCl$_2$, 0.5 mM EGTA, 10 μM ATP, 2 mM NaN$_3$, 5 μM GTPγS, and 100 μg/ml *p*-amidinophenyl methanesulfonyl fluoride hydrochloride.

6. Cell-free assay buffer: 1.0 mM FAD, 1.0 mM EGTA, 1.0 mM MgCl$_2$, 100 μM GTPγS, 1.0 mM NaN$_3$, and 100 mM potassium phosphate, pH 7.0.

7. 100 mM Potassium phosphate, pH 7.0.

8. 100 mM Potassium phosphate, pH 7.0, with 0.5 M KCl.

9. HBS-BSA: HBS containing 0.04% bovine serum albumin (BSA).

10. SDS-sample buffer: 5% SDS, 20% glycerol, 0.2% bromophenol blue, and 125 mM Tris–HCl, pH 6.8.

2.2. Chemicals and Reagents

1. Expression vectors: pGEX-6P (GE Healthcare) for expression of GST–fusion proteins in *Escherichia coli*; pcDNA3 (Invitrogen) and pEF-BOS (28) for expression of FLAG-, Myc-, or HA-tagged proteins in mammalian cells.

2. Glutathione-Sepharose 4B (GE Healthcare).

3. PMA (Sigma-Aldrich): prepare a 10 μg/ml stock solution in DMSO and freeze at –80°C.

4. Modified DIOGENES solution: dissolve the DIOGENES reagent and the DIOGENES activator (National Diagnostics) in 12 ml of 25 mM luminol. Divide the solution into 40-μl aliquots and freeze at –80°C.

5. Superoxide dismutase (SOD) (Wako): prepare a 1 mg/ml stock solution and freeze at –80°C.

6. Ferricytochrome *c*: prepare a 5 mM stock solution in 100 mM potassium phosphate, pH 7.0, and freeze at –80°C.

7. Polyvinylidene difluoride (PVDF) membrane (Millipore).

8. Coomassie stain and destain solutions. Stain solution: 0.25% Coomassie brilliant blue-R250, 5% acetic acid, and 50% methanol:destain solution:10% acetic acid and 10% methanol.

9. Human blood: obtain from healthy volunteers using blood collection bags containing sodium citrate.

10. Dextran 2000 solution: 3% dextran in 0.9% NaCl solution, sterile. Prepare freshly.

11. 0.9% NaCl solution, sterile.

12. 10% NaCl solution, sterile.

13. Lymphocyte Separation Medium 1077 (Wako) or Histopaque 1077 (Sigma-Aldrich).

14. Protease inhibitor cocktail for use with mammalian cell and tissue extracts (Sigma-Aldrich).

15. GTPγS in 100 mM potassium phosphate, pH 7.0: prepare a 100 mM stock solution, and freeze at –80°C.

16. 10 mM MgCl$_2$.

17. SDS: freshly prepare a 10 mM solution in distilled water.

18. NADPH: freshly prepare a 100 mM solution in 100 mM potassium phosphate, pH 7.0.

19. Arachidonic acid: prepare a 10 mM solution in ethanol and stored at –20°C.

20. Cell lines: CHO, HeLa, HEK, and COS.

21. Media for cell culture: Ham's F12 medium and Dulbecco's modified Eagle's medium (DMEM), both supplemented with 10% fetal calf serum (FCS), Opti-MEM (Invitrogen).

22. Transfection reagents: FuGENE (Roche), Lipofectamine (Invitrogen).

23. Trypsin–EDTA (Invitrogen).

24. Discontinuous sucrose gradient: 2.0 ml of 40% (w/v) sucrose and 1.5 ml of 15% (w/v) sucrose in 40 mM NaCl, 1.0 mM MgCl$_2$, and 0.5 mM EGTA.

25. Antibodies: anti-Rac monoclonal antibody (BD Biosciences); anti-FLAG monoclonal antibody (Sigma-Aldrich); anti-Myc monoclonal antibody (Roche Diagnostics); anti-HA monoclonal antibody (Covance); and horseradish peroxidase-conjugated anti-mouse antibody (GE Healthcare Biosciences).

26. ECL plus detection kit (GE Healthcare Biosciences) for immunoblot analysis.

27. Double-strand small-interfering RNAs (siRNAs) targeting human Rac1: for the coding region, 5′-UCCGUGCAAAGU GGUAUCCUGAGGU-3′ (sense) and 5′-ACCUCAGGAUAC CACUUUGCACGGA-3′ (antisense); for the 3′-untranslated region, 5′-CUUGGAACCUUUGUACGCUUUGCUC-3′ (sense) and 5′-GAGCAAAGCGUACAAAGGUUCCAAG-3′ (antisense) (17).

28. Medium GC Duplex of Stealth RNAi Negative Control Duplexes (Invitrogen) as a negative control.

2.3. Equipment

1. Protein electrophoresis (BE-280, Bio Craft) and transfer equipment (Trans-Blot SD semidry transfer cell, BioRad).

2. Sonicator (Astrason XL, Misonix).

3. Spectrophotometer (Hitachi 557 dual-wavelength spectrophotometer, Hitachi).

4. Luminometer (Auto Lumat LB953, EG&G Berthold).

2.4. Plasmid Construction

The cDNA encoding a C-terminally truncated p67^{phox} (amino acids 1–242) is prepared by PCR using the cDNA of full-length p67^{phox} (526 amino acids) as a template. The cDNA encoding a C-terminally truncated Noxa1 (amino acids 1–243) is prepared by PCR using the cDNA of full-length Noxa1 (476 amino acids). Mutations leading to the indicated amino acid substitutions are introduced by PCR-mediated site-directed mutagenesis. For expression in mammalian cells, the cDNAs for Nox1, Nox2, and Nox3 are ligated to pcDNA3, whereas the cDNAs encoding p22^{phox}, Rac1, p67^{phox}, Noxa1, p47^{phox}, and Noxo1 are ligated to pEF-BOS: p22^{phox} is constructed for expression without a tag; Rac as a Myc-tagged protein; p67^{phox} and Noxa1 as a FLAG- or Myc-tagged protein; p47^{phox} and Noxo1 as an HA-tagged protein. The cDNAs encoding Rac1, p67^{phox}, Noxa1, and p47^{phox} are also ligated to pGEX-6P (GE Healthcare Biosciences) for expression in *E. coli*.

2.5. Preparation of Recombinant Rac1, p67^{phox}, Noxa1, and p47^{phox}

GST–Rac1 (Q61L/C189S), GST–Rac1 (C189S), GST–p67^{phox}-(1–526), GST–p67^{phox}-(1–242), GST–Noxa1-(1–460), GST–Noxa1-(1–243), GST–p47^{phox}-(1–390), and GST–p47^{phox}-(1–286) are expressed in the *E. coli* strain BL21 and affinity purified with glutathione-Sepharose-4B (20, 27) using standard procedures. Purified GST–Rac1 (Q61L/C189S) and GST–Rac1 (C189S) are digested with PreScission protease (GE Healthcare) to remove the GST-tag following the manufacturer's instructions (20, 27). Recombinant proteins at about 1 mg/ml are frozen in liquid nitrogen and stored at –80°C.

3. Methods

3.1. Estimation of Binding of Rac to p67^{phox} or Noxa1

1. Incubate GST-tagged p67^{phox} or Noxa1 (0.2 nmol) for 10 min at 4°C with 1.0 nmol of Rac1 (Q61L/C189S) (see Note 1) in 400 μl of binding buffer (see Note 2).

2. Add 30 μl of a 50% slurry of glutathione-Sepharose-4B beads, and incubate the mixture for 30 min at 4°C.

3. Centrifuge the sample at 400×*g* for 2 min at 4°C, and remove the supernatant using a plastic transfer pipette and discard.

4. Wash the beads by resuspending them in 1.0 ml of washing buffer and centrifuging at 400×*g* for 2 min at 4°C.

5. Wash the beads twice more by repeating step 4.

6. Remove the supernatant, and elute proteins from the beads with 30 μl of elution buffer.

7. Resolve the eluted proteins by 12% SDS–PAGE in two gels using standard procedures.

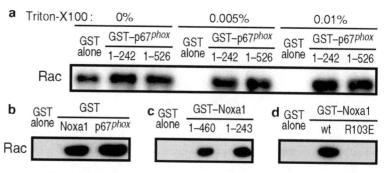

Fig. 2. Direct binding of Rac to p67*phox* and Noxa1. (**a**) Recombinant Rac1 (Q61L) was incubated with GST-fused of the N-terminus p67*phox* (amino acids 1–242) or full-length p67*phox* (amino acids 1–526) in binding containing the indicated concentration of Triton X-100. (**b–d**) Rac1 (Q61L) was incubated with GST–p67*phox*-(1–242) or GST–Noxa1-(1–243) (**b**); GST–Noxa1-(1–460) or GST–Noxa1-(1–243) (**c**); or GST–Noxa1-(1–243) with or without the R103E substitution (**d**). Proteins were pulled down with glutathione-Sepharose 4B beads. The precipitated proteins were analyzed by immunoblot with an anti-Rac monoclonal antibody.

8. Stain one gel with Comassie for estimation of the recovery of p67*phox* or Noxa1.

9. Transfer proteins on the second gel to a PVDF membrane, probe the membrane with the anti-Rac1 monoclonal antibody, and detect Rac using the horseradish peroxidase-conjugated anti-mouse antibody and the ECL plus detection kit.

A representative example of the outcome of this method is depicted in Fig. 2. p67*phox* and Noxa1 are capable of binding to Rac to a similar extent (Fig. 2a, b). The binding ability of C-terminally truncated p67*phox* and Noxa1 is similar to that of full-length proteins (Fig. 2a, c). The R103E substitution in Noxa1 results in a loss of interaction with Rac (Fig. 2d). Note that the presence of at least 0.005% Triton X-100 in the binding solution is crucial for the assay, as without the detergent, Rac1 binds not only to GST–p67*phox* but also to GST alone (Fig. 2a).

3.2. Estimation of the Role of Rac in Nox2 Activation in a Cell-Free System

3.2.1. Preparation of Human Neutrophil Membrane Fraction as a Source of Nox2

Following protocol is designed for 200 ml of blood. If different blood volumes are required, adjust the indicated volumes proportionally.

1. Collect human blood into blood collection bags containing sodium citrate (for collecting 200 ml of blood).

2. Combine the equal volume (200 ml) of dextran 2000 solution with 200 ml of blood in a large polypropylene tube. Mix by gently inverting the tube.

3. Allow the blood–dextran mixture to sediment for 30–45 min at room temperature.

4. Transfer the clear, red blood cell-depleted layer to two 250-ml polypropylene centrifuge tubes with a plastic transfer pipette.

5. Centrifuge the sample at $800 \times g$ for 10 min with low brake at room temperature.

6. Remove the supernatant using a plastic transfer pipette and discard.

7. Resuspend the pellet in 40 ml of sterile 0.9% NaCl solution.

8. Lyse erythrocytes by adding 140 ml of sterile H_2O. Mix by gently stirring for 30–40 s at room temperature.

9. Immediately add 13.8 ml of 10% NaCl solution, and mix well by gently stirring.

10. Centrifuge the sample at $400 \times g$ for 10 min at room temperature.

11. Resuspend the white blood cell pellet in 10 ml of sterile 0.9% NaCl solution.

12. Dispense 10 ml of Lymphocyte Separation Medium 1077 or Histopaque 1077 into a conical 50-ml polypropylene tube, and carefully layer the cell suspension on the top of the separation medium.

13. Centrifuge the sample at $400 \times g$ for 10 min with no brake at 4°C.

14. Remove the supernatant using a plastic transfer pipette and discard.

15. Resuspend the neutrophil pellet in sterile 0.9% NaCl solution.

16. Wash the cells by resuspending them in 40 ml of ice-cold, sterile 0.9% NaCl solution and centrifuging at $400 \times g$ for 10 min at room temperature.

17. Wash the cells twice more by repeating step 16.

18. Resuspend purified cells at the concentration of 2×10^8 cells in ice-cold HBS containing 1% (v/v) protease inhibitor cocktail.

19. Sonicate the cells four times for 30 min on ice without foaming.

20. Clear the lysate by centrifugation at $1,000 \times g$ for 10 min.

21. Centrifuge the postnuclear supernatant at $20,000 \times g$ for 10 min at 4°C.

22. Ultracentrifuge the resultant supernatant at $150,000 \times g$ for 1 h at 4°C.

23. Remove the cytosolic fraction using a glass transfer pipette.

24. Wash the pellet by resuspending it in 100 mM potassium phosphate, pH 7.0, containing 0.5 M KCl, and by centrifuging again at $150,000 \times g$ for 1 h at 4°C (see Note 3).

25. Resuspend the pellet in 100 mM potassium phosphate, pH 7.0, and use it as the membrane fraction rich in Nox2 (which is tightly dimerized with p22^{phox} as cytochrome b_{558}) (see Note 4). Determine protein concentration using a standard method.

1. Preload recombinant Rac1 (see Note 5) with 100 μM GTPγS in 100 mM potassium phosphate, pH 7.0, and stop the GDP/GTP exchange by the addition of 10 mM $MgCl_2$ (see Note 6).

2. Mix GTPγS-loaded Rac1 with the membrane fraction of human neutrophils (3.0–6.0 μg/ml), 100 nM recombinant p47^{phox}, and 100 nM recombinant p67^{phox} in the cell-free assay buffer containing 75 μM ferricytochrome c (see Note 7).

3. Add SDS (100 μM final concentration) or arachidonic acid (50 μM final concentration) to the mixture, and incubate the mixture for 2.5 min at room temperature.

4. Immediately add NADPH at a final concentration of 1.0 mM to initiate the reaction (see Note 8).

5. Measure reduction of ferricytochrome c at 550 nm or at 550-minus-540 nm for 5–15 min using a usual or dual-wavelength spectrophotometer, respectively. The molar extinction coefficients of reduced-minus-oxidized cytochrome c at 550 nm and at 550–540 nm are 21.0×10^3 and 19.1×10^3/M/cm, respectively. Determine the NADPH-dependent superoxide-producing activity by subtracting the rate of ferricytochrome c reduction measured in the presence of 10 μg/ml of SOD from that measured in the absence of SOD.

A representative example of this methods is shown in Fig. 3. See also Note 3.

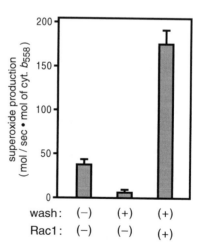

Fig. 3. Role for Rac in a cell-free system for Nox2 activation. Human neutrophil membranes (8 μg/ml) as a source of flavocytochrome b_{558} (the Nox2–p22^{phox} dimer), washed or unwashed with a high salt buffer, were mixed with 100 nM GST–p67^{phox}-(1–242) and 100 nM GST–p47^{phox}-(1–286) in the presence or absence of 800 nM GTPγS-loaded Rac1. The mixture was preincubated for 2.5 min at 25°C with 100 μM SDS, and the reaction was initiated by the addition of 1.0 mM NADPH. Superoxide-producing activity was measured by determining the rate of SOD-inhibitable ferricytochrome c reduction. Each *bar* represents the mean ± SD of data from three independent experiments.

**3.3. Estimation
of the Role of Rac
in Whole-Cell
Activation Systems
of Nox1–3**

*3.3.1. Reconstitution
of Nox Activation
in Whole-Cell Systems*

1. Culture CHO cells in Ham's F12 medium supplemented with 10% FCS; or HeLa, COS-7, or HEK293 cells in DMEM supplemented with 10% FCS.

2. Transfect CHO cells (see Note 9) in a 60-mm dish with 1–4 μg of plasmid cDNAs using 15 μl of FuGENE6 reagent (see Note 10) in 200 μl of Ham's F12 medium; HeLa cells with 1–4 μg of plasmid cDNAs using 12 μl of Lipofectamine in 200 μl Opti-MEM (see Note 11); or COS-7 cells with 1–4 μg of plasmid cDNAs using 12 μl of Lipofectamine reagent and 8 μl of Plus reagent in 200 μl of Opti-MEM (see Notes 12 and 13). For transfection, follow the manufactures' instructions.

3. Incubate the cells at 37°C in a CO_2 incubator for 48 h.

4. Harvest adherent cells by treatment with trypsin–EDTA for 1 min at 37°C, and wash them with HBS.

5. Centrifuge the cells at $300 \times g$ for 3 min and resuspend the cell pellet in HBS-BSA at a density of 8×10^5 cells/ml.

6. Mix 1.0 ml of the cell suspension with 40 μl of the DIOGENES-luminol solution in a polystyrene tube.

7. Place the tube in a luminometer, and monitor chemilumines-cence change continuously at 37°C using 100 μl of the original DIOGENES solution or 40 μl of the modified DIOGENES solution (29).

8. After preincubation for 5–10 min at 37°C, add 20 μl of PMA (10 μg/ml final concentration) to the cell suspension and incubate for 30 min.

9. Add 50 μl of SOD (1 mg/ml) to confirm that chemilumines-cence change is due to the superoxide produced. The change occurs immediately after the addition.

A representative example is depicted in Fig. 4. Rac1 (Q61L) induces Nox2 activation in PMA-stimulated cells (Fig. 4a). Rac is also involved in activation of Nox1 (Fig. 4b, c), whereas neither Cdc42 nor RhoA can replace Rac (Fig. 4b). Both p67[phox] and Noxa1 enhance Nox3 activity in a Rac-dependent manner (Fig. 4d).

*3.3.2. Estimation
of Oxidase Proteins
Ectopically Expressed
in Cultured Cells*

1. Resuspended transfected cells (4×10^5 cells/ml) (see Subheading 3.3.1, steps 1–4) in 200 μl of HBS.

2. Add 200 μl of 2× SDS-sample buffer.

3. Incubate the cell lysate for 5 min at 98°C.

4. Load 5–20 μl of the cell lysate and resolve the proteins on an SDS–PAGE gel using standard procedures.

5. Transfer proteins to a PVDF membrane, probe the membrane with antibodies against tagged proteins, and detect proteins using the horseradish peroxidase-conjugated anti-mouse anti-body and the ECL plus detection kit (see Fig. 4a).

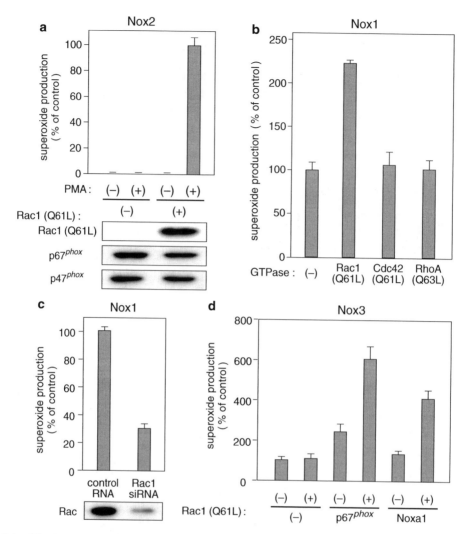

Fig. 4. Role of Rac in a whole-cell system for activation of Nox1–Nox3. (**a**) Requirement of Rac for Nox2 activation. Nox2, Myc–p67^{phox}, and HA–p47^{phox} were coexpressed in HeLa cells with or without a constitutively active Rac1 (Q61L). After cells were stimulated with or without PMA, superoxide-producing activity was determined by SOD-inhibitable chemilumines-cence change using DIOGENES. Expression of Myc–p67^{phox} and HA–p47^{phox} was estimated by immunoblot analysis with anti-Myc and anti-HA monoclonal antibodies, respectively. (**b, c**) Involvement of Rac in Nox1 activation. Nox1, Noxa1, and Noxo1 were coexpressed in HeLa cells with or without a constitutively active form of Rac1 (Q61L), Cdc42 (Q61L), or RhoA (Q63L) (**b**). Nox1, Noxa1, and Noxo1 were coexpressed in HeLa cells where endogenous Rac was deprived by siRNA-mediated RNA interference (**c**). The content of endogenous Rac was estimated by immunoblot analysis with anti-Rac monoclonal antibody. Superoxide-producing activity was determined by SOD-inhibitable chemiluminescence change using DIOGENES. (**d**) Involvement of Rac in Nox3 activation. Nox3, p22^{phox}, and p67^{phox} or Noxa1 were coexpressed in HEK293 cells with or without Rac1 (Q61L). Superoxide-producing activity was determined by SOD-inhibitable chemiluminescence change using DIOGENES. Each *graph* represents the mean ± SD of data from three independent experiments.

3.3.3. RNA Interference (RNAi) for Knockdown of Rac1 in HeLa Cells

1. Culture HeLa cells in a 60-mm dish (see Subheading 3.3.1).

2. Transfect cells with 10 µl of 20 µM siRNAs using 10 µl of Oligofectamine in 200 µl of Opti-MEM following the manufacturer's instructions.

3. Incubate the cells at 37°C in a CO_2 incubator for 24 h.

4. Transfect the cells with cDNAs for expression of Nox proteins using Lipofectamine (see Subheading 3.3.1, step 2).

5. Incubate the cells at 37°C in a CO_2 incubator for 48 h.

6. Estimate the amount of Rac1 in the lysate of HeLa cells by immunoblot analysis (see Subheading 3.3.2), and measure superoxide-producing activity of the cells by the chemiluminescence method (see Subheading 3.3.1). A representative example of this method is shown in Fig. 4c.

3.3.4. Estimation of the Role for Rac in Membrane Localization of p67^{phox} in HEK293 Cells

1. Culture HEK293 cells in a 150-mm dish (see Subheading 3.3.1).

2. Transfect cells with 1–4 µg of the cDNAs for Nox3, p67^{phox}, and Rac1 (Q61L) using 80 µl of Lipofectamine 2000 reagent in 1.6 ml of Opti-MEM (see Subheading 3.3.1, step 2).

3. Incubate the cells at 37°C in a CO_2 incubator for 48 h.

4. Harvest adherent cells by treatment with trypsin–EDTA for 1 min at 37°C, and wash them with HBS.

5. Centrifuge the cells at $300 \times g$ for 3 min at 4°C.

6. Resuspend the cell pellet in 1.0 ml of sonication buffer.

7. Sonicate the suspended cells four times for 30 s on ice without foaming.

8. Layer 1.0 ml of the sonicate onto a discontinuous sucrose gradient.

9. Ultracentrifuge the sample at $200,000 \times g$ for 1 h at 4°C.

10. Collect the layer between the 15 and 40% sucrose fractions and use it as the membrane fraction.

11. Estimate proteins in the membrane fraction (equivalent to 10^6 cells) by immunoblot analysis (see Subheading 3.3.2).

A representative example of this method is depicted in Fig. 5. Rac1 (Q61L) induces membrane recruitment of p67^{phox} in HEK293 cells, which enhances superoxide production by Nox3. p67^{phox} with the R102E substitution, defective in binding to Rac, fails to translocate to the membrane, and to facilitate Nox3 activation.

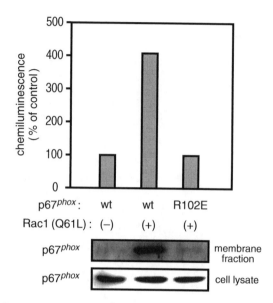

Fig. 5. Role of Rac in membrane recruitment of p67^{phox} during Nox3 activation. Nox3 and Myc–p67^{phox} (wt) or Myc–p67^{phox} (R102E) were expressed with or without Rac1 (Q61L) in HEK293 cells. Superoxide-producing activity was determined by SOD-inhibitable chemiluminescence change using DIOGENES. The membrane fraction of HEK293 cells (equivalent to 4.5×10^5 cells) or the whole-cell lysate (equivalent to 1.0×10^5 cells) was analyzed by immunoblot with an anti-Myc monoclonal antibody.

4. Notes

1. Rac1 (Q61L/C189S) is constitutively active due to the Q61L substitution, whereas the C189S substitution prevents this protein from being aggregated under aerobic conditions.

2. The presence of 0.005% Triton X-100 in the binding solution is crucial for the assay. Without the detergent, Rac1 (Q61L/C189S) binds not only to GST–p67^{phox} but also to GST alone (see Fig. 2).

3. Washing the neutrophil membrane fraction with a solution with a high salt concentration is required to remove contaminated Rac protein. This treatment results in an almost complete dependence on Rac in activation of Nox2 in a cell-free system (see Fig. 3).

4. The content of heme of cytochrome b_{558} (the Nox2–p22^{phox} dimer) is calculated from reduced-minus-oxidized difference spectra using the extinction coefficient at 427–411 nm of 200×10^3/M/cm or that of at 558/559–540 nm of 21.6×10^3/M/cm (30). Chemical reduction of cytochrome is performed by the addition of a few grains of solid sodium dithionite to the sample.

5. Human Rac paralogues (Rac1, Rac2, and Rac3) undergo geranylgeranylation at the C-terminal region when expressed in mammalian cells. This modification does not occur in prokaryotes such as *E. coli*. Membrane translocation of nonprenylated Rac is considered to largely depend on the polybasic region N-terminal to the site for geranylgeranylation. Rac1 contains six contiguous basic residues in this region, whereas Rac2 and Rac3 harbor three and four basic residues, respectively. This likely explains why Rac1 most potently activates Nox2 in a cell-free system among the three homologous proteins expressed as a nonprenylated form in *E. coli*; on the other hand, Rac2 is the least potent one (31). In contrast, when geranylgeranylated, Rac2 is as active as Rac1 in cell-free activation of Nox2 (32).

6. When Rac1 (Q61L/C189S), a constitutively active Rac1, is used for cell-free activation of NADPH oxidase instead of Rac1 (C189S), loading with GTPγS is not required.

7. A pair of GST–p67^{phox}-(1–212) and GST–p47^{phox}-(1–286), each lacking the C-terminal region, is capable of supporting superoxide production by Nox2 to the same extent as a pair of the full-length proteins p67^{phox}-(1–526) and p47^{phox}-(1–390) (33). The GST-tag on p67^{phox} or p47^{phox} does not affect the ability to support Nox2 activation.

8. It is very important that NADPH is added to the reaction mixture exactly 2–3 min after the addition of SDS or arachidonic acid. Either longer or shorter incubation results in a markedly reduced activation of Nox2.

9. In CHO cells, ectopic expression of p22^{phox} is absolutely required for superoxide production by Nox2 or Nox3; it effectively enhances Nox1 activation, albeit dispensable (14). CHO cells in a 60-mm dish (about 80% confluent) are transfected using 15 μl of FuGENE6 or X-tremeGENE9 with 1.0 μg of pcDNA3–Nox1, pcDNA3–Nox2, or pcDNA3–Nox3; 0.1 μg of pEF-BOS–p22^{phox}; and 0.5–1.0 μg of pEF-BOS–p67^{phox}, pEF-BOS–p47^{phox}, pEF-BOS–Noxa1, and/or pEF-BOS–Noxo1.

10. FuGENE6 can be successfully substituted with X-tremeGENE9 (Roche Diagnostics) in the experiments described here.

11. In HeLa cells, endogenous p22^{phox} is capable of fully supporting superoxide production by Nox1, Nox2, and Nox3; thus transfection with the p22^{phox} cDNA is not required for reconstitution of Nox activity in HeLa cells. On the other hand, reconstitution of Nox2 but not Nox1 or Nox3 requires expression of a constitutively active Rac1 such as Rac1 (Q61L). HeLa cells in a 60-mm dish (about 80% confluent) are transfected using 12 μl of LipofectAMINE with 1.0 μg of pcDNA3–Nox1, pcDNA3–Nox2, or pcDNA3–Nox3; 0.5 μg of pEF-BOS–Rac1 (Q61L);

and 0.5–1.0 µg of pEF-BOS–p67^{phox}, pEF-BOS–p47^{phox}, pEF-BOS–Noxa1, and/or pEF-BOS–Noxo1.

12. In COS-7 cells, reconstitution of Nox activity does not require ectopic expression of p22^{phox} or Rac1 (Q61L). COS-7 cells in a 60-mm dish (about 80% confluent) are transfected using 8 µl of Plus Reagent and 12 µl of LipofectAMINE with 1.0 µg of pcDNA3–Nox1, pcDNA3–Nox2, or pcDNA3–Nox3; and 0.5–1.0 µg of pEF-BOS–p67^{phox}, pEF-BOS–p47^{phox}, pEF-BOS–Noxa1, and/or pEF-BOS–Noxo1.

13. The extent of Rac requirement for activation of Nox oxidases, especially for Nox3, depends in part on the cell line used (19).

Acknowledgments

This work was supported in part by Grants-in-Aid for Scientific Research and Targeted Proteins Research Program (TPRP) from the Ministry of Education, Culture, Sports, Science and Technology (MEXT), Japan, and by Japan Foundation for Applied Enzymology.

References

1. Sumimoto, H. (2008) Structure, regulation and evolution of Nox-family NADPH oxidases that produce reactive oxygen species. *FEBS J* 275, 3249–3277.

2. Abo, A., Pick, E., Hall, A., Totty, N., Teahan, C. G., and Segal, A. W. (1991) Activation of the NADPH oxidase involves the small GTP-binding protein p21rac1. *Nature* 353, 668–670.

3. Knaus, U. G., Heyworth, P. G., Evans, T., Curnutte, J. T., and Bokoch, G. M. (1991) Regulation of phagocyte oxygen radical production by the GTP-binding protein Rac 2. *Science* 254, 1512–1515.

4. Mizuno, T., Kaibuchi, K., Ando, S., Musha, T., Hiraoka, K., Takaishi, K., Asada, M., Nunoi, H., Matsuda, I., and Takai, Y. (1992) Regulation of the superoxide-generating NADPH oxidase by a small GTP-binding protein and its stimulatory and inhibitory GDP/GTP exchange proteins. *J Biol Chem* 267, 10215–10218.

5. Molshanski-Mor, S., Mizrahi, A., Ugolev, Y., Dahan, I., Berdichevsky, Y., and Pick, E. (2007) Cell-free assays: the reductionist approach to the study of NADPH oxidase assembly, or "all you wanted to know about cell-free assays but did not dare to ask". *Methods Mol Biol* 412, 385–428.

6. Ambruso, D. R., Knall, C., Abell, A. N., Panepinto, J., Kurkchubasche, A., Thurman, G., Gonzalez-Aller, C., Hiester, A., deBoer, M., Harbeck, R. J., Oyer, R., Johnson, G. L., and Roos, D. (2000) Human neutrophil immunodeficiency syndrome is associated with an inhibitory Rac2 mutation. *Proc Natl Acad Sci USA* 97, 4654–4659.

7. Williams, D. A., Tao, W., Yang, F., Kim, C., Gu, Y., Mansfield, P., Levine, J. E., Petryniak, B., Derrow, C. W., Harris, C., Jia, B., Zheng, Y., Ambruso, D. R., Lowe, J. B., Atkinson, S. J., Dinauer, M. C., and Boxer, L. (2000) Dominant negative mutation of the hematopoietic-specific Rho GTPase, Rac2, is associated with a human phagocyte immunodeficiency. *Blood* 96, 1646–1454.

8. Gu, Y., Filippi, M. D., Cancelas, J. A., Siefring, J. E., Williams, E. P., Jasti, A. C., Harris, C. E., Lee, A. W., Prabhakar, R., Atkinson, S. J., Kwiatkowski, D. J., and Williams, D. A. (2003) Hematopoietic cell regulation by Rac1 and Rac2 guanosine triphosphatases. *Science* 302, 445–449.

9. Yamauchi, A., Marchal, C. C., Molitoris, J., Pech, N., Knaus, U., Towe, J., Atkinson, S. J., and Dinauer, M. C. (2005) Rac GTPase isoform-specific regulation of NADPH oxidase and chemotaxis in murine neutrophils *in vivo*.

Role of the C-terminal polybasic domain. *J Biol Chem* 280, 953–964.

10. Shiose, A., and Sumimoto, H. (2000) Arachidonic acid and phosphorylation synergistically induce a conformational change of p47^{phox} to activate the phagocyte NADPH oxidase. *J Biol Chem* 275, 13793–13801.

11. Akasaki, T., Koga, H., and Sumimoto, H. (1999) Phosphoinositide 3-kinase-dependent and -independent activation of the small GTPase Rac2 in human neutrophils. *J Biol Chem* 274, 18055–18059.

12. Koga, H., Terasawa, H., Nunoi, H., Takeshige, K., Inagaki, F., and Sumimoto, H. (1999) Tetratricopeptide repeat (TPR) motifs of p67^{phox} participate in interaction with the small GTPase Rac and activation of the phagocyte NADPH oxidase. *J Biol Chem* 274, 25051–25060.

13. Lapouge, K., Smith, S. J., Walker, P. A., Gamblin, S. J., Smerdon, S. J., and Rittinger, K. (2000) Structure of the TPR domain of p67^{phox} in complex with Rac.GTP. *Mol Cell* 6, 899–907.

14. Takeya, R., Ueno, N., Kami, K., Taura, M., Kohjima, M., Izaki, T., Nunoi, H., and Sumimoto, H. (2003) Novel human homologues of p47^{phox} and p67^{phox} participate in activation of superoxide-producing NADPH oxidases. *J Biol Chem* 278, 25234–25246.

15. Ueyama, T., Geiszt, M., and Leto, T. L. (2006) Involvement of Rac1 in activation of multicomponent Nox1- and Nox3-based NADPH oxidases. *Mol Cell Biol* 26, 2160–2174.

16. Cheng, G., Diebold, B. A., Hughes, Y., and Lambeth, J. D. (2006) Nox1-dependent reactive oxygen generation is regulated by Rac1. *J Biol Chem* 281, 17718–17726.

17. Miyano, K., Ueno, N., Takeya, R., and Sumimoto, H. (2006) Direct involvement of the small GTPase Rac in activation of the superoxide-producing NADPH oxidase Nox1. *J Biol Chem* 281, 21857–21868.

18. Ueno, N., Takeya, R., Miyano, K., Kikuchi, H., and Sumimoto, H. (2005) The NADPH oxidase Nox3 constitutively produces superoxide in a p22^{phox}-dependent manner: its regulation by oxidase organizers and activators. *J Biol Chem* 280, 23328–23339.

19. Miyano, K., and Sumimoto, H. (2007) Role of the small GTPase Rac in p22^{phox}-dependent NADPH oxidases. *Biochimie* 89, 1133–1144.

20. Miyano, K., Koga, H., Minakami, R., and Sumimoto, H. (2009) The insert region of the Rac GTPases is dispensable for activation of superoxide-producing NADPH oxidases. *Biochem J* 422, 373–382.

21. Heyworth, P. G., Knaus, U. G., Settleman, J., Curnutte, J. T., and Bokoch, G. M. (1993) Regulation of NADPH oxidase activity by Rac GTPase activating protein(s). *Mol Biol Cell* 4, 1217–1223.

22. Kwong, C. H., Adams, A. G., and Leto, T. L. (1995) Characterization of the effector-specifying domain of Rac involved in NADPH oxidase activation. *J Biol Chem* 270, 19868–19872.

23. Diekmann, D., Abo, A., Johnston, C., Segal, A. W., and Hall, A. (1994) Interaction of Rac with p67^{phox} and regulation of phagocytic NADPH oxidase activity. *Science* 265, 531–533.

24. Nisimoto, Y., Freeman, J. L., Motalebi, S. A., Hirshberg, M., and Lambeth, J. D. (1997) Rac binding to p67^{phox}. Structural basis for interactions of the Rac1 effector region and insert region with components of the respiratory burst oxidase. *J Biol Chem* 272, 18834–18841.

25. Sarfstein, R., Gorzalczany, Y., Mizrahi, A., Berdichevsky, Y., Molshanski-Mor, S., Weinbaum, C., Hirshberg, M., Dagher, M. C., and Pick, E. (2004) Dual role of Rac in the assembly of NADPH oxidase, tethering to the membrane and activation of p67^{phox}: a study based on mutagenesis of p67^{phox}-Rac1 chimeras. *J Biol Chem* 279, 16007–16016.

26. Mizrahi, A., Berdichevsky, Y., Casey, P. J., and Pick, E. (2010) A prenylated p47^{phox}-p67^{phox}-Rac1 chimera is a quintessential NADPH oxidase activator: membrane association and functional capacity. *J Biol Chem* 285, 25485–25499.

27. Maehara, Y., Miyano, K., Yuzawa, S., Akimoto, R., Takeya, R., and Sumimoto, H. (2010) A conserved region between the TPR and activation domains of p67^{phox} participates in activation of the phagocyte NADPH oxidase. *J Biol Chem* 285, 31435–31445.

28. Mizushima, S., and Nagata, S. (1990) pEF-BOS, a powerful mammalian expression vector. *Nucleic Acids Res* 18, 5322.

29. Takeya, R., Ueno, N., and Sumimoto, H. (2006) Regulation of superoxide-producing NADPH oxidases in nonphagocytic cells. *Methods Enzymol* 406, 456–468.

30. Pick, E., Bromberg, Y., Shpungin, S., and Gadba, R. (1987) Activation of the superoxide forming NADPH oxidase in a cell-free system by sodium dodecyl sulfate. Characterization of the membrane-associated component. *J Biol Chem* 262, 16476–16483.

31. Ueyama, T., Eto, M., Kami, K., Tatsuno, T., Kobayashi, T., Shirai, Y., Lennartz, M. R., Takeya, R., Sumimoto, H., and Saito, N. (2005) Isoform-specific membrane targeting

mechanism of Rac during FcγR-mediated phagocytosis: positive charge-dependent and independent targeting mechanism of Rac to the phagosome. *J Immunol* 175, 2381–2390.

32. Ando, S., Kaibuchi, K., Sasaki, T., Hiraoka, K., Nishiyama, T., Mizuno, T., Asada, M., Nunoi, H., Matsuda, I., Matsuura, Y., Polakis, P., McCormick, F., and Takai, Y. (1992) Post-translational processing of *rac* p21s is important both for their interaction with the GDP/GTP exchange proteins and for their activation of NADPH oxidase. *J Biol Chem* 267, 25709–25713.

33. Hata, K., Takeshige, K., and Sumimoto, H. (1997) Roles for proline-rich regions of p47[phox] and p67[phox] in the phagocyte NADPH oxidase activation in vitro. *Biochem Biophys Res Commun* 241, 226–231.

Part IV

Imaging and High Throughput Methods

Chapter 15

Multiplex Imaging of Rho Family GTPase Activities in Living Cells

Désirée Spiering and Louis Hodgson

Abstract

Here, we provide procedures for imaging the Rho GTPase biosensors in both single and multiplex acquisition modes. The multiplex approach enables the direct visualization of two biosensor readouts from a single living cell. Here, we take as an example a combination of the RhoA biosensor based on a CFP/YFP FRET modality and the Cdc42 biosensor based on organic dyes that change fluorescence as a function of the local solvent polarity. We list the required optical components as well as cellular manipulation techniques necessary to successfully image these two ratiometric biosensors in a single living cell.

Key words: Rho GTPase, Dynamics, Live-cell imaging, FRET, Biosensors, Multiplex

1. Introduction

The fluorescent biosensors for detecting the activation of the Rho family of p21 small GTPases have opened a new window into the spatiotemporal regulation of these important molecular switches in living cells (1–7). The ability to observe the effector-binding events of these proteins in their native environment has significantly changed the way in which we understand the biology of Rho GTPases (1–5). While the early imaging attempts of these sensors were limited to the single-mode observations wherein only one biosensor was imaged in a single living cell at a time, we have now capabilities to observe two distinct Rho GTPase activities in a single living cell (4). This latter approach has enabled the direct characterization of the interrelationships between the activation states of different Rho GTPases to better understand the balance and coordination of Rho GTPase activations that govern the leading edge dynamics (4). Here, we provide methods for both the single-mode and the multiplex-mode imaging approaches.

Francisco Rivero (ed.), *Rho GTPases: Methods and Protocols*, Methods in Molecular Biology, vol. 827,
DOI 10.1007/978-1-61779-442-1_15, © Springer Science+Business Media, LLC 2012

1.1. General Considerations for the Use of Rho GTPase Biosensors

The biosensors based on CFP/YFP fluorescence resonance energy transfer (FRET) such as the Rho GTPase biosensors (3) or those that are based on solvent-sensitive dyes (2, 8) require ratiometric calculations. Methods for observing single protein activity in living cells using these types of biosensors are described in detail elsewhere (9). Here, we summarize some key issues, specific caveats, and pitfalls encountered during this mode of imaging. There are several major considerations: (1) expression levels of the biosensor and the dominant negative effect; (2) motion artifacts; (3) lateral chromatic aberrations within the field of view and dual-camera alignment; and (4) general imaging considerations for optimal live-cell microscopy. These issues are addressed in the following sections.

The expression level of the genetically encoded biosensors requires careful control in order to maintain a minimal level of biosensors needed to obtain a sufficient signal-to-noise ratio (SNR) during imaging. Typically, the SNR for the widefield epifluorescence mode we employ is 2:1 at the dimmest part of the cell periphery compared to the background fluorescence signals outside of the cell (9). We have found that in order to achieve this SNR, we must optimize the microscope optics by custom designing the bandpass filters and dichromatic mirrors. The specifications of the microscope optics that we employ for this mode of imaging are shown in Table 1. Using these custom components, we are able to maintain the biosensor expression levels at 20–30% of the endogenous material (see Fig. 1) (4) and obtain sufficient SNR during imaging. To achieve these low levels of expressions, we routinely utilize an inducible gene expression system (tet-OFF) coupled to a retroviral transduction system where we can tightly regulate the induction levels of biosensors by titrating the amount of doxycyclin (3, 10, 11), and also to control the relative copy numbers of the biosensor cDNA incorporated into the genome by optimizing the number of retroviral transduction cycles.

1.2. General Considerations for the Single-Mode Imaging of Rho GTPase Biosensors

The widefield imaging of live cells expressing the genetically encoded biosensor can be performed in one of two ways. The Sequential Imaging mode utilizes a single camera attached to the microscope imaging port via a filterwheel which can be used to switch the emission filters between the CFP and YFP FRET emissions. Two images of the biosensor are taken sequentially and later ratiometric calculations are performed. The major advantage of this approach is the simplicity of the hardware design where only a single high-sensitivity cooled charge-coupled device (CCD) camera is needed (see Fig. 2). A minor advantage could be that because a single CCD is used, it will be highly unlikely for any significant image misalignment to be present in the two fields of views acquired. However, the lateral chromatic aberration will be present, which will depend on the extent of the wavelength differences of the two acquisition wavelengths and on the relative location of the placement

Table 1
List of bandpass filters and dichromatic mirrors required for the single-biosensor mode imaging

Single-camera/single-mode imaging

Bandpass filters		
ET436/20X	CFP excitation	Filterwheel 1
ET500/20X	YFP excitation	Filterwheel 1
ET480/40M	CFP emission	Filterwheel 2
ET535/30M	YFP (FRET) emission	Filterwheel 2
Dichromatic mirrors		
94:6 mirror	All fluororphores	Main turret

Dual-camera/single-mode imaging (optional third camera)

Bandpass filters		
ET436/20X	CFP excitation	Filterwheel 1
ET500/20X	YFP excitation	Filterwheel 1
ET480/40M	CFP emission	At camera 2
ET535/30M	YFP(FRET) emission	At camera 1
Linear polarizer	DIC camera 3	Filterwheel 2
Dichromatic mirrors		
94:6 mirror	All fluororphores	Main turret
T555LPXXR	CFP/FRET to side port	At side port prism
T505LPXR	Separate CFP/FRET	Side port beam-splitter

The bandpass filters and the dichromatic mirrors are from Chroma Technology. The 94:6 mirror is from Olympus. The linear polarizer is from Chroma Technology. All filters and mirrors are antireflection coated

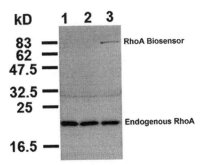

Fig. 1. Expression levels of the genetically encoded single-chain RhoA biosensor in MEFs, taken from ref. 4. *Lane 1*: doxycyclin at 1 μg/ml; *Lane 2*: doxycyclin at 0.1 μg/ml; and *Lane 3*: no doxycyclin. The expression level of the biosensor is approximately 20% of the endogenous RhoA under these conditions (4).

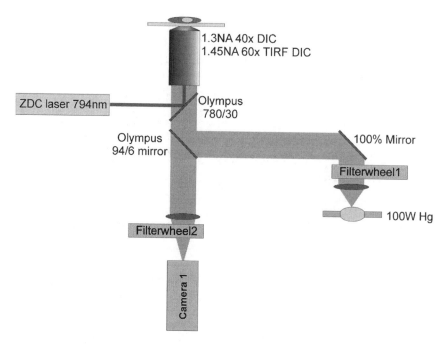

Fig. 2. The light path diagram for the single-camera, single-biosensor mode imaging. The list of bandpass filters and dichromatic mirrors required in this setup is shown in Table 1.

of the CCD within the field of view. Furthermore, because the filterwheel is normally positioned within the focused optical space, the differences in the thicknesses of the emission bandpass filters as well as their slight imperfections in the mounting of these filters could cause misalignments and defocusing effects at the image plane. These latter issues can be fully corrected via the calibration and computational correction methods developed previously (9), along with the use of a priori determined focus offsets for each of the channels. One of the major drawbacks of the sequential mode of acquisition is that, because it takes a finite amount of time to switch between the two emission wavelengths (together with the appropriate exposure times at each wavelength), the cell in the field of view could move in between the two sequential acquisitions. This will produce "motion artifacts" during the ratiometric calculations (2, 9), and will make the interpretation of the resulting ratiometric data set difficult. Approaches to roughly estimate the extent of such motion artifacts are described elsewhere (2, 9). In order to sidestep this critical problem, it is possible to set up two identical cameras for a simultaneous image acquisition. This requires an additional beam-splitter in the light path in place of the filterwheel in the emission optical train as well as a second identical CCD camera (see Fig. 3). The advantage of this approach is that it removes the motion artifacts through eliminating the need to switch the wavelengths. However, because the alignment of the

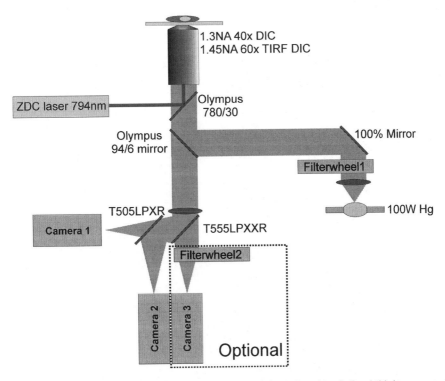

Fig. 3. The light path diagram for the dual-camera, single-biosensor mode imaging. Optional third camera path is also shown. The third camera will enable acquisitions of DIC and any other red and far-red fluorescence beyond the cut-off point of the long-pass filter to be placed in the optical path. The list of bandpass filters and dichromatic mirrors required in this setup is shown in Table 1.

cameras and the beam-splitter mirrors are not always perfect, it introduces additional sources of image misalignments including the field shear or otherwise nonlinear displacements in the field of view. These issues are corrected using a priori calibration and morphing-based computational image matching technologies allowing to register the two ratiometric images to within one-twentieth of a pixel accuracy (9).

1.3. General Considerations for the Multiplex-Mode Imaging of Rho GTPase Biosensors

Using a combination of two biosensor systems with compatible fluorescence characteristics, it is now possible to perform the wide-field imaging of two protein activities in single living cells (4, 12). This necessitates separations of four wavelengths for two ratiometric determinations of protein activities. One previous approach was to use a combination of a CFP/YFP FRET biosensor and another biosensor that was based on organic dyes (4). In this approach, dye fluorescence wavelengths and their placements on the biosensor molecule were carefully adjusted to be complementary to the genetically encoded biosensor in a single living cell. More recently, an approach was developed where a set of two genetically encoded biosensors (one using CFP/YFP FRET, another using the monomeric orange fluorescent protein 2 and monomeric Cherry) have

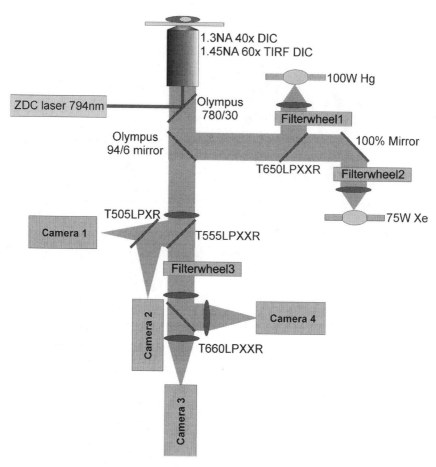

Fig. 4. The light path diagram for the four-channel, dual-biosensor mode imaging. The list of bandpass filters and dichromatic mirrors required in this setup is shown in Table 2.

been used in single living cells (12). However, this work, while enabling the measurements of two distinct readouts, utilized one biosensor that was intracellular and another as an extracellular expressed protein, segregating the compartments occupied by these two sensors, making the separation of wavelengths simpler (12). Naturally, if measurements of two protein activities from components that existed in the same compartment (such as the case of many Rho GTPases) are required, separation of fluorescence wavelengths is more challenging using this latter approach.

The optical system that enables separation of four wavelengths is shown in Fig. 4. The associated bandpass filters and dichromatic mirror specifications are shown in Table 2. In this specific configuration, the system is capable of imaging CFP/YFP FRET together with the dye-based biosensor utilizing the ratiometric dye combination of mero87 (13) and a far-red dye, Atto-700-NHS with an excitation/emission wavelengths of 681/713 nm. Here, we will

Table 2
List of bandpass filters and dichromatic mirrors required for the dual-biosensor mode imaging

Four-channels/dual-mode imaging

Bandpass filters		
ET436/20X	CFP excitation	Filterwheel 1
ET500/20X	YFP excitation	Filterwheel 1
ET480/40M	CFP emission	At camera 2
ET535/30M	YFP(FRET) emission	At camera 1
Linear polarizer	DIC camera 4	Filterwheel 3
FF585/29	mero87 excitation	Filterwheel 1
FF684/24	Atto700 excitation	Filterwheel 2
FF628/32	mero87 emission	At camera 4
FF794/160	Atto700 emission	At camera 3
Dichromatic mirrors		
94:6 mirror	All fluororphores	Main turret
T650LPXXR	Combine Xe and Hg	Arc lamp combiner
T555LPXXR	CFP/FRET to side port	At side port prism
T505LPXR	Separate CFP/FRET	Side port beam-splitter
T660LPXXR	Separate mero87/Atto700	Side port beam-splitter

The bandpass filters and the dichromatic mirrors starting with the "ET" and "T" specifications, respectively, are from Chroma Technology. The bandpass filters with the "FF" specifications are from Semrock. The 94:6 mirror is from Olympus. The linear polarizer is from Chroma Technology. All filters and mirrors are antireflection coated

focus our discussions on the RhoA biosensor based on CFP/YFP FRET (3) and a modified version of the MeroCBD biosensor for detecting the activity of endogenous Cdc42 GTPase (2, 4).

2. Materials

2.1. Cell Culture, Transfection, and Viral Transduction

1. Cell lines: GP2-293 (Clontech) and mouse embryonic fibroblasts (MEF) (Clontech).

2. DMEM (4.5 g/ml glucose with 110 mg/L sodium pyruvate; Mediatech) supplemented with 10% fetal calf serum (FCS), penicillin (100 IU)/streptomycin (100 μg/ml), and 2 mM Glutamax (Invitrogen).

3. Opti-MEM (Invitrogen).

4. Trypsin (Mediatech).

5. Plasmids: pRetro-X-tet-OFF system including the pRetro-X-Tight-Pur or the pRetro-X-Tight-Hygro backbone system

(Clontech). The biosensor cDNA is restricted from the cloning vector and ligated into the multiple cloning site of the pRetro-X backbone at the *Not*I restriction site. Both the vector backbone and the insert must be blunt-ended using Klenow end-filling. The viral expression constructs for the RhoA biosensor can be obtained from Addgene (http://www.addgene.org) in the previous generation format (pBabe-sin-puro-tet-CMV-RhoA Biosensor) or the current configuration in the pRetro-X system by writing to the authors.

6. Lipofectamine 2000 reagent (Invitrogen).

7. DPBS, calcium and magnesium free (Mediatech): 0.2 g/L KCl; 0.2 g/L KH_2PO_4; 8 g/L NaCl; 1.15 g/L Na_2HPO_4 (anhydrous).

8. Poly-L-lysine (Sigma P4707) diluted 1:10 in DPBS. For coating, treat plates with this solution for 5 min at 24°C and aspirate prior to plating cells.

9. Fibronectin (Sigma) diluted in DPBS at a final concentration of 10 μg/ml. For coating, treat coverslips with this solution for 1 h at 24°C and aspirate prior to plating cells.

10. Retro-X concentrator (Clontech).

11. Polybrene (Sigma), 8 μg/μl stock.

12. Antibiotics: G-418/neomycin (100 μg/μl stock), puromycin (10 μg/μl stock), hygromycin (50 μg/μl stock), and doxycyclin (10 μg/ml stock).

13. Beckman Coulter MoFlo XDP FACS system.

2.2. Dye Labeling of the Multiplex-Comptible Biosensor for Cdc42

1. *Escherichia coli* strain expressing a Cdc42-binding domain fused to maltose-binding protein (CBD-MBP) (available from the author's laboratory).

2. Atto-700-NHS-succinimidylester (Sigma).

3. Mero87-Iodoacetamide (13). This reactive dye is available from the Hahn laboratory; University of North Carolina at Chapel Hill.

4. Centrifugal concentrators (Amicon Ultra 10 kDa cut-off; Millipore).

5. Sephadex G-15 (Sigma).

6. 50 mM NaH_2PO_4 at pH 7.5 and 5.5.

7. 2-Mercaptoethanol.

8. 1.5 M Hydroxylamine (Sigma) at pH 5.5, prepared fresh.

2.3. Imaging

1. Refer to Tables 1 and 2 for the required optical filters and mirrors for the imaging setup. The placement of the CCD cameras for the different imaging modes is shown in Figs. 2–4.

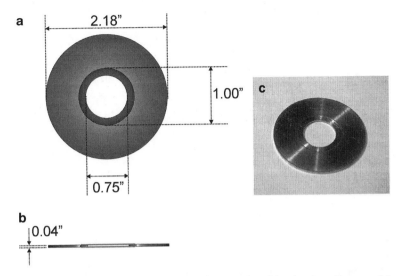

Fig. 5. Design of the custom imaging chamber. (**a**) Top view of the chamber with appropriate measurements. (**b**) Side view of the chamber. The two 25-mm round coverslips are assembled in a sandwich-like manner using silicone vacuum grease. Cells are cultured onto these coverslips and the small void created by the chamber and the two coverslips contains the imaging medium (approximately 280 µl). (**c**) Photograph of an actual chamber. The chamber assembly was designed to fit into the QE-2 heated stage assembly of Warner Instruments.

2. Custom-made chamber for imaging (see Fig. 5). This chamber is designed to be air-tight and is temperature regulated to 37°C when used in combination with a temperature regulator (QE-2; Warner Instruments).

3. 25-mm Round coverslips (# 1.5; Warner Instruments).

4. Imaging medium: 6 ml of Ham's F-12K without phenol-red (Biosource) (see Note 1) is warmed to 37°C to release any dissolved gases. Argon gas is then bubbled into the medium for 1 min to displace the oxygen. FCS is added at 2% and the medium is aliquoted into 2-ml tubes together with Oxyfluor reagent (Oxyrase) at 1:100 dilution and with 5 mM DL-lactate. The mixture is then incubated at 37°C for 1 h and spun for 1 min at 24°C, $20,000 \times g$ to remove any debris from the Oxyfluor treatment prior to imaging (see Note 2).

5. Inverted tissue culture microscope (Olympus CKX31) equipped with a microinjection system (Femtojet system; Eppendorf).

6. Microinjection needles are pulled using a Sutter P-97 puller (Sutter Instruments) following the manufacturer's protocols. We use borosilicate glass capillary tubes with filament, 10 cm length with 1.0 mm outer diameter and 0.58 mm inner diameter (Sutter Instruments: BF100-58-10). Ramp test should be performed on the puller to determine the proper melting point of the glass.

7. Loading pipette tips (Eppendorf).

8. Multispectral beads for calibrations (Bangs Laboratory, custom design courtesy of Dr. Robert H. Singer; Cat#: FS02F; Lot#: 8227; Inventory#: L070619A).

9. Metamorph ver.7.6 software (Molecular Devices).

10. Light-intensity meter (ThorLabs).

3. Methods

3.1. Dye Labeling of the Multiplex-Compatible Biosensor for Cdc42

The CBD-MBP is produced in bacteria essentially as detailed elsewhere for the original MeroCBD biosensor (9). For dual-labeling of the CBD-MBP, we first perform the cysteine-targeted site-specific labeling using mero87-iodoacetamide (2, 4) at pH 7.5, followed by a change of pH to an acidic condition (pH 5.5) and perform the N-terminal amine-targeted site-specific labeling using Atto-700-NHS-succinimidylester.

1. The mero87 labeling of the CBD-MBP is based on the cysteine-targeted chemistry we used previously (2). We normally use 600–800 μl of the CBD-MBP protein stock solution at 200 μM in 50 mM NaH_2PO_4 at pH 7.5 to take into account the later loss of protein through precipitation at lower pH conditions. One milligram of Mero87-iodoacetamide is dispersed into 20 μl of anhydrous DMSO which will yield a concentration of approximately 40 mM (measure precisely using the extinction coefficient of 135,000 in DMSO at 596 nm absorption). This stock dye solution is then added into the protein solution to yield a 6:1 dye to protein ratio in the final reaction mix. This mixture is quickly vortexed, and put on a rocking mixer for 1 h at room temperature (see Note 3).

2. At the end of 1 h, the reaction is quenched by the addition of 3 μl of 2-mercaptoethanol, and allowed to further proceed the quenching reaction at room temperature on the rocking mixer for additional 5 min. The mixture is then briefly centrifuged at $20,000 \times g$ for 1 min at room temperature.

3. For the purification of the labeled material, use a Sephadex G-15 column equilibrated with 50 mM NaH_2PO_4 at pH 7.5.

4. Following a successful labeling of the CBD-MBP, the protein is concentrated using a centrifugal concentrator to the final concentration of 200 μM, and it can be frozen at –86°C for storage.

5. The N-terminal labeling of the CBD-MBP-mero87 requires the pH of the buffer to be reduced to 5.5, rendering the intra-chain lysines nonreactive. At this condition, the stability of the protein is slightly compromised and we have found that 3 h is the maximum we can maintain the protein at this reduced pH

and still obtain workable yields. To exchange the pH, dialyze 600–800 μl of the mero87-labeled CBD-MBP against a 4 L volume of 50 mM NaH_2PO_4 at pH 5.5 for 1 h at 24°C (see Note 4).

6. Immediately following the dialysis, protein concentration is measured spectrophotometrically and the protein is then concentrated as required to 200 μM using a centrifugal concentrator.

7. One milligram of Atto 700 NHS is first dissolved in 40 μl of anhydrous DMSO and the concentration is measured spectrophotometrically based on the extinction coefficient provided by the manufacturer. Here, we aim at a 20–40 mM stock concentration. The optimal dye to protein ratio should be empirically determined, but a starting point should be approximately 10:1 (see Note 5). Allow the labeling reaction to proceed at 24°C for 2 h on a rocking mixer.

8. At the end of 2 h, quench the reaction by addition of 1/3 reaction volume of 1.5 M hydroxylamine at pH 5.5 and then allow the quenching reaction to proceed for 5 min prior to a brief centrifugation at $20,000 \times g$ for 1 min at room temperature and purification using a size exclusion column.

9. For the purification of the labeled material, use a Sephadex G-15 column equilibrated with 50 mM NaH_2PO_4 at pH 5.5 (see Note 6). Upon collection, the labeled fraction is immediately subjected to dialysis against 50 mM NaH_2PO_4 at pH 7.5 at 4°C overnight.

10. Following the dialysis, the protein concentration is determined spectrophotometrically. Here, care must be taken to properly account for the multiple labeling effects. The correction factors for the absorption measurements at 280 nm must first be determined. In order to calculate these, make dilutions of the dyes in 50 mM NaH_2PO_4 buffer and measure the spectra from 280 nm to the dye absorption maxima for both the mero87 and Atto 700. Take the absorption value at 280 nm from these solutions in the absence of any proteins, and divide that value by the absorption value of the major dye absorbance peak. The resulting correction factor is the percentage absorption of the dye at 280 nm when the dye-conjugated protein concentration is measured. We have determined that the correction factor at 280 nm is 0.076 for mero87, and 0.4166 for Atto 700.

11. Adjust the protein concentration to approximately 50 μM and flash freeze 40 μl aliquots of CBD-MBP-mero87-Atto700 (Cdc42 biosensor) using liquid N_2, and store at –86°C.

3.2. Viral Transduction and Stable tet-OFF Cell Line Production

1. Plate GP2-293 cells at 8×10^6 cells/plate overnight in a 10-cm tissue culture dish coated with poly-L-lysine.

2. Prepare transfection mix. The optimal amounts of DNA plasmids for the transfection are 22 μg of pRetro-X-tet-OFF

and 2 μg of pVSVg together with 60 μl of Lipofectamine 2000 reagent in a total transfection mix volume of 1 ml following the manufacturer's instructions. Allow complexes to form for approximately 20 min at 24°C (see Note 7).

3. Add the mix dropwise to the cell monolayer to result in a total volume of 5 ml.

4. Exchange medium for 7 ml of fresh growth medium containing 5% FCS and normal antibiotic concentrations 24 h posttransfection, and transfer plates to a 32°C incubator.

5. Collect viral supernatant 48 h following the medium exchange. Concentrate the viral supernatant using a Retro-X concentrator following the manufacturer's protocols to a 7:1 ratio. Store at –80°C for later use (see Note 8).

6. Plate MEFs (or the cells of interest) at 1×10^5 cells/plate overnight in a 10-cm tissue culture dish.

7. Add supernatant stock (0.5 ml of 7:1 concentrated supernatant) together with 8 μg/ml polybrene in a total medium volume of 4 ml. We normally infect four consecutive times, 12 h apart, before starting the selection.

8. Select for the stable incorporation of the tet-OFF tetracyclin trans-activator (tTA) using G-418 at the final concentration of 1 mg/ml.

3.3. Viral Transduction and Stable Biosensor Cell Line Production

1. Perform production of the virus containing the biosensor expression construct as described in Subheading 3.2, steps 1 through 5.

2. Perform transduction as described in step 7 of Subheading 3.2. However, it is necessary to monitor the cell fluorescence every 24 h (every second consecutive transduction) to ascertain that a sufficient level of biosensor expression is achieved prior to repressing the biosensor expression and then initiating the selection process.

3. The selection for stable incorporation of the biosensor must be performed in a stepwise manner, by gradually increasing the puromycin or hygromycin concentration. We routinely start at 1 μg/ml for the puromycin selection and 10 μg/ml for the hygromycin selection. The final concentrations for the stable selection for MEFs are 10 μg/ml puromycin and 50 μg/ml hygromycin (see Note 9).

4. Once the stable, inducible cell lines are established, cells are FACS sorted based on the fluorescence intensity of the expressed biosensors to obtain populations of low, medium, and high expressors.

5. The induction procedure for the biosensors is to remove the doxycyclin from the growth medium. Cells cultured in medium

containing doxycyclin are detached when subconfluent, pelleted at $300 \times g$, and washed once with fresh medium without doxycyclin. The cells are then replated onto fresh 10-cm tissue culture dishes at 5×10^4 cells/plate without doxycyclin. Imaging is typically performed 48–72 h postinduction depending on the efficiency of the induction in different cell types (see Note 10).

3.4. A Typical Procedure and Key Considerations for the Single-Mode Imaging of the Rho GTPase Biosensor

1. Induced cells are plated onto sterile, precleaned, 25-mm round coverslips at 4.5×10^4 cells/coverslip in normal growth medium. We find it convenient to place the coverslips in 6-well plates. The coverslips are usually treated with fibronectin prior to plating.

2. Allowed cells to attach and spread for approximately 2–3 h prior to imaging. The imaging medium used is Ham's F-12K without phenol-red (Biosource).

3. Assemble the imaging chamber (see Fig. 5) (see Note 11).

4. For the sequential acquisition mode, follow the flow diagram of Fig. 6a. Control the image acquisition using the Metamorph

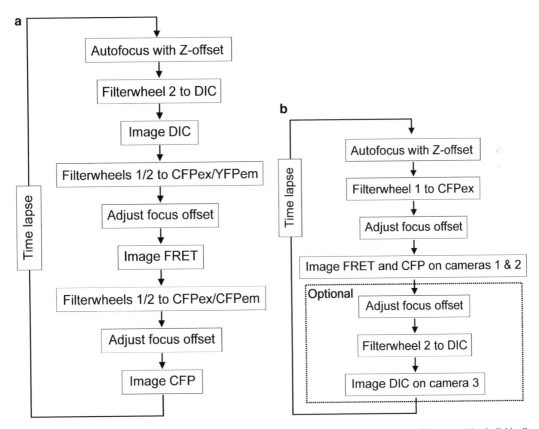

Fig. 6. (a) Flow diagram for the single-camera, single-biosensor mode imaging. The focus offsets must be individually determined empirically for each cell type and wavelength. (b) Flow diagram for the dual-camera, single-biosensor mode imaging. The optional DIC acquisition using a *third camera* is also shown.

ver.7.6 software and use the macro programs that automate the majority of acquisition functions together with the multi-dimensional acquisition application. In order to ascertain the extent of potential motion artifacts, we acquire first a CFP image followed by the FRET image, and then one additional CFP image at every acquisition time point. The ratio between the first CFP image and the last CFP image constitutes the extent of motion artifacts at each time point.

5. For the simultaneous acquisition mode, follow the flow diagram of Fig. 6b. As in the sequential acquisition mode, use the Metamorph software with the automation macro programs.

Concerning beam-splitting at the side port for CFP and FRET on two cameras, if a beam-splitter that does not require a relay lens cannot be placed in the focused distance of the tube lens, there are two possible solutions: (a) use the commercially available Dual-View/Dual-Cam module (Optical Insights) which would reduce the light throughput due to inclusions of additional relay lens components (16–25% loss as measured in our hands) or (b) remove the tube lens from the microscope and perform all of the beam-splitting and filtering in the infinity space outside of the microscope body and then bring to focus the two channels at some distance away from the final beam-splitter. The second solution requires additional thoughts, in that not all manufacturers fully correct the objective lens and some rely on the tube lens to perform a portion of the lateral chromatic correction (excepting the Nikon CF optics). Therefore, if more than one tube lens is used or if nonoriginal equipment manufacturer lenses are used, this issue will become problematic.

One of the key points is to choose the dual-camera system so that the cameras are identical. We carefully test the CCD readout noise characteristics of two identical camera units and select based on similar noise characteristics. Typically, readout noise of less than six electrons at 1×1 binning is desired and the two CCDs should have near identical noise characteristics. The beam-splitter that reflects the image onto the second CCD camera requires some consideration. The ideal approach would be to place the dichromatic mirror directly within the emission optical train without additional relay lenses. There are several approaches to achieving this. If bottom and side ports of the microscope can be dedicated to the CFP/YFP FRET acquisition modality, the simplest solution is to replace the internal side port prism to an appropriate long-pass mirror and then attach the two cameras par-focal at the two ports, placing appropriate emission bandpass filters in front of each camera. If only one camera port is available, it is usually possible to attach a beam-splitter module within the focused distance of the optical train without the need for an additional relay lens by using a dual-camera module such as the U-DPCAD from Olympus for the IX2 series of microscopes. For the image acquisition, we take simultaneously

FRET and CFP emissions upon a single CFP excitation using the two-camera setup at each time point during a timelapse experiment. Following image acquisitions, it is necessary to take a calibration image set using a coverslip with mounted multispectral beads. Because the mounting of cameras and the dichromatic mirrors are never perfect and are susceptible to fluctuations, it is necessary to correct for any potential misalignment between the FRET and CFP channels prior to ratiometric calculations. In order to achieve the required pixel-by-pixel register, the calibration image set is used to obtain the nonlinear field morphing parameters using a computer algorithm available elsewhere (9). Using this approach, it is possible to register two images to an accuracy of one-twentieth of a pixel, allowing for corrections of chromatic aberrations as well as translational and shear effects within the FRET/CFP image set.

3.5. A Typical Procedure and Key Considerations for the Multiplex-Mode Imaging of Rho GTPase Biosensors

1. Plate induced cells onto fibronectin-coated 25-mm round coverslips as described in step 1 of Subheading 3.4. The cells to be imaged must be healthy and should have been passaged at optimal confluence. The microinjection of the Cdc42 biosensor is performed ~4 h postplating.

2. Thaw a frozen 40-μl aliquot of the Cdc42 biosensor over ice, adjust to 50 μM using 50 mM NaH_2PO_4 pH 7.5, and centrifuge at $15,000 \times g$ at 4°C for 20 min. Carefully transfer the supernatant to a fresh tube, wrapped in aluminum foil and keep on ice until microinjection.

3. Transfer coverslips with cells from 6-well plates to a 35-mm round tissue culture dish with approximately 4 ml of regular growth medium. This is mounted atop the inverted tissue culture microscope equipped with a microinjection system.

4. Using a loading pipette tip back-fill 5 μl of the Cdc42 biosensor solution into the pulled needles and attached a needle tip to the microinjection system. Typically, we inject 20–40 cells per coverslip in approximately 10 min at ambient conditions. MEFs appear to lose integrity much past this 10 min window and this affects the survival rate following the injection and recovery. Cells are allowed to recover for 30 min in the incubator prior to imaging.

5. Assemble the imaging chamber (see Fig. 5) (see Note 11).

6. Follow the flow diagram of Fig. 7. We use the Metamorph software with the automation macro programs. The image acquisition sequence is such that we first autofocus to find the focused plane, followed by CFP/YFP FRET acquisition using the side-mounted two-camera setup, then acquire the mero87/Atto700 using the bottom-mounted two-camera setup, and then acquire a DIC image using one of the channels of the bottom-mounted cameras. Here, we need to place a linear polarizer in and out of

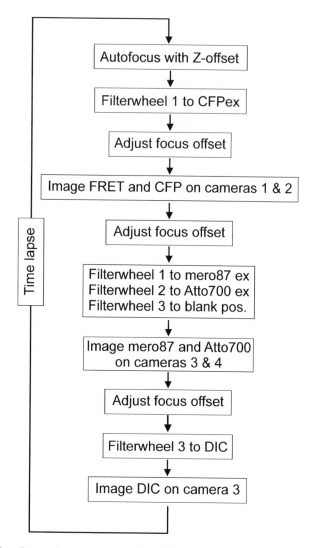

Fig. 7. Flow diagram for the four-channel, dual-biosensor mode imaging.

the light path for the DIC imaging; therefore we have designed a filterwheel within the focused distance of the bottom port prior to the beam-splitter (see Fig. 4). For the beam-splitter, we use the DualCam module (Optical Insights) with the appropriate set of filters (Table 2). Because of the complexity of imaging paths (see Fig. 4), we have opted for using this beam splitting solution even with the loss of light associated with the internal relay optics within the DualCam. In doing so, we have retained the original lateral chromatic correction characteristics of the tube lens and enough distance to place a single filterwheel within the optical path, greatly expanding the versatility of the imaging system. Again, it is possible to achieve a system that is more light efficient through the placement of a tube lens at some distant position away from the original design of the microscope.

Similar image processing considerations as in the single-mode, simultaneous imaging described in Subheading 3.4 apply. The primary concern is therefore the channel-to-channel alignment on a pixel-by-pixel basis prior to ratiometric calculations. Although we maintain the types of CCDs identical on all four channels, we have noticed significant magnification differences between the side port versus the bottom port of the microscope. This should also be corrected using the morphing-based image registration approaches described elsewhere (9).

Since the excitations of mero87 and Atto700 are achieved using two different light sources, the light intensity of each excitation source must be carefully monitored. We use a light-intensity meter to measure the milliWatt power of light at each excitation wavelength at the specimen plane prior to imaging. Usually, we measure approximately 0.5 mW of light power at each excitation wavelength, ±5%. If the light intensities are significantly different, neutral density filters must be used to adjust the relative exposure intensities at the two cameras.

Imaging conditions are carefully selected so as to minimize the production of the reactive oxygen species. It appears that cells tolerate much better longer exposure times at low light intensities compared to a short exposure at a high excitation light intensity. To achieve this, the multiplex imaging system (see Fig. 4 and Table 2) has been optimized to deliver a low intensity excitation through a combination of the neutral density filters and the 94:6 reflective mirror in the main turret. The use of this mirror allows one to sidestep the need for a custom-designed dichromatic mirror and then use the bandpass filters that are designed very close in wavelength separation to one another. This approach enables the system to: (1) tightly pack the bandpass filters allowing separation of more colors within the visible spectrum and (2) separate multiple fluorescence wavelengths at a minimal loss of light compared to using the traditional dichromatic mirrors. Using this system, we routinely achieve ~1,000 ms exposure at ND1.0 (10% incident light) to fill approximately 80% of the dynamic range of the cameras.

4. Notes

1. This formulation (14, 15) reduces the background autofluorescence in the channels used for the CFP/YFP FRET; however, it is no longer available from this manufacturer. We list the formulation in Table 3.

2. The production of oxygen radicals during the fluorescence excitation of the fluorophores must be rigorously controlled, particularly when having two biosensor systems in a single living cell, as this necessitates a greater dose of exposure to the

Table 3
Formulation of the Ham's F-12K medium without phenol-red and without glutamine, pH 7.4–7.5

Inorganic salts	mg/L	Other compounds	mg/L
NaCl	7,530	Glucose	1,260
KCl	285	Linoleic acid	0
$MgCl_2-6H_2O$	106	Hypoxanthine–Na	4
$MgSO_4-7H_2O$	393	Phenol-red	0
$CaCl_2$	135	Putresine–HCl	0.3
$Na_2HPO_4-7H_2O$	218	Sodium pyruvate	220
KH_2PO_4	59	Thymidine	0.7
$NaHCO_3$	2,500		
$FeSO_4-7H_2O$	0.8		
$CuSO_4-5H_2O$	0.002		
$ZnSO_4-7H_2O$	0.14		

Amino acids	mg/L	Vitamins	mg/L
L-Alanine	17.8	L-Ascorbic acid	0
L-Arginine–HCl	421.3	Biotin	0.07
L-Asparagine–H_2O	30	D-Calcium pantothenate	0.48
L-Aspartic acid	26.6	Choline chloride	13.96
L-Cysteine–HCl–H_2O	70.04	Cyanocobalamin	1.36
L-Cystine	0	Folic acid	1.32
L-Glutamic acid	29.4	Inositol	18
Glycine	15	Nicotinamide	0.04
L-Histidine–HCl–H_2O	41.9	Pyridoxine–HCl	0.06
L-Isoleucine	7.9	Riboflavin	0.04
L-Leucine	26.2	Thiamine–HCl	0.21
L-Lysine–HCl	73.1	DL-Thioctic acid	0.21
L-Methionine	8.9		
L-Phenylalanine	9.9		
L-Proline	69.1		
L-Serine	21		
L-Threonine	23.8		
L-Tryptophan	4.1		
L-Tyrosine	10.9		
L-Valine	23.4		

fluorescence excitation light. We do this through pretreatment of the imaging media with Ar gas followed by the Oxyfluor reagent treatment.

3. There could be batch-to-batch variations in the reactivity of the dye. The proposed 6:1 dye to protein ratio and the reaction time of 1 h at 24°C should be optimized as required to achieve 95–100% efficiency of labeling.

4. We find it critical that this pH exchange be a dialysis-based rather than using an ion/salt exchange column. It appears that using these ion/salt exchange columns does not result in a proper protonation of the intrachain lysines in our hands and always results in excessive conjugation of the succinimidylester dye.

5. We have found that the reactivity of this dye is somewhat low and have therefore required a relatively large dye to protein ratio for the optimal reaction.

6. The separation of the labeled material from the unlabeled free dye is usually straightforward. Only caution here is a potential overloading of the column resulting in an incomplete separation and smearing.

7. Transfection is best performed by reducing the DNA suspension media and Lipofectamine 2000 suspension volumes to 500 μl each. Here, it is critical to remove any antibiotics from the media during transfection.

8. Because VSVg is toxic, it has been possible to reliably collect only one round of the supernatant in our hands. Therefore, we normally produce multiple tissue culture dishes of viral supernatant, pool them, and then concentrate the virus at a ratio of 7:1.

9. We usually transduce 4–6 consecutive times, and then repress the biosensor expression by applying doxycycline at the final concentration of 2 μg/ml prior to starting the selection process.

10. We have seen the timing for the optimal imaging postinduction to vary depending on the particular cell type; for example, MEFs appear to be optimal at 48 h, while the rat mammary carcinoma MtLn3 cell line requires 72 h for an optimal induction. It is possible to slightly boost the total expression level of the biosensor by simply trypsinizing and replating directly back onto the same culture dish at 24-h intervals, though we find this to be rarely necessary.

11. For stimulation experiments, the glass coverslip comprising the top part of the heated chamber is excluded during the experiment. In this case, 25 mM HEPES buffer should be added to the medium, and stimulation experiments should be designed in such a way to ensure a good mixing upon addition of the stimulant. We have found that having 500 μl of medium in this open chamber and an addition of another 500 L with the stimulant at 2× concentration works effectively.

Acknowledgments

This work was funded by GM093121 (D.S. and L.H.) and "Sinsheimer Foundation Young Investigator Award" (L.H.).

References

1. Kraynov, V.S., Chamberlain, C., Bokoch, G.M., Schwartz, M.A., Slabaugh, S., and Hahn, K.M. (2000) Localized Rac activation dynamics visualized in living cells, *Science* 290, 333–337.

2. Nalbant, P., Hodgson, L., Kraynov, V., Toutchkine, A., and Hahn, K.M. (2004) Activation of endogenous Cdc42 visualized in living cells, *Science* 305, 1615–1619.

3. Pertz, O., Hodgson, L., Klemke, R.L., and Hahn, K.M. (2006) Spatiotemporal dynamics of RhoA activity in migrating cells, *Nature* 440, 1069–1072.

4. Machacek, M., Hodgson, L., Welch, C., Elliott, H., Pertz, O., Nalbant, P., Abell, A., Johnson, G.L., Hahn, K.M., and Danuser, G. (2009) Coordination of Rho GTPase activities during cell protrusion, *Nature* 461, 99–103.

5. Seth, A., Otomo, T., Yin, H.L., and Rosen, M.K. (2003) Rational design of genetically encoded fluorescence resonance energy transfer-based sensors of cellular Cdc42 signaling, *Biochemistry* 42, 3997–4008.

6. Ridley, A.J., and Hall, A. (1992) The small GTP-binding protein rho regulates the assembly of focal adhesions and actin stress fibers in response to growth factors, *Cell* 70, 389–399.

7. Ridley, A.J., Paterson, H.F., Johnston, C.L., Diekmann, D., and Hall, A. (1992) The small GTP-binding protein rac regulates growth factor-induced membrane ruffling, *Cell* 70, 401–410.

8. Garrett, S.C., Hodgson, L., Rybin, A., Toutchkine, A., Hahn, K.M., Lawrence, D.S., and Bresnick, A.R. (2008) A biosensor of S100A4 metastasis factor activation: inhibitor screening and cellular activation dynamics, *Biochemistry* 47, 986–996.

9. Hodgson, L., Shen, F., and Hahn, K.M. (2010) Biosensors for characterizing the dynamics of rho family GTPases in living cells In: Bonifacino, J.S, Dasso, M., Harford, J.B., Lippincott-Schwartz, J. and Yamada, K.M. (eds) Current Protocols in Cell Biology, Unit 14.11. Wiley, New York.

10. Gossen, M., and Bujard, H. (1992) Tight control of gene expression in mammalian cells by tetracycline-responsive promoters, *Proc Natl Acad Sci USA* 89, 5547–5551.

11. Hodgson, L., Pertz, O., and Hahn, K. M. (2008) Design and optimization of genetically encoded fluorescent biosensors: GTPase biosensors, *Methods Cell Biol* 85, 63–81.

12. Ouyang, M., Huang, H., Shaner, N. C., Remacle, A.G., Shiryaev, S.A., Strongin, A.Y., Tsien, R.Y., and Wang, Y. (2010) Simultaneous visualization of protumorigenic Src and MT1-MMP activities with fluorescence resonance energy transfer, *Cancer Res* 70, 2204–2212.

13. Toutchkine, A., Nguyen, D.V., and Hahn, K.M. (2007) Merocyanine dyes with improved photostability, *Org Lett* 9, 2775–2777.

14. Kaighn, M.E. (1973) Tissue Culture Methods and Applications. Academy Press, New York.

15. Robey, P.G., and Termine, J.D. (1985) Human bone cells in vitro. *Calcif Tissue Int* 37, 453–460.

Chapter 16

FRET-Based Imaging of Rac and Cdc42 Activation During Fc-Receptor-Mediated Phagocytosis in Macrophages

Adam D. Hoppe

Abstract

Fluorescence resonance energy transfer (FRET) imaging can measure the spatial and temporal distributions of activated Rho GTPases within living cells. This information is essential for understanding how signaling networks influence Rho-GTPase switching and for elucidating the mechanisms of Rho GTPase control of the cytoskeleton. This chapter describes FRET microscopy methods to image the distribution of GTP-bound Rac and Cdc42 during the well-defined morphological transitions of phagocytosis by macrophages. Specifically, we describe the use of FRET microscopy to detect the binding of genetically encoded fluorescent protein fusions to Rac1 or Cdc42 with a fluorescent protein fusion to a p21-binding domain (PBD) that recognizes their GTP-bound states. We focus on quantifying the kinetics and activation levels of Rac and Cdc42 during Fc receptor-mediated phagocytosis by macrophages. This process is a Rac1, Cdc42, and actin-dependent process, by which macrophages engulf particles ranging in size from 0.5 to 20 μm and is an ideal model system for studying the spatial and temporal control of these GTPases. Quantitative FRET analysis for measuring the fractions of activated GTPase to allow comparison between cells, independent of the relative expression levels of the fluorescent fusions is also discussed.

Key words: Fluorescence resonance energy transfer, FRET, Microscopy, Imaging, Phagocytosis, Rac1, Cdc42, GTPase, Fc receptor, Macrophage, Fluorescent proteins

1. Introduction

Since their discovery as cell-shape regulating proteins, much has been learned about how the Rho-GTPases regulate the cytoskeleton. Numerous guanine nucleotide exchange factors (GEFs), GTPase activating proteins (GAPs) and effector proteins have been identified (1). Yet, there is much to learn about the spatial and temporal orchestration of signaling to and from the Rho GTPases during the reorganization of the cytoskeleton (2). The growing number of imaging methods that target this research question hint at its importance and suggest that with the right tools, new insights into

Francisco Rivero (ed.), *Rho GTPases: Methods and Protocols*, Methods in Molecular Biology, vol. 827,
DOI 10.1007/978-1-61779-442-1_16, © Springer Science+Business Media, LLC 2012

Rho GTPases function will impact topics ranging from metastatic cancer to phagocyte function in the immune system.

Fc-receptor γ (FcR)-mediated phagocytosis is the uptake of antibody (Immunoglobulin G) coated particles by macrophages through Rho-family GTPase-coordinated rearrangements of the actin cytoskeleton (3–6). During FcR phagocytosis, macrophages reorganize actin to extend plasma membrane from their surface to form a structure called the pseudopod. As the actin-rich pseudopod envelopes the particle, forming a cup, actin depolymerizes from the base of the cup. Contractile activities close the cup and move the newly formed phagosome through the cortical actin layer into the cytoplasm. This tightly organized series of spatial and temporal events provides a well-defined framework on which to interpret signal transduction and function of the Rho GTPases.

FRET microscopy provides a powerful tool for quantifying the spatio-temporal distribution of GTP-bound GTPases within living cells. This is accomplished by detecting the association of fluorescently labeled GTPase with a fluorescently labeled effector domain. Provided that this interaction brings the donor and acceptor fluorophores to within approximately 5 nm, then nonradiative energy transfer can occur resulting in a decrease in donor fluorescence and a corresponding increase in acceptor fluorescence, called the sensitized emission. FRET microscopy is particularly effective for localizing Rho-GTPase interactions with effector domains since their small size (<25 kDa) and small effectors domains (~5 kDa) readily bring fluorescent protein donor and acceptors close enough for FRET to occur.

This chapter describes the use of FRET Stoichiometry (sFRET) (7) to image GTP-bound Rac or Cdc42 during phagocytosis. This analysis requires that a yellow fluorescent protein (YFP)-labeled Rac or Cdc42, (the FRET acceptor) be co-expressed with a cyan fluorescent protein (CFP)-labeled binding domain (the FRET donor). Fluorescence images of the cell are captured with distinct combinations of excitation and emission wavelengths corresponding to the YFP and CFP excitation and emission spectra. sFRET uses a predetermined calibration to convert these raw data into apparent FRET efficiencies. These computed images reflect the product of the fraction of donors or acceptors in complex and the amount of energy transferred from the donor to the acceptor (true FRET efficiency). Additionally, sFRET produces an image of the molar ratio of acceptors to donors in the cell. This ratio image allows correction of the apparent FRET efficiency image for differences in acceptor/donor expression levels. Simpler FRET-based methods are possible, however, they provide skewed intensity values depending upon the ratios at which the two proteins are expressed (7, 8). Here, we describe the use of sFRET to generate images that reflect fractions of activated Rac or Cdc42 within the living cell (4). These data allow comparison of Rac or Cdc42 activation between cells and signaling events despite variations in expression levels of YFP-tagged GTPase and CFP-tagged effector domains.

sFRET uses three fluorescence images and an instrument calibration to recover the apparent FRET efficiencies and the molar ratio of acceptors to donors. These images include an image of the fluorescence from the acceptor (I_A) only, fluorescence from the donor (I_D) only, and an image of the "FRET fluorescence (I_F)." The I_F image is rarely, if ever, a pure image of the sensitized emission, and is usually contaminated by directly emitted fluorescence from the donor as well as fluorescence from the direct (non-FRET) excitation of the acceptor. sFRET uses a calibration scheme to isolate the sensitized emission from I_F and to compute images of the apparent FRET efficiencies and FRET-corrected molar ratio. The sFRET calibration is obtained by expressing the acceptor fluorophore alone (typically YFP), the donor fluorophore alone (typically CFP), and a linked CFP–YFP fusion construct of known FRET efficiency. Once the instrument is calibrated, timelapse sFRET movies allow visualization of the fraction of activated Rho-GTPase. Here, we provide the methods for sFRET, including calibration and application to imaging Rac and Cdc42 activity during phagocytosis.

2. Materials

2.1. Cells, Media, and Opsonin

1. COS7 cells (ATCC) are easy to transfect and are used for calibration of the FRET microscope and testing of preliminary FRET constructs.

2. RAW 264.7 cells (ATCC) are a moderately transfectable murine macrophage cell line suitable for expression of fluorescent protein chimeras for FRET analysis. We prefer this cell line over the frequently encountered J774 cell line (see Note 1).

3. Dulbecco's Modified Eagle Medium supplemented with 10% heat-inactivated fetal bovine serum, 4 mM L-glutamine, 20 U/mL penicillin, and 20 U/mL streptomycin is used to culture both lines.

4. Ringer's buffer: 155 mM NaCl, 5 mM KCl, 2 mM CaCl$_2$, 1 mM MgCl$_2$, 2 mM NaH$_2$PO$_4$, 10 mM HEPES, and 10 mM glucose. It is used for imaging, as it provides low autofluorescence and good pH stability in room air.

5. PBS: 137 mM NaCl, 1.47 mM KH$_2$PO$_4$, 2.68 mM KCl, pH 7.4.

6. PBS–EDTA: 152 mM, 1.63 mM KH$_2$PO$_4$, 2.98 mM KCl, 60 mM EDTA, pH 7.4.

7. Sheep red blood cells (SRBCs) are the target particles for phagocytosis. They are prepared by centrifuging 2 mL of sheep blood plus 12 mL of PBS–EDTA for 15 min at $1,200 \times g$

at 4°C followed by two additional washes with 12 mL of PBS–EDTA. Count SRBCs with a hemocytometer to determine cell density. Dilute stock cells to 10^9 SRBCs/mL in PBS–EDTA and store at 4°C for up to 1 month.

8. Polyclonal, rabbit anti-SRBC IgG (MP Biomedicals) is used to opsonize SRBCs for phagocytosis.

9. FuGene6 transfection reagent (Roche) is used to transfect the RAW 264.7 cell line. Typical transfection efficiency is ~10%.

10. TE: 1 M Tris–HCl pH 7.3, 250 mM EDTA.

11. Calcium buffer: 2.5 M $CaCl_2$ in 10 mM HEPES pH 7.2.

12. HEPES-buffered saline (HBS) solution: 50 mM HEPES, 1.5 mM Na_2HPO_4, 0.28 M NaCl, 10 mM glucose, pH 7.2.

2.2. Plasmids and Fluorophores

Constructs were built from pECFP-C1/N1 or pEYFP-C1/N1 (Clontech) (4).

1. The donor/acceptor pair for FRET is mECFP. We have also used the CFP variant mCerulean (9) and mCitrine (a YFP variant with reduced environmental sensitivity and improved Förster distance) (7, 10). Both CFP and YFP variants used in these studies also have the A206K mutation to reduce dimerization tendencies and are denoted in the plasmid name by "m" for monomeric (11). Accurate FRET microscopy requires that the same fluorescent protein variants be used for both experiment and calibration. We refer to these variants simply as CFP and YFP (see Table 1).

2. Calibration of the FRET microscope requires expression of CFP alone, YFP alone, and a YFP–CFP fusion protein with a known FRET efficiency (Table 1). These constructs are available from the Center for Live Cell Imaging (CLCI) at the University of Michigan (http://sitemaker.umich.edu/4dimagingcenter/center_for_live-cell_imaging_home).

3. Expression constructs for YFP–Cdc42, YFP–Rac1, and YFP–Rac2 and the CFP-tagged p21-binding domain (CFP–PBD) from hPAK1, are available from Addgene.org. The constitutively active (V12 or L61) and dominant negative (N17) mutants of these YFP–Rac and YFP–Cdc42 fusions are also available from the author's laboratory (4).

4. Small fluorescent beads (~100 nm) (PS-Speck Microscope Point Source Kit, Invitrogen) are needed for verification and correction of image alignment.

2.3. Microscope for FRET Microscopy and Leiden Chamber

1. Microscope. Typical sFRET microscopes use filter wheels for selecting excitation and emission wavelengths (7, 12, 13). In addition to these platforms, sFRET imaging has been achieved using two custom dual-camera microscopes; one built on a Nikon TE2000 (14) and one around a fully

Table 1
Plasmids needed for the calibration and analysis of Rac1 and Cdc42 activity by FRET microscopy

Application	Construct	Features	Source
Calibration	pmCitrine-N1	Expresses YFP only	Center for Live Cell Imaging
	pmECFP-N1	Expresses CFP only	Center for Live Cell Imaging
	pCit-CFP	Linked construct of known FRET efficiency	Center for Live Cell Imaging
Effector domain	pmCFP-PBD	CFP labeled effector domain that binds Cdc42, Rac1, or Rac2 with comparable affinity	Addgene.org
	pmCFP-PBD(L83,L86)	Mutant defective for binding activated Cdc42 and Rac	Addgene.org
Rac1	pmCit-Rac1	Expression of YFP–Rac1	Addgene.org
	pmCit-Rac1(L61)	Constitutively active	Addgene.org
	pmCit-Rac1 (L61,H40)	Constitutively active and partially defective in binding PBD	Addgene.org
	pmCit-Rac1(N17)	Dominant negative YFP–Rac1	Addgene.org
Cdc42	pmCit-Cdc42	YFP–Cdc42	Addgene.org
	pmCit-Cdc42(V12)	Constitutively active	Addgene.org
	pmCit-Cdc42(N17)	Dominant negative	Addgene.org

automated iMIC (Till Photonics/Toptica). We describe the filter-based microscope since this is the type of instrument most users have access to.

The sFRET microscope must be capable of collecting three images: donor fluorescence, I_D (e.g., CFPex, CFPem), acceptor fluorescence, I_A (e.g., YFPex, YFPem), and the sensitized emission, I_F (e.g., CFPex, YFPem).

2. Filters. Most commercial microscopes can be configured for sFRET by outfitting them with filter wheels to select bandpasses of excitation and emission light. Typically, we have used filter wheels by Sutter Instruments and Prior Scientific in combination with matched bandpass filter sets such as the 86006 filter set from Chroma Technology and CFP/YFP sets from Omega Optical.

3. Objective Lenses. High-resolution images with good light collection efficiency necessitate the use of high numerical aperture (NA) objective lenses: 60× oil immersion lens, NA 1.4 or 60× water immersion lens, NA 1.2.

4. Cameras. High-sensitivity cameras are desired, as they will minimize photobleaching of the sample and photodamage to the cell. Cooled charge-coupled devices (CCD) cameras such

as the Photometrics Cool Snap are sufficient for sFRET. In addition, we have seen that some gains can be made for weakly fluorescent samples by using electron-multiplying CCD (emCCD) cameras such as the iXon 885 (Andor Technology) or the Photometrics Cascade II (Roper Scientific/Photometrics). However, additional noise associated with the EM-signal amplification and loss of dynamic range can be detrimental to the signal processing for sFRET. Ultimately, a balance must be struck between EM amplification of weak fluorescence signals and loss of dynamic range and additional noise created by EM amplification.

5. An environmental chamber for keeping cells near physiological temperature is an essential part of the sFRET microscope. This can be a simple heated microscope stage such as those available from Warner Instruments or microscope enclosures that allow for heating of the entire sample and all or part of the microscope. A major limitation with the stage heater approach is focal drift during heating and cooling cycles; which makes capturing movies of phagocytosis difficult. Recently, we constructed a simple enclosure for our iMIC platform consisting of a commercially available Plexiglas cylinder, small heater (Cirrus 40, Genesis Automation), and a programmable temperature controller (Cole Palmer). This system displays good thermal stability and is ideal for observation of phagocytosis.

6. We employ a robotic stage for imaging five different fields on the coverglass during a single sFRET acquisition.

7. Cells are cultured on acid-washed 25-mm round No. 1.5 coverglasses in the bottom of a six-well culture dishes. These coverglasses are mounted into Leiden-type chambers such as the Attofluor for imaging (Invitrogen/Molecular Probes).

2.4. Data Analysis Software and the FRET Calculator

In principle, any software platform capable of performing image arithmetic can perform sFRET. Given the number of computations required for sFRET and the need for image refinements such as image co-registration, preconfigured software greatly eases implementation of this method. We have constructed a freely available sFRET calculator built in Matlab (The Mathworks) and DIPImage (www.diplib.org, Delft University of Technology, Netherlands). This calculator has an interface that guides the user through image alignment, sFRET calibration, and the sFRET analysis. This chapter focuses on the use of the sFRET calculator for data analysis. The FRET calculator can be obtained for Matlab/Dipimage or as a stand-alone distribution from the University of Michigan CLCI (http://sitemaker.umich.edu/4dimagingcenter/fret_calculator).

Prism (GraphPad Software) or equivalent fitting software is needed to fit the mass-action binding data required for converting sFRET data into fractions of activated GTPases.

3. Methods

3.1. Cell Transfection

Calcium phosphate transfection provides an inexpensive way to transfect COS7 cells without complicating fluorescence background that is often observed with lipid-based reagents (see Note 2). To transfect one well of a six-well dish:

1. Mix in a microcentrifuge tube 300 μL of TE, 7 μg of plasmid DNA (500 ng/μL), and 17 μL of calcium buffer.

2. Use a pipette to blow air bubbles into 167 μL of HBS. Add the mix of step 1 dropwise to the HBS.

3. Incubate at room temperature for 30 min.

4. Transfect one six-well dish with the mixture by dropwise addition of the mixture to the well.

5. Replace medium after 4 h.

3.2. Expression of Fluorescent Proteins in COS7 Cells for Calibration

1. Place cleaned and ethanol flamed 25-mm, No. 1.5 circular coverglasses into each well of a six-well dish.

2. Seed COS7 cells into the dish at $\sim 1 \times 10^5$ cells/well. The lower density used here will allow space between cells making individual cell recording easier.

3. Transfect two wells each with the calibration constructs (CFP, YFP, and CFP–YFP) using calcium phosphate transfection (see Subheading 3.1 and Note 2).

4. Incubate cells overnight.

3.3. Image Alignment

Prior to calibrating and using the FRET microscope, corrections must be applied when image shift and rotation is observed (see Notes 3 and 4).

1. Measure the image shift by capturing images of pointspread function beads. Begin by making a 1:100 dilution of the beads.

2. Pipette 20 μL of diluted beads onto a cleaned 25-mm No. 1.5 coverglass then dry by setting the coverglass onto a 37°C aluminum block.

3. Mount the dry coverglass into a Leiden chamber and add Ringer's buffer. It may be necessary to rinse to remove beads that detach from the surface of the glass.

4. Mount the Leiden chamber on the microscope and capture I_A, I_D, I_F for 2–3 fields-of-view. Try to capture images where the PSF beads are uniformly spaced over the majority of the field-of-view.

5. Measure the transform mapping the I_D image onto the I_A and I_F images using the image alignment tool in sFRET calculator or Matlab.

6. Record the image transform for future use.

3.4. Calibration of the FRET Microscope

The calibration parameters and required samples are given in Table 2.

1. Mount coverglasses of COS7 cells transfected with CFP, YFP, or YFP–CFP into a Leiden chamber and cover with 0.5–1 mL of warm Ringer's buffer.

2. Place the Leiden chamber on the microscope and capture I_A, I_D, and I_F for ~20 fields-of-view for each condition. The exposure times for the I_A, I_D, and I_F images will need to be adjusted for each sample. For example, when imaging the YFP only sample, the fluorescence in I_F will be low, necessitating a greater exposure time to determine α. Different exposures for I_A, I_D, and I_F are accommodated in the sFRET calculator. Additionally, excessive exposure to the cells should be avoided during data acquisition (see Note 5).

3. Load data into the sFRET calculator and run the sFRET calibration routine. The sFRET calculator provides a number of options for correcting for camera offset (bias), background fluorescence, and variable illumination patterns (shading). For a well-adjusted microscope, bias/background can be corrected by subtracting a cell-free region. Often no shading correction is necessary when a liquid light guide or fiber is used to deliver the excitation light to the microscope. The characteristic FRET efficiency of the linked YFP–CFP construct is needed to finish the calibration. These values are preloaded in the FRET calculator for the linked constructs from the CLCI. The output of the calibration consists of four parameters: α, β, γ, and ξ (see Table 2).

4. Verify the sFRET calibration by running the images of the linked construct through the FRET calculator (see Subheading 3.7). These images should return $E_A = E_D = E_C$ and $R_M = 1$.

Table 2
Calibration Parameters for sFRET

Parameter	Sample	Equation	Meaning
α	Acceptor only (YFP)	$\alpha = \dfrac{I_F}{I_A}$	Direct acceptor fluorescence in I_F versus I_A
β	Donor only (CFP)	$\beta = \dfrac{I_F}{I_D}$	Donor emission in I_F versus I_D
γ	Linked construct of known FRET efficiency (YFP–CFP)	$\gamma = \dfrac{E_C \alpha I_A}{I_F - \alpha I_A - \beta I_D}$	Ratio of acceptor absorbance to donor absorbance at the donor excitation wavelength
ξ	Linked construct of known FRET efficiency (YFP–CFP)	$\xi = \dfrac{E_C I_D}{(I_F - \alpha I_A - \beta I_D)(E_C - 1)}$	Ratio of donor emission to acceptor emission at their respective emission wavelengths

3.5. Preparation of Samples

1. Two days prior to the experiment, plate RAW 264.7 cells at a density of $\sim 2 \times 10^5$ cells/well of a six-well dish containing clean 25 mm, No 1.5 coverslips.

2. One day prior to the experiment, transfect RAW cells with YFP–Rac1 and CFP–PBD using FuGene6.

3. On the day of the experiment, opsonize SRBCs by diluting 2–3 μL of anti-SRBC IgG into 750 μL of PBS and mix with 250 μL of SRBCs (2.5×10^8 cells). Incubate 30 min at 37°C. Pellet the SRBCs at $3,000 \times g$ for 1 min and resuspend in 1 mL of PBS. Repeat the wash three additional times with PBS. Make a working stock of cells by diluting the SRBCs 1:10 in PBS; keep on ice.

3.6. Data Acquisition and Particle Bombardment

1. Load a coverslip with transfected RAW cells into a Leiden chamber and cover with warm Ringer's buffer.

2. Place the sample on the microscope and incubate at 37°C. Some time (5–10 min) maybe required for the cells to adapt to their new environment and Rac/Cdc42 activity to subside.

3. Record a baseline sFRET timelapse by recording I_A, I_D, and I_F at 10–30 s intervals with exposure times of 50–200 ms for each image (also see Notes 6–10).

4. After ten acquisitions, pause the acquisition and place 10–100 μL of opsonized SRBCs into the medium directly above the point where the objective lens contact the coverglass.

5. Observe the cells under brightfield, phase contrast, or differential interference contrast until the first SRBC approaches the macrophage. Restart the timelapse acquisition.

6. Continue imaging until all phagocytic events are completed or stop after 10–15 min.

3.7. Data Analysis of Rac/Cdc42 Activity During Phagocytosis

1. Load image data (I_A, I_D, and I_F) into the sFRET calculator and input calibration parameters (α, β, γ, and ξ) and exposure times.

2. Select method for correction of camera bias level and excitation shading if needed, otherwise use the region of interest (ROI) in a cell-free region to correct the bias/background (see Notes 11 and 12).

3. Input the image transform (see Subheading 3.2) and perform image alignment if needed.

4. Select masking method and set masking parameters. The purpose of this step is to exclude all pixels with fluorescence near zero (e.g., areas in the image not covered by the cell) to eliminate artifacts arising from the division operations used to create E_A, E_D, and R_M. A good starting point here is to create a mask for I_A, I_D, and I_F separately by setting a threshold on the minimum

Table 3
sFRET calculations and output images

Data output	Calculation	Meaning
$E[DA]$	$E[DA] = SE = I_F - \alpha I_A - \beta I_D$	The sensitized emission (SE) provides the distribution of donor and acceptor molecules in complex scaled by the FRET efficiency ($E[DA]$) (e.g., the concentration of YFP–Rac and CFP–PBD complexes times the FRET efficiency).
$[D]$	$[D] = SE + \dfrac{I_D}{\xi}$	The FRET-corrected distribution of total donors (e.g., CFP–PBD).
$[A]$	$[A] = \dfrac{\alpha I_A}{\gamma}$	The FRET-corrected image of total acceptors (e.g., YFP–Rac).
E_A	$E_A = \dfrac{E[DA]}{[A]} = \dfrac{\gamma SE}{\alpha I_A}$	Fraction of acceptors in complex times the FRET efficiency (e.g., the fraction of YFP–Rac bound to CFP–PBD). Pathlength independent.
E_D	$E_D = \dfrac{E[DA]}{[D]} = \dfrac{SE}{SE + \dfrac{I_D}{\xi}}$	Fraction donors in complex times the FRET efficiency (e.g., the fraction of CFP–PBD bound to YFP–Rac). Pathlength independent.
R_M	$R_M = \dfrac{[A]}{[D]} = \dfrac{\alpha I_A}{\gamma \left(SE + \dfrac{I_D}{\xi} \right)}$	Image showing the molar ratio of acceptors to donors (e.g., YFP–Rac/CFP–PBD). Pathlength independent.
E_{AVE}	$E_{AVE} = \dfrac{E_A + E_D}{2}$	The average of E_A and E_D. This parameter is a simple way to describe the FRET efficiency for an interaction while minimizing the impact of differences in the ratio of expressed donor and acceptor.

intensity included in the calculation. This is accomplished in the FRET calculator by moving a slider to select the edge of the cell. The mask is then created from the union of the user-defined minimum for I_A, I_D, and I_F.

5. Perform the sFRET calculation. The calculation will produce the images described in Table 3. Figure 1 illustrates the output of a sFRET calculation for a macrophage phagocytosing an opsonized SRBC.

3.8. Interpreting sFRET Data of Rac and Cdc42 Activation and Conversion of these Images to Represent the Fractions of Active GTPase

Which image do we turn to, to learn when and where our Rac or Cdc42 was active during phagocytosis? For a cell expressing YFP–Rac and CFP–PBD the [A] and [D] images provide the distributions of total GTPase and binding domain while accounting for FRET processes (Fig. 1). The $E[DA]$ image provides the distribution of YFP–Rac/CFP–PBD complexes and is the first answer to our question. This image, however, does not correct for the changes in cell thickness that occur during phagocytosis (Fig. 2a). Thus, increases

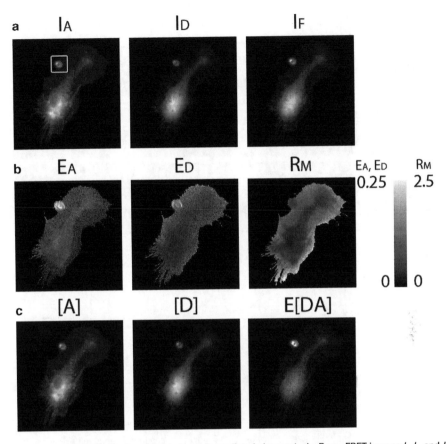

Fig. 1. sFRET microscopy of YFP–Rac1 activity during FcR-mediated phagocytosis. Raw sFRET images I_A, I_D, and I_F (**a**) show a macrophage during phagocytosis of an IgG-coated erythrocyte (*boxed region* in I_A). E_A, E_D, and R_M images produced by the sFRET calculator (**b**) reflect the fraction of YFP–Rac1 in complex (E_A), the fraction of CFP–PBD in complex (E_D), and the molar ratio of YFP–Rac1/CFP–PBD (R_M). (**c**) sFRET calculator images showing distributions of total (bound and free) YFP–Rac1 [A], CFP–PBD [D], and the YFP–Rac1/CFP–PBD complex times the FRET efficiency E[DA]. The *box* in (**a**) is 6 μm.

in fluorescence could either originate from the recruitment of molecules to the forming phagosome or from an increase in the number of YFP–Rac1/CFP–PBD complexes. To uncouple changes in cell thickness from changes in complexes, three-dimensional image reconstruction methods can be used (14). Alternatively, the interpretation of GTPase activation can be simplified by using the E_A image. E_A shows the fraction of acceptor bound to donor (times the FRET efficiency) and is therefore a ratiometric image that mitigates the impact of variable cell thickness. For example, in Fig. 1, E_A shows that the fraction of YFP–Rac1 bound to CFP–PBD is much greater on the phagosome than in the rest of the cell, whereas the E[DA] and [A] images are ambiguous since they both show bright signals associated with the phagosome. Additionally, the E_A signal can be quantified via the scale bar (Fig. 1) with an absolute range of 0–1, with 1 being 100% FRET efficiency and 100% of the acceptors bound by donors (e.g., all of the YFP–Rac bound by

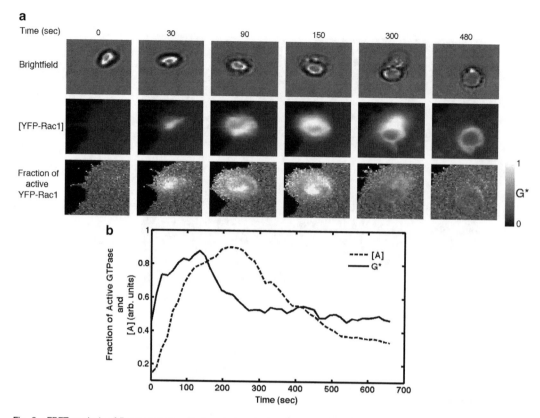

Fig. 2. sFRET analysis of Rac1 activation during phagocytosis of an opsonized SRBC. (**a**) Time lapse images of the *boxed region* in Fig. 1 show the kinetics of YFP–Rac1 activation on a forming phagosome. Pseudopod extension (30–150 s) and phagosome closure (300 s) can be observed in the brightfield and [A] (YFP–Rac1) images. The E_A image from Fig. 1 has been converted to display the fraction of active YFP–Rac1 (note the intensity bar). (**b**) Particle tracking on the center of the SRBC in brightfield provided a region of interest for measuring the fraction of active YFP–Rac1 (G^*) and total YFP–Rac1 ([A]) during phagocytosis.

CFP–PBD). Thus, the E_A image resolves the ambiguity in determining where YFP–Rac1 is bound to CFP–PBD and provides a quantitative measure that can be used to compare the amplitudes of Rac or Cdc42 signaling within and between cells.

Interpreting Rac/Cdc42 activation from sFRET data is simplified by assuming that each YFP–GTPase bound to CFP–PBD produces the same three-dimensional molecular structure. Then, the FRET efficiency between YFP and CFP will be a fixed characteristic (a constant value) of that structure (termed the characteristic FRET efficiency E_C) (7). If E_C is determined then the E_A image can be scaled to display the fraction of YFP–GTPase bound to CFP-binding domain. This analysis, however, does not account for mass-action effects owing to the relative concentrations of YFP–GTPase and CFP-binding domain present in the cell. If for example, R_M for a cell is high, indicating that YFP–Rac1 is in excess, then the concentration of CFP–PBD will limit the maximum values that can be observed in the E_A image. Conversely, if R_M is low for the cell then the E_A values will be high and E_D values will be low. In fact, one can imagine elevating the concentration of CFP–PBD to saturate binding to YFP–Rac1, thereby leading to E_A values that approach E_C as the fraction of acceptors in complex with donors approaches 1. At first glance, this may sound attractive for imaging GTPase activity. However, this case will lead to saturation of both the labeled and unlabeled GTPase and will inhibit the transmission of signals through the GTPase (see Note 8). A better approach, takes advantage of activating mutations in the small GTPases to define mass action binding curves for various expression ratios of GTPase and binding domain (4). Once measured, these binding curves can be used to express the E_A image in terms of the fraction of activated GTPase (relative to the constitutively active mutants) for cells of various expression ratios of GTPase and binding domain. Figure 2 illustrates the measurement of the fraction of activated YFP–Rac1 on the phagosome from the sFRET data in Fig. 1. To convert E_A into fraction of activated GTPase:

1. Plate RAW cells onto coverglasses as described in Subheading 3.5 and transfect with varying ratios (20:1, 1:1, 1:20) of constitutively active YFP–GTPase and CFP–PBD (e.g., YFP–Rac1(Q61L) and CFP–PBD).

2. Collect snapshots of sFRET microscopy data for ~30 cells with expression ratios ranging from ~20:1 to ~1:20 (CFP:YFP, the wider this range the better). This will require changing the exposure times to compensate for weak fluorescence signals and maintain good signal to noise in each of the sFRET images. For example, when collecting the 20:1 data the I_D (CFP) exposure time may have to be increased 20×.

3. Process the sFRET data as described in Subheading 3.7 and record the E_A and R_M values for each cell using the ROI tool in the FRET calculator or other software.

4. Create plots of E_A versus $1/R_M$. Fit these plots to the single site binding equation:

$$E_A^*(R_M^{-1}) = \frac{E_C R_M^{-1}}{K_{RA} + R_M^{-1}}$$

for K_{RD}, K_{RA}, and E_C using Prism or equivalent fitting software.

5. Estimate the fraction of active GTPase image (G^*) for the experimental E_A data by determining the R_M values for the entire cell and using that value to determine E_A^* from the fitted equation above. G^* is then obtained by dividing the E_A image by this value:

$$G^* = \frac{E_A}{E_A^*\left(R_M^{-1}\right)}.$$

4. Notes

1. In some immortalized cell lines, Rac1 is hyperactivated, thereby making imaging of its activity on forming phagosomes difficult. An example is the J774 macrophage cell line. Here, we observed that YFP–Rac1 and CFP–PBD nearly always gave high FRET signals in the absence of stimulant (unpublished result). This phenomenon has been observed by biochemical pull down assays of Rac1 activity (15).

2. One of the major challenges for analyzing Rac or Cdc42 activity in transfected macrophage cell lines is that transfection reagents such as FuGene6 often create spurious fluorescent aggregates. These aggregates appear primarily in the I_F image as strong FRET signals in the E_A and E_D images. A tell-tale signature of these aggregates is that they appear both inside the cell and attached to the coverglass, outside of the cell. The appearance of these aggregates is worst when the ratio of DNA to FuGene6 in the transfection mix is too high. Optimization of plasmid DNA and transfection reagent ratio can minimize these aggregates. Interestingly, these aggregates are not unique to FuGene6, but we have also observed this effect with Lipofectamine and TransitLT reagents. Calcium phosphate transduction methods do not have this problem.

3. Poor image alignment usually arises from misalignment of dual-view or dual camera systems or from imperfect matching of emission filters on single camera systems. Perfect mechanical correction can be difficult to achieve, necessitating computational imaging alignment. Since the I_A and I_F images are typically recorded by the same detector/filter combination, it is sufficient to align I_D with I_A and I_F using alignment tools such as those available in the FRET calculator and the Matlab Image Toolbox.

4. E_A, E_D, and R_M images that look like a three-dimensional surface are indicative of poor image alignment. Correct this using the alignment tool in the sFRET calculator as in Subheading 3.3.

5. Photobleaching while finding a cell and focusing can be detrimental to the FRET signal. Molecules engaged in FRET will show accelerated photobleaching (8, 16) that can skew the results if excessive exposure occurs prior to capturing your data. This problem can be overcome by focusing in brightfield or by using a small number of single snapshots to "bracket" the focal plane.

6. Focusing just above where the lamella contact the coverglass typically captures the best midplane view of phagocytosis. Additionally, a piezo z-drive can be used to capture images in multiple z-planes to improve visualization of phagocytosis.

7. Opsonized SRBCs can sometimes be clumpy making the observation of phagocytic events of individual particles difficult. If this occurs, opsonize SRBCs with a range of antibody concentrations to find a subagglutinating concentration of antibody.

8. Photobleaching during movie acquisition can result from excessive exposure times and illumination intensities. Photobleaching can be observed as changes in R_M measured for the entire cell during the movie. We typically will not use data that displays greater than a 10% change in R_M during the entire timelapse. Photobleaching corrections are possible, however, they are complicated and can alter interpretation since each species free acceptor, free donor, acceptor in a FRET complex and donor in a FRET complex will all bleach at different rates. Work to minimize photobleaching by minimizing exposure of the cell. Use the shortest exposure times and lowest excitation light intensities possible while still maintaining a workable signal-to-noise ratio. Use on-camera binning of pixels to improve signal intensity while minimizing the contributions of CCD read noise.

9. Overexpression of CFP–PBD will inhibit phagocytosis. Measure phagocytic uptake of internalized and surface bound particles as a function of CFP–PBD intensity to determine the workable range of CFP–PBD expression.

10. Overexpression of YFP–Rac1 or YFP–Cdc42 can lead to saturation of the cellular RhoGDI pool (17). By observing cells as a function of brightness, one will see that brighter cells show a larger fraction of YFP–Rac1 or YFP–Cdc42 on plasma and internal membranes. These cells are likely expressing the GTPases at a concentration that exceeds the cellular pool of RhoGDIs. Observe the dimmest cells possible and then search for a range of YFP-fluorescence that provides a similar subcellular distribution.

11. Poor background correction and a masking that is too loose cause false-positive signals appearing as a halo around the cell in E_A and E_D images. This problem can readily be detected in control experiments such as imaging phagocytosis in cells expressing YFP–Rac1 with CFP in place of CFP–PBD. To correct this problem, improve the background correction (say by measuring it from a cell free region, or by using cleaner coverglasses and media) and apply more stringent masking.

12. For some microscope systems, the excitation light is not uniform across the field of view and can vary with wavelength. To correct the illumination field, capture images of a thin solution of fluorophore trapped between a slide and a coverglass suspended by a small spacer, such as tape or broken coverglass fragments (8). Any fluorophore that can be imaged in I_A, I_D, and I_F will

work (we typically use purified fluorescent proteins). Start by capturing images of these solutions and then adjust the microscope excitation optics to obtain the most uniform excitation field possible. Next, record 10–20 images of I_A, I_D, and I_F at different locations on the coverslip. These images can then be averaged in the FRET calculator to provide a correction. One can also compare the excitation uniformity by dividing these images. If the differences are small (<3%) the excitation field of the microscope is uniform and can be neglected. If there are differences greater than this, then use these images in the "shading correction" function of the sFRET calculator to correct all calibration and raw data. To apply this correction, the camera bias will need to be obtained by collecting a series of 10–20 images while blocking all light to the camera. Excitation shading problems can be seen as a gradient across the field of view that is evident in the E_A, E_D, and R_M images.

Acknowledgments

The author thanks Brandon Scott for creating Fig. 1 and editing. The author also thanks Jason Kerkvliet, John Robinson, and Joel Swanson for edits and helpful suggestions. This work was supported in part by NSF grant #0953561, the South Dakota Governor's 2010 Initiative and the South Dakota Board of Reagents Competitive Research Grant Program.

References

1. Bishop, A.L., and Hall, A. (2000) Rho GTPases and their effector proteins. *Biochem J* 348, 241–255.
2. Pertz, O. (2010) Spatio-temporal Rho GTPase signaling - where are we now? *J Cell Sci* 123, 1841–1850.
3. Allen, W.E., Jones, G.E., Pollard, J.W., and Ridley, A.J. (1997) Rho, Rac and Cdc42 regulate actin organization and cell adhesion in macrophages. *J Cell Sci* 110, 707–720.
4. Hoppe, A.D., and Swanson, J.A. (2004) Cdc42, Rac1, and Rac2 display distinct patterns of activation during phagocytosis. *Mol Biol Cell* 15, 3509–3519.
5. Beemiller, P., Zhang, Y., Mohan, S., Levinsohn, E., Gaeta, I., Hoppe, A.D., and Swanson, J.A. (2010) A Cdc42 activation cycle coordinated by PI 3-kinase during Fc receptor-mediated phagocytosis. *Mol Biol Cell* 21, 470–480.
6. Swanson, J.A., and Hoppe, A.D. (2004) The coordination of signaling during Fc receptor-mediated phagocytosis. *J Leukoc Biol* 76, 1093–1103.
7. Hoppe, A., Christensen, K., and Swanson, J.A. (2002) Fluorescence resonance energy transfer-based stoichiometry in living cells. *Biophys J* 83, 3652–3664.
8. Hoppe, A.D. (2007) Quantitative FRET microscopy in live cells. In Shorte, S. L., Frischknecht, F. (eds) Imaging cellular and molecular biological function. Springer, New York.
9. Rizzo, M.A., Springer, G.H., Granada, B., and Piston, D.W. (2004) An improved cyan fluorescent protein variant useful for FRET. *Nat Biotechnol* 22, 445–449.
10. Griesbeck, O., Baird, G.S., Campbell, R.E., Zacharias, D.A., and Tsien, R.Y. (2001) Reducing the environmental sensitivity of

yellow fluorescent protein. Mechanism and applications. *J Biol Chem* 276, 29188–29194.

11. Zacharias, D.A., Violin, J.D., Newton, A.C., and Tsien, R.Y. (2002) Partitioning of lipid-modified monomeric GFPs into membrane microdomains of live cells. *Science* 296, 913–916.

12. Fontaine, J.M., Sun, X., Hoppe, A.D., Simon, S., Vicart, P., Welsh, M.J., and Benndorf, R. (2006) Abnormal small heat shock protein interactions involving neuropathy-associated HSP22 (HSPB8) mutants. *FASEB J* 20, 2168–2170.

13. Liu, J., Ernst, S.A., Gladycheva, S.E., Lee, Y. Y., Lentz, S.I., Ho, C.S., Li, Q., and Stuenkel, E.L. (2004) Fluorescence resonance energy transfer reports properties of syntaxin1a interaction with Munc18-1 in vivo. *J Biol Chem* 279, 55924–55936.

14. Hoppe, A.D., Shorte, S.L., Swanson, J.A., and Heintzmann, R. (2008) Three-dimensional FRET reconstruction microscopy for analysis of dynamic molecular interactions in live cells. *Biophys J* 95, 400–418.

15. Patel, J.C., Hall, A., and Caron, E. (2002) Vav regulates activation of Rac but not Cdc42 during FcgammaR-mediated phagocytosis. *Mol Biol Cell* 13, 1215–1226.

16. Jares-Erijman, E.A., and Jovin, T.M. (2003) FRET imaging. *Nat Biotechnol* 21, 1387–1395.

17. Michaelson, D., Silletti, J., Murphy, G., D'Eustachio, P., Rush, M., and Philips, M. R. (2001) Differential localization of Rho GTPases in live cells: regulation by hypervariable regions and RhoGDI binding. *J Cell Biol* 152, 111–126.

Chapter 17

High-Throughput Flow Cytometry Bead-Based Multiplex Assay for Identification of Rho GTPase Inhibitors

Zurab Surviladze, Susan M. Young, and Larry A. Sklar

Abstract

Rho family GTPases and their effector proteins regulate a wide range of cell signaling pathways. In normal physiological conditions, their activity is tightly controlled and it is not surprising that their aberrant activation contributes to tumorigenesis or other diseases. For this reason, the identification of small, cell permeable molecules capable of inhibition of Rho GTPases can be extraordinarily useful, particularly if they are specific and act reversibly. Herein, we describe a flow cytometric assay, which allows us to measure the activity of six small GTPases simultaneously. GST-tagged small GTPases are bound to six glutathione bead sets each set having a different intensity of red fluorescence at a fixed wavelength. The coated bead sets were washed, combined, and dispensed into 384-well plates with test compounds, and fluorescent-GTP binding was used as the read-out. This multiplex bead-based assay was successfully used for to identify both general and selective inhibitors of Rho family GTPases.

Key words: Rho GTPases, Cdc42, Bead-based multiplex assay, Fluorescent GTP binding, Screen, Flow cytometry

1. Introduction

1.1. Rho GTPase Inhibitors

During the past few years, Rho GTPases and their effector proteins have been recognized as major regulators of a wide range of signaling pathways that control a number of biological processes such as cell-cycle progression and gene transcription. These molecules have also been implicated in cellular processes such as adhesion, migration, phagocytosis, cytokinesis, neurite extension and retraction, cellular morphogenesis and polarization, growth and cell survival (1–3).

Francisco Rivero (ed.), *Rho GTPases: Methods and Protocols*, Methods in Molecular Biology, vol. 827,
DOI 10.1007/978-1-61779-442-1_17, © Springer Science+Business Media, LLC 2012

The function of Rho-family GTPases in disease pathogenesis has been reviewed extensively (4–7). For these reasons, the identification of small, cell permeable molecules capable of regulating Rho GTPases activity can be extraordinarily useful, particularly if they are specific and act reversibly. There has been limited success in identifying inhibitors that specifically interact with individual small Rho GTPases. Most previous efforts have been focused on inhibiting post-translational GTPase modification by lipids, which are necessary for their membrane localization and activation. Unfortunately, these inhibitors and drugs are not specific to GTPases and affect other cell signaling pathways, which complicate the interpretation of results and create toxicity issues. Recently, one small molecule inhibitor of Cdc42 activation (secramine) (8) and several low molecular weight inhibitors for Rac1 have been described (9, 10). Here, we describe a bead-based multiplex flow cytometry high-throughput screening (HTS) assay, which allowed simultaneous screening of six GTPase targets against ~200,000 compounds in the Molecular Libraries Small Molecule Respository (MLSMR). This approach resulted in the publication of two probe reports, one describing the discovery of novel small molecule probes for a selective Cdc42 inhibitor, and the other describing a pan-activator family of probes for Ras-related GTPases (http://mli.nih.gov/mli/mlp-probes/) as well as a general inhibitor of Rho family GTPases (11).

1.2. Fluorescence-Based Assay

Small GTPases exist in two interconvertible forms: GDP-bound inactive and GTP-bound active forms. GTP/GDP exchange studies frequently use guanine nucleotide analogs, which behave similarly to the native species and have been modified such that they can be sensitively detected. Radiolabeled GTP analogs, such as $[\gamma\text{-}^{32}P)$ GTP and $[\gamma\text{-}^{35}S]$ GTPγS have been most commonly used. While these analogs are very sensitive and nonperturbing, their use has obvious drawbacks associated with radioactivity including the requirement for measurement at fixed time points. Recently developed BODIPY-labeled nucleotides are therefore increasingly being adopted for characterizing of GTPase nucleotide binding activities over time (12–14). The fluorescence emission of BODIPY-guanine nucleotides is directly affected by protein binding. Free BODIPY-nucleotides in solution exhibit quenched fluorescence, which is unquenched upon protein binding. The resulting two- to tenfold fluorescence enhancement allows real-time detection of protein–nucleotide interactions. Fluorescent polarization assay has also been used in these experiments; however, flow cytometry measurements have several advantages over the polarization assay (15), and could be successfully used for HTS of chemical libraries for identification of Rho GTPase inhibitors.

Recently developed technology dramatically accelerates the speed of analysis. Whereas previous methods using flow cytometry only allowed the measurement of ten samples per minute, now HyperCyt technology allows analysis of up to one sample per second. Methods using flow cytometry as the read-out in screening assays and general principles for applying such systems in multiwell plates have been described previously (15, 16).

Recently, assays using beads as a solid support matrix for molecular interactions, have become an industry standard of flow cytometry fluorescence-based measurements (15, 17). This approach has been used to develop a bead-based flow cytometric, fluorescent GTP-binding assay which is highly sensitive and allows real-time measurements (18). This bead-based format was subsequently extended to a multiplexed target assay and successfully utilized for HTS to identify regulators of small GTPases (11). The unique multiplexing capabilities of flow cytometry enabled the simultaneous quantitative analysis of the activation or inhibition of six or more GTPases immobilized on color-coded beads. The high surface density of the fusion protein on the beads is believed to promote GST dimerization and to promote stable attachment of the fusion protein to the bead during the course of analysis. Since the beads were obtained in quantity at discount (~1 penny per color per well), the approach resulted in a low-cost, time saving, and highly efficient method for the simultaneous measurements of the activity for a subset of proteins in the family. Moreover, it simplifies the high-content analyses involved with comparison and detection of the specific inhibitors of individual GTPases.

1.3. High-Throughput Screening

Identification of inhibitors by multiplex bead-based HTS is a multistep process. The first step usually includes determination of conditions that allow simultaneous measurement of several bead-attached enzymes. The second step of inhibitor discovery, so-called primary screening, is performed with one fixed concentration of compound. This is often a fast, cost-efficient assay, to screen many compounds for identification of potential "hits" (i.e., active compounds). The third step would be further analysis of potential hits (compounds initially identified in the primary screen) in a dose–response series. And finally, the fourth step would use molecular- or cell-based assays to validate the particular function of hit compounds.

Finding the right conditions for multiplex assay is a potentially difficult, and time-consuming process requiring some preliminary data from pilot studies. Initially, optimal conditions are determined for individual enzyme activities in a single-plex format. Often, these conditions vary in accordance to individual GTPases. For example, BODIPY-GTPγNH is identified as a better substrate for some of the small GTPases such as RhoA wt, Rac1wt, and Cdc42, while others (Rab proteins and Ras) prefer binding to BODIPY-FL-GTP. Mg^{2+} ions are crucial for the Rho enzyme activity, while Rab proteins and H-Ras more effectively bound fluorescent GTP in the presence of EDTA. The binding activity of H-Ras is completely inhibited in the presence of 0.01% dodecyl maltoside in the assay, while this detergent has negligible effect on GTP binding by other GTPases. Based on these preliminary results, we chose specific conditions that were not optimal for all GTPases, but allowed us to measure simultaneously GTP binding to several small GTPases. High-affinity binding of GTP to GTPases in the presence of Mg^{2+} ions complicates screening of competitive inhibitors ($K_d = 0.6$ nM for Cdc42 binding BODIPY-FL-GTP). For this reason, during primary screening Mg^{2+} ions were excluded from the assay system, which significantly decreased nucleotide affinity to GTPases by adding 1 mM EDTA ($K_d \sim 30$ nM). This approach was used for the multiplex analysis of GTPase activity and, therefore, the HTS of the MLSMR.

2. Materials

Prepare all solutions using deionized water and analytical grade reagents. Prepare and store all reagents at 4°C or on ice (unless indicated otherwise).

2.1. Reagents and Buffers

1. Test compounds are provided in 384-well plates and screened using 10^{-3} M stocks in dimethylsulfoxide (DMSO). Obtained from MLSMR (http://mlsmr.glpg.com/MLSMR_HomePage/).

2. 10% Bovine serum albumin (BSA). Freeze in aliquots; do not use a microwave oven for thawing.

3. 100 mM Dithiothreitol (DTT). Freeze at –20°C in aliquots.

4. 5 mM BODIPY-FL-GTP stock. (G-12411, Invitrogen Molecular Probes). Store at –20°C. 1 M HEPES, pH 7.5: Weigh 23.83 g of HEPES, add water to a volume of 90 mL. Mix and adjust pH with KOH. Make up to 100 mL. Store at 4°C.

5. 5 M NaCl: Dissolve 292.2 g of NaCl in 1 L of water.

6. 1 M KCl: Dissolve 75 g of KCl in 1 L of water.

7. 1 M $MgCl_2$: Dissolve 20.3 g of $MgCl_2$ $6H_2O$ in 100 mL of water.

8. 100 mM EDTA: Dissolve 3.7 g of EDTA in 80 mL of water. Adjust pH to 7.5 with NaOH. Then bring volume to 100 mL.

9. 10% Nonidet P-40 (NP-40): Dilute 1 mL of NP40 in water. Bring volume up to 10 mL.

10. NP-HPS buffer: 30 mM HEPES, pH 7.5, 100 mM KCl, 20 mM NaCl, 0.01% (vol/vol) NP-40. Mix 30 mL of 1 M

HEPES, 100 mL of 1 M KCl, 4 mL of 5 mM NaCl, 1 mL of 10% NP-40, and bring volume up to 1 L with water. Store at 4°C.

11. NP-HSP buffer containing 1 mM EDTA (NP-HPSE): Add to 100 mL NP-HPS buffer 1 mL 100 mM EDTA.

12. NP-HSP buffer containing 1 mM $MgCl_2$ (NP-HPSM): Add to 100 mL of NP-HPS buffer 100 µL of 1 M $MgCl_2$.

13. 200 nM BODIPY-FL-GTP in NP-HPSE buffer: Add in 50 mL of NP-HPSE 2 µL of 5 mM BODIPY-FL-GTP. Prepare fresh and keep on ice.

14. 2 nM BODIPY-FL-GTP in NP-HPSM buffer for dose–response experiments: Prepare a 2-µM solution of BODIPY-FL-GTP by adding 1 µL of the 5 mM stock in 2.5 mL of NP-HPSM. Use a 1:1,000 dilution of this solution for preparation of 2 nM BODIPY-FL-GTP. Keep on ice.

15. GST-GTPases: wild type (wt) Cdc42, Rac1, RhoA, and H-Ras and constitutively active mutants Cdc42Q61L, Rac1Q61L, RhoAQ63L, and H-RasG12V are purchased from Cytoskeleton, Inc. Rab2 and Rab7 are purified from *Escherichia coli* as described (18). Store as 1 mg/mL stocks at –80°C.

16. 4 µm Diameter glutathione-bead (GSH-beads) sets for multiplex assays, distinguished by seven different intensities of red fluorescence (representing several orders of magnitude variation of emission at 665 ± 10 nm with excitation at 635 nm) are obtained from Duke Scientific Corp. (but may now be ordered from Thermo Fisher). Each polystyrene bead set is supplied at 1.4×10^5 beads/µL with about 1.2×10^6 glutathione sites per bead as determined by using GST–green fluorescent protein (GFP).

17. Fluorescence standard beads (Bangs Laboratories, cat. No. 825B). This kit contains five sets of beads, with a measured green fluorescence for each set in the FITC, or fluorescein, channel, using a 488-nm laser for excitation and (in our instrument) a 530 ± 40 nm emission filter. The fluorescence is given in mean equivalents of soluble fluorophores (MESF) ranging from 40,000 soluble fluorescein equivalents to 1,100,000 soluble fluorescein equivalents, and is used to calibrate the instrument response.

18. 384-Well assay plates (Greiner Bio-One), 30 µL maximum volume.

19. V-bottom 96-well PCR plates (ISC Bioexpress).

20. Sealing covers for plates (Gene Mate). A roller seals the cover onto the plate.

2.2. Equipment

1. Biomek FX^P (Beckman-Coulter) multitip dispensing instrument, or robot, with a pin tool device (V&P Scientific).

2. Computer with Microsoft Windows 2000 or Windows XP, 512 MB or more RAM, 500 MB or more of free disk space, and a USB port.

3. HyperView™ program (IntelliCyt).

4. GraphPad Prism 4 or 5 software.

5. Flow cytometer (CyAn ADP Dako, now Beckman-Coulter) or LSRII (Becton-Dickinson) and an Accuri C6 (Accuri). For multiplex assay, both 488- and 635-nm lasers are required. The data acquisition software must include a time parameter capable of binning data at 100-ms intervals continuously for 15 min or more.

6. HyperCyt™ instrument (IntelliCyt). This instrument includes an autosampler, a peristaltic pump, 25-G stainless steel tube inlet probes, and PVC tubing. HyperCyt is set up as described earlier (16). Briefly, the peristaltic pump speed is set to 15 rpm to result in a flow rate of about 2 μL/s. Faster or slower speed is typically suboptimal and can also result in increased particle carryover. Peristaltic pump clamping pressure: when adjusted properly, there should be uniform air bubbles on both sides of the pump. If the bubbles are broken up on the flow cytometer side of the pump, the tension on the tubing is too great and can be appropriately adjusted.

7. Peltier cooler for standard size plates (Inheco, TEC Control 96 and CPAC Ultra Flat). The cooling device is placed on the autosampler deck of the HyperCyt.

8. Software for HyperCyt™ (IntelliCyt). Includes two programs that are needed to run the HyperCyt™ platform: HyperCytSampler controls the autosampler, while HyperCytDataAnalysis is used to bin the time-resolved files stored in flow cytometry standard (FCS) 2.0 or 3.0 formats.

3. Methods

3.1. Primary Screening of 384-Well Plates

A set of color-coded glutathione-microspheres, having different intensities of red fluorescence, is coated with an individual low molecular weight GST–GTPase on each microsphere (Fig. 1a). After washing, individual GTPase coupled beads are combined and 5-μL aliquots of the resulting suspension are added into each well of a 384-well plate. A green fluorescent-GTP is used as a binding ligand to look for molecules that could regulate the binding of GTP to small GTPases.

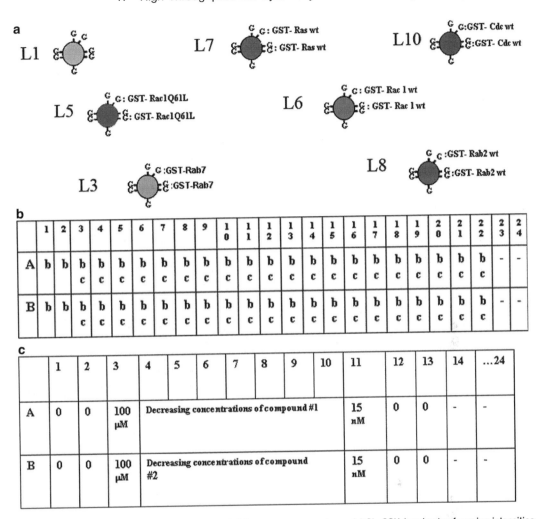

Fig. 1. Experimental setup for primary screening and dose–response analyses. (a) Six GSH-bead sets of varying intensities of red fluorescence are individually coated with GST-Ras family GTPases, and the seventh set of blank beads serves as a scavenger. (b) Setup of 384-well plates for primary screening. The columns are labeled with numbers 1–24, and the rows are labeled with letters A–P. Wells with a symbol "b" have the multiplex (seven different bead sets) in each well. Wells with a symbol "c" have compounds in them to be screened, a total of 320 different compounds per plate. Wells in the first two columns have no compounds, and serve as positive controls. Wells with a "-" symbol in the last two columns have no beads or compounds, and are used to mark the end of each row when binning the data. (c) Setup of 384-well plates for dose–response analyses. Wells in columns 14 and 24 are marked with a "-" symbol, do not have beads and compounds in them, they contain only BODIPY-FL-GTP in assay buffer and are used as markers for the end of the dose–response series for a given compound. (These wells wash the tube for traces of compound to reduce contamination). Each well in columns 1–13 has seven different kinds of beads. Wells in columns 1, 2, 12, and 13 do not have compounds in them, and serve as positive controls. Compound 1 is added to wells A3–A11 at concentrations of 100 μM, 33 μM, ..., 15 nM, and the other compounds are dosed in the same way (starting from B3, C3, and so on). The HyperCyt autosampler picks up ~2 μL of suspension from each well across each row in turn: A1–A24, B1–B24, ending at P24.

3.1.1. Coating of the Bead Sets with Small GTPases and Preparation of 384-Well Plates for Primary Screening

In primary screening, each well contains seven bead types in different intensities of red fluorescence, coupled with individual GST–GTPases. This mixture is incubated for 45 min with 200 nM fluorescent GTP in the presence of 10 μM compound, and the protein binding fluorescence is determined using a flow cytometer with the beads delivered by a HyperCyt system (11).

1. Mix the 4-μm glutathione bead slurries, and load 250 μL of each bead type into seven individual tubes (see Note 1).

2. Collect beads by centrifugation for 30 s at maximum speed.

3. Aspirate the supernatant and wash beads with NP-HPSE buffer.

4. Block nonspecific binding in each bead set individually with 0.1% BSA in NP-HPSE buffer for 30 min at room temperature (see Note 2).

5. Collect bead sets (repeat step 2).

6. Resuspend the beads in 100 μL of NP-HPSE buffer.

7. Thaw aliquots of six recombinant GST-GTPases.

8. Add ~20 μg of each GTPase to the corresponding tube with the individual bead set. Pipet gently and incubate the separate suspensions on a rotator overnight at 4°C. One bead set consists of GSH-beads without any bound protein and serves as a "scavenger" for GST-proteins that might dissociate during the assay to minimize cross-contamination of protein-bound bead sets (see Note 3).

9. To remove unbound GST–GTPase, collect beads by centrifugation (see step 2). Wash individual GTPase-coupled beads twice with 100 μL of ice-cold NP-HPSE buffer supplemented with 0.1% BSA and 1 mM DTT and keep in separate tubes on ice.

10. To minimize the dissociation of protein from the beads, split equal volumes of protein–bead suspension into four tubes. This volume is enough for ten 384-well plates.

11. Resuspend the contents of the first sets of tubes in 600 μL NP-HPSE/BSA/DTT buffer. Keep the protein–bead mixture in the remaining tubes on ice with a minimal amount of buffer (~10 μL) before use.

12. Load 60 μL aliquots of each GTPase-bead sets in a new tube, dilute up to 2,100 μL with NP-HPSE/DTT/BSA buffer, and load 130 μL of this suspension in the first two columns of a 96-well plate for distribution to an assay plate by the Biomek FX[P]. Use eight-tip pod to deliver 5 μL of coated bead suspension from 96-well plate in wells of column 1–22 of a 384-well plate. Wells 23 and 24 in the last two columns have no beads, which results in a large temporal gap in data acquisition at the end of each row that allows row separation during data analysis (Fig. 1b).

13. The Beckman Coulter Biomek FX[P] robot is programmed to deliver 0.1 μL of compound (wells #2–22 of each row) or control in DMSO (wells #1, 2) to 384-well plates using the pin tool device (see Note 4).

14. Next, the robot adds 5 μL of 200 nM BODIPY-GTP to each well (using the 384-tip pod), and mixes them.

15. Cover sample-loaded 384-well plate with foil cover and rotate slowly to maintain suspension at 4°C for 45–60 min. We use a rotating mixer that rotates the plate at 10 rpm. The plate is hold on with rubber bands.

16. As a negative control, prepare a plate with the same GTPase-bead mixture with fluorescent GTP and 0.5 mM unlabeled GTP as a competitor. Measure separately.

3.1.2. Flow Cytometry Measurement

Sample delivery from a multiwell plate to a flow cytometer occurs in a continuous stream with an autosampler, which is connected with flexible tubing and a peristaltic pump to a flow cytometer. The autosampler sips ~2 µL suspension from each of the wells of a multiwell plate, leaving an air gap between samples. Data acquisition is obtained in a single time-resolved data file (one file per 384-well plate).

If the laboratory is equipped with all the necessary equipment (robots and Flow cytometry) a single person could screen ~30–35 384-well plates per day. Screening of one plate takes ~15 min.

1. Use following conditions from the outset to measure fluorescence of the standard calibration beads with the CyAn flow cytometer: use ~550–650 V voltage setting of the photomultiplier tube for the fluorescein channel (488-nm excitation, 530-nm emission). Adjust the voltage/gain of the photomultiplier tube for the red channel (635-nm excitation, 665-nm emission) so that the bead sets are well separated (Fig. 2a, b) (see Note 5).

2. Fill sample tubing with 100 nM BODIPY-FL-GTP solution in NP-HPSE/DTT/BSA buffer, and allow it to coat the interior tubing wall for several seconds before running the assay plate (see Note 6).

3. Move the plate from the rotator (at 4°C) and place the plate-cooler on the HyperCyt™ autosampler deck for high-throughput sampling and flow cytometric measurements.

4. Start the plate run. The autosampler will move from well to well, under the control of HyperCytSampler software, sampling for ~1.1 s from each well and pausing 0.4 s in the air before sampling the next well. The series of 384 bubble-separated samples are delivered to the flow cytometer and measured as a single data file (see Note 7).

3.1.3. Data Analysis

Data analyses use HyperView™ software to merge the raw instrument FCS files and work lists associated with the compound library. Results from analyses of the compound activity in each well, corresponding to single compounds, are then processed through a Microsoft Excel spreadsheet template file segregating target specific data from each well and automatically calculating assay quality statistics (Z and Z' analyses) (19, 20). Data is then normalized with

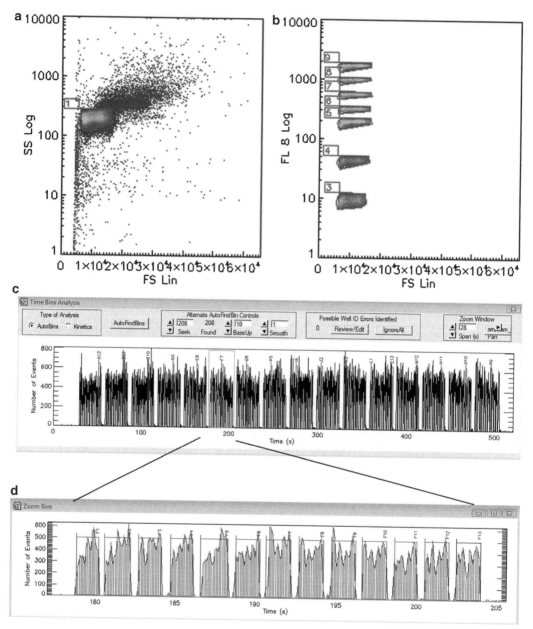

Fig. 2. Data analysis with HyperView™ software. (a) In a plot of forward scatter versus side scatter, a gate is drawn around the singlet beads. (b) Gates are drawn around each bead set using their red fluorescence. (c) After gating of individual bead sets a time versus number of events histogram is created to identify time bins. A plot of events (beads per 0.1 s) versus time for a plate is displayed. There are no beads in columns 23 and 24, leaving 352 bins with beads. (d) An enlarged view of row F. Software allows merging of the FCS files and work lists associated with the compound library. Results from analyses of the compound activity in each well, corresponding to single compounds, are then processed through a Microsoft Excel spreadsheet.

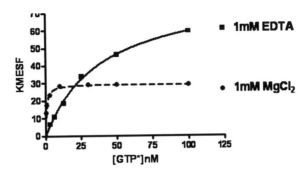

Fig. 3. Equilibrium binding assay. Cdc42 wt coupled on GSH-beads was depleted of nucleotide, and equilibrium binding of various concentration of fluorescent GTP in the presence of 1 mM EDTA or 1 mM MgCl$_2$ was measured with a flow cytometer.

respect to control wells and corrected for systematic error trends across plates (see Note 8).

Data gating is performed as follows:

1. Open a data file with HyperView™ software.

2. Choose forward scatter (FCS linear) on the *X*-axis versus side scatter (SSC log) on the *Y*-axis. The beads on the screen will appear in a cluster based on particle size. As Fig. 2a shows, there may be a cluster of aggregated beads (increased FSC) in addition to singlet beads. Draw a gate around the singlet beads.

3. Select to display only the beads inside the gate, and in a two parameter plot of side scatter versus log red fluorescence intensity (FL8). The beads will be in seven clusters. Create a gate around each of the bead sets as shown in Fig. 2b.

4. These gates for the first plate become a template for subsequent plates measured that day under the same conditions. This is the benefit of using MESF beads to ensure the appropriate voltages for the entire library analysis.

5. Create a time versus number of events histogram to identify time bins. Time bins will be created in the HyperView™ program. Columns 23 and 24 of a 384-well plate have no beads; therefore, there will be 352 bins of data to analyze (Fig. 2c, d).

6. These data are correlated with the compound list used for that plate and are exported to a Microsoft Excel spreadsheet. A decrease in the median fluorescence intensity (MFI) demonstrates that the compound in a particular well inhibits the binding of the GTPase to the fluorescent ligand.

3.2. Dose–Response Measurements

As mentioned above (see Subheading 1.3) Mg^{2+} ions dramatically increase the affinity of GTP to GTPases. Equilibrium binding assay of single-plex Cdc42 binding to BODIPY-FL-GTP revealed that the $K_d \sim 36$ nM (in the presence of EDTA) drops up to 0.6 nM in the presence of Mg^{2+} (Fig. 3).

1. Bind the wild-type GST-Cdc42 (4 μM) to GSH-beads overnight at 4°C. Any type of GSH-beads suitable for flow cytometry can be used (see Subheading 3.1.1, steps 1 and 2).

2. Deplete Cdc42 on GSH-beads of nucleotide by incubating them with NP-HPS buffer containing 10 mM EDTA for 20 min at 30°C.

3. Wash twice with NP-HPS buffer, then resuspend in the same buffer containing 1 mM EDTA/or 1 mM MgCl$_2$, 1 mM DTT, and 0.1% BSA.

4. Block Cdc42 binding to unbound sites by incubating the protein–bead complex for 15 min at room temperature.

5. Collect beads by centrifugation and resuspend them in 30 μL of the same buffer and add 30 μL of appropriate dilutions (3–200 nM) of ice-cold BODIPY-FL-GTP.

6. Incubate samples at 4°C for 45 min and measure the binding of fluorescent nucleotide to the enzyme using a flow cytometer (see Note 9).

7. Export raw data and plot using GraphPad Prism software.

3.2.2. Dose–Response
Assay

Figure 4a, panel A demonstrates a previously published example of a GTPase pan-inhibitor. The mean channel fluorescence of all individual GTPases is reduced. Panel B demonstrates an excerpt from the Excel file of the primary screen, in which one compound specifically inhibits the activity of only one GTPase (arrow) (11). In the next stage of evaluation, the inhibitory effect of active compounds obtained from the primary screening should be confirmed with dose–response analyses in the presence of (a) 1 mM EDTA and (b) 1 mM MgCl$_2$. For assays with EDTA use primary screening conditions. As detected, with equilibrium binding K_d of Cdc42 to BODIPY-FL-GTP in the presence of Mg^{2+} ions changed, for this reason, in dose–response assays concentration of fluorescent GTP was decreased up to 1 nM. Six Rho family GST–GTPases were assayed simultaneously in a single multiplex (Rac1 wt, Rac1Q61L, RhoA wt, RhoAL63, Cdc42 wt and Cdc42Q61L) and 1 nM BODIPY-FL-GTP binding was measured in the presence or absence of the serial drug dilution series. Both of these measurements confirmed that compound MLS000693334 (Fig. 4b) specifically inhibits GTP binding of Cdc42 (Fig. 5a, b).

The inhibitory effect of the compound could be the result of dissociation of GST-tagged GTPase from the beads. To eliminate the possibility that the compound affects GST–protein binding to beads, additional single-plex analysis with GST–GFP was examined in the presence of EDTA and Mg^{2+} ions (Fig. 5c) (see Note 10).

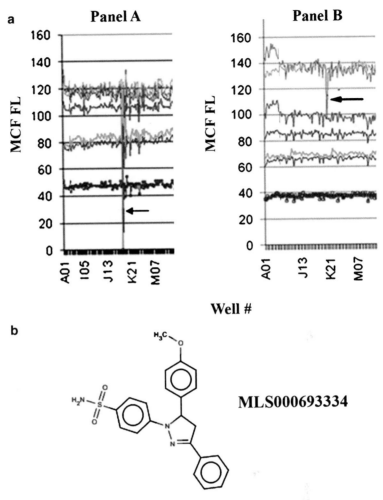

Fig. 4. High-throughput screen identifies small-molecule inhibitors of small GTPases. (a) The raw data were parsed in HyperView to produce annotated fluorescence summary data for each well. The parsed data were then processed through an Excel template file constructed specifically for the assay to segregate data for each target and the fluorescence scavenger in the multiplex. The Excel file shows one compound (*arrow*) that inhibits all GTPases (Panel A). Panel B Shows a specific inhibitor of one GTPase. (b) Structure of Cdc42 specific inhibitor MLS000693334.

For dose–response assays compounds are diluted serially 1:3, a total of eight times from a starting concentration of 10 mM giving a nine-point dilution series in DMSO (Fig. 1c).

1. Load 10 µL of 100% DMSO in columns 1, 2, and 12, 13 of a 384-well plate. The wells in these columns do not have compounds in them, and serve as positive controls.

2. Dispense 9 µL of 10 mM compound in DMSO in column 3 and 6 µL of DMSO in columns 4–11.

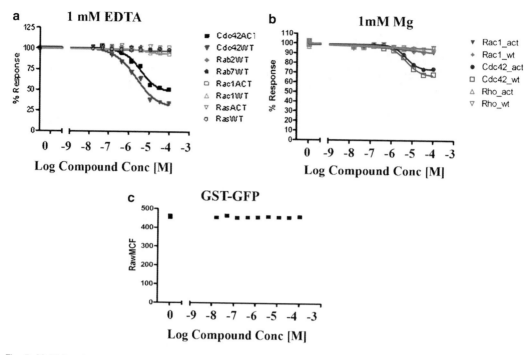

Fig. 5. Multiplex dose–response assay. (**a**, **b**) Confirmation of active compounds identified in HTS multiplexed primary screening with dose–response analysis of multiplex in the presence of EDTA (**a**) and Mg^{2+} (**b**). Graphs show percent of activity versus compound concentration. (**c**) In order to evaluate the effect of compounds on the binding of GST fusion proteins to beads, bead sets were coated with GST–GFP and evaluated with the compounds of interest (active wells from HTS) in a dose–response assay. The *graph* represents the raw mean channel fluorescence (MCF) value versus concentration of test compound for a nine level dose response. Analysis of the data demonstrates no interference between the compounds of interest and GST–GFP binding (see Note 11).

3. Transfer 3 μL of 10 mM compound (from column 3) in the adjacent well and mix with pipetting.

4. Continue sequentially for eight wells. The concentration of the final well will be 15 nM.

5. The Beckman Coulter Biomek FX^P robot is programmed to deliver 5 μL of bead suspension into the wells of columns 1–13 of the 384-well plate.

6. Add 0.1 μL of compound (wells in columns 3–11) or control DMSO (wells in columns 1, 2, 12, and 13) using the pin tool device.

7. Add 5 μL of 200 nM BODIPY-FL-GTP (or 2 nM in assays with Mg^{2+}) to each well, and mix them.

8. Cover the sample-loaded 384-well plate with a foil cover and rotate slowly to maintain suspension at 4°C for 45–60 min.

9. Measure fluorescence as described above (see Subheading 3.1.2) (Fig. 6).

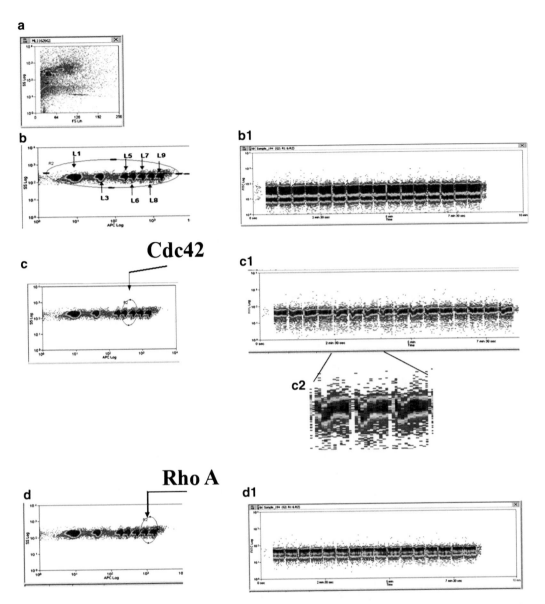

Fig. 6. Multiplex dose–response analysis with CyAn. After incubation with fluorescent-GTP for ~45 min, the mixed bead set is analyzed as FCS versus SSC (**a**), a gate is drawn around the singlet beads discriminated by SSC versus FL8 (uniformly sized bead populations discriminated by fluorescence intensity) (**b**). Each bead set is associated with a unique optical address. Electronic gates allow for separate analysis of each bead population. b1 shows green fluorescence (on the Y-axis) versus time when all seven beads sets are gated (**b**). Gating Cdc42 containing beads (**c**) shows that test compounds affect Cdc42 binding to GTP in a dose-dependent manner (c1). c2 shows an enlarged data fragment of c1; Gating RhoA wt and RhoA constitutively active mutant on coated beads (**d**) indicates no inhibition with test compounds, confirming that these compounds are specific to Cdc42 (d1).

4. Notes

1. It is common to keep the beads suspended during coating with protein. We usually use overnight rotation at 4°C. Longer incubation times result in increased protein binding to beads. Occasionally, some fusion proteins lose about 50% of their binding activity while rotating. We let the proteins settle overnight, since all of them exhibited good binding in the process.

2. Blocking of beads with 0.1% BSA containing buffer can be done after coupling GST–GTPases with beads.

3. Most GST–GTPase fusion proteins will bind to the beads well, but some of these fusion proteins may have unstable binding to the glutathione beads. This binding stability depends on chosen conditions. For example, we found that GST-Rac1 dissociates from the beads in the presence of NP-HPSE, but not in the presence of Mg^{2+} ions. To minimize cross-contamination of beads with dissociated proteins we retain the "scavenger beads" (GSH-beads without any protein added) in the system. Do not use the particular GST-tagged protein that dissociates from the beads in multiplex sets during dose–response assays. Use a single-plex assay for such proteins. To minimize adsorption of beads to pipette tips and HyperCyt tube walls we use of 0.01% detergent in the assay buffer.

4. Using the robot (Biomek FX^P) helps greatly in the preparation of 40 or more 384-well plates for large libraries. But if it is planned to investigate the effect of only a few compounds for GTPase ligand interaction, plates can easily be prepared manually.

5. Red laser alignment will help to solve problems with poor resolution of the separate bead sets.

6. Reduced fluorescent intensity at the start of plate screening could be due to the adsorption of fluorophore to the HyperCyt sample tubing walls. Running more fluorophore through the tubing to equilibrate fluorophore adsorption (just prior to sampling the plate) will solve this problem.

7. Sometimes the bead number in wells may be low. Usually increasing the sampling time from 0.9 to 1.3 s doubles the number of beads sampled per well. But there may be other reasons, for example, HyperCyt tubing or probe clogging, and replacing these parts will solve the problem.

8. The quality control Z' statistics were very good for each target. Beads are more homogeneous than cells, and therefore Z' scores are better for bead-based assays than for cells. In particular, the average Z' was 0.87 ± 0.04 for Cdc42, 0.85 ± 0.04

for Rac1 wt, and 0.90 ± 0.03 for Rac1Q61L. The complete results from the multiplex screen are available on PubChem (http://pubchem.ncbi.nlm.nih.gov/, AID#: 757, 758, 759, 760, 761, 764, 1333–1337, 1339–1341).

9. For CyAn flow cytometer use settings of Subheading 3.1.2. New generation Accuri cytometers have pre-optimized detector settings.

10. An observed inhibitory effect of a compound could be the result of inhibition of GSH-bead GST–protein interaction. In this case, the compounds would give the identical dose–response curve for all checked enzymes. To eliminate these false-positive compounds, an additional assay should be done. Incubate 2 nM GST–GFP overnight with the beads, wash unbound GST–GFP, and measure fluorescence with the addition of the suspect compounds. In our dose–response assays, we have used GST–GFP screening for all compounds tested.

11. Some compounds may be fluorescent by themselves and will show higher fluorescence intensity than the positive control. For this reason, it is recommended to check all your interesting compounds in an auto-fluorescence assay. In an auto-fluorescence assay, use the same bead–protein sets, add compound, but instead of fluorescent GTP, add assay buffer (NP-HPSE, or NP-HPSM), and then sample the plate. Some highly fluorescent compounds can carryover into the next wells. In this case, run the row or entire plate backwards (to designate the first highly fluorescent well as the last well sampled for that plate).

Acknowledgments

This work was supported by NIH grants U54 MH074425 and U54 MH084690.

References

1. Chimini, G., and Chavrier, P. (2000) Function of Rho family proteins in actin dynamics during phagocytosis and engulfment. *Nature Cell Biol* 2, E191–196.

2. Etienne-Manneville, S., and Hall, A. (2002) Rho GTPases in cell biology. *Nature* 420, 629–635.

3. Raftopoulou, M., and Hall, A. (2004) Cell migration: Rho GTPases lead the way. *Dev Biol* 265, 23–32.

4. Boettner, B., and Van Aelst, L. (2002) The role of Rho GTPases in disease development. *Gene* 286, 155–174.

5. Gomez del Pulgar, T., Benitah, S.A., Valeron, P.F., Espina, C., and Lacal, J.C. (2005) Rho GTPase expression in tumourigenesis: evidence for a significant link. *Bioessays* 27, 602–613.

6. Sahai, E., and Marshall, C.J. (2002) RHO-GTPases and cancer. *Nat Rev Cancer* 2, 133–142.

7. Vega, F.M., and Ridley, A.J. (2008) Rho GTPases in cancer cell biology. *FEBS Lett* 582, 2093–2101.

8. Pelish, H.E., Peterson, J.R., Salvarezza, S.B., Rodriguez-Boulan, E., Chen, J.L., Stamnes, M., Macia, E., Feng, Y., Shair, M.D., and Kirchhausen, T. (2006) Secramine inhibits Cdc42-dependent functions in cells and Cdc42 activation in vitro. *Nature Chem Biol* 2, 39–46.

9. Gao, Y., Dickerson, J.B., Guo, F., Zheng, J., and Zheng, Y. (2004) Rational design and characterization of a Rac GTPase-specific small molecule inhibitor. *Proc Natl Acad Sci USA* 101, 7618–7623.

10. Shutes, A., Onesto, C., Picard, V., Leblond, B., Schweighoffer, F., and Der, C.J. (2007) Specificity and mechanism of action of EHT 1864, a novel small molecule inhibitor of Rac family small GTPases. *J Biol Chem* 282, 35666–35678.

11. Surviladze, Z., Waller, A., Wu, Y., Romero, E., Edwards, B.S., Wandinger-Ness, A., and Sklar, L.A. (2010) Identification of a small GTPase inhibitor using a high-throughput flow cytometry bead-based multiplex assay. *J Biomol Screen* 15, 10–20.

12. McEwen, D.P., Gee, K.R., Kang, H.C., and Neubig, R.R. (2001) Fluorescent BODIPY-GTP analogs: real-time measurement of nucleotide binding to G proteins. *Anal Biochem* 291, 109–117.

13. Jameson, E.E., Roof, R.A., Whorton, M.R., Mosberg, H.I., Sunahara, R.K., Neubig, R.R., and Kennedy, R.T. (2005) Real-time detection of basal and stimulated G protein GTPase activity using fluorescent GTP analogues. *J Biol Chem* 280, 7712–7719.

14. Evelyn, C.R., Ferng, T., Rojas, R.J., Larsen, M.J., Sondek, J., and Neubig, R.R. (2009) High-throughput screening for small-molecule inhibitors of LARG-stimulated RhoA nucleotide binding via a novel fluorescence polarization assay. *J Biomol Screen* 14, 161–172.

15. Salas, V.M., Edwards, B.S., and Sklar, L.A. (2008) Advances in multiple analyte profiling. *Adv Clin Chem* 45, 47–74.

16. Edwards, B.S., Young, S.M., Oprea, T.I., Bologa, C.G., Prossnitz, E.R., and Sklar, L.A. (2006) Biomolecular screening of formylpeptide receptor ligands with a sensitive, quantitative, high-throughput flow cytometry platform. *Nature Protocols* 1, 59–66.

17. Saunders, M.J., Edwards, B.S., Zhu, J., Sklar, L.A., and Graves, S.W. (2010) Microsphere-based flow cytometry protease assays for use in protease activity detection and high-throughput screening. *Curr Protoc in Cytom* Unit 13.12.1–17.

18. Schwartz, S.L., Tessema, M., Buranda, T., Pylypenko, O., Rak, A., Simons, P.C., Surviladze, Z., Sklar, L.A., and Wandinger-Ness, A. (2008) Flow cytometry for real-time measurement of guanine nucleotide binding and exchange by Ras-like GTPases. *Anal Biochem* 381, 258–266.

19. Zhang, J. H., Chung, T. D., and Oldenburg, K. R. (1999) A simple statistical parameter for use in evaluation and validation of high throughput screening assays. *J Biomol Screen* 4, 67–73.

20. Malo, N., Hanley, J.A., Cerquozzi, S., Pelletier, J., and Nadon, R. (2006) Statistical practice in high-throughput screening data analysis. *Nature Biotech* 24, 167–175.

Chapter 18

Functional Analysis of Rho GTPase Activation and Inhibition in a Bead-Based Miniaturized Format

Michael Schmohl, Stefanie Rimmele, Peter Gierschik, Thomas O. Joos, and Nicole Schneiderhan-Marra

Abstract

Extensive knowledge about protein–protein interactions is fundamental to fully understand signaling pathways and for the development of new drugs. Rho GTPases are key molecules in cellular signaling processes and their deregulation is implicated in the development of a variety of diseases such as neurofibromatosis type 2 and cancer. Here, we describe a bead-based protein–protein interaction assay for overexpressed HA-tagged Rho GTPases to study the GTPγS-dependent interaction with the regulatory protein RhoGDIα. This assay provides a useful tool for the analysis of both macromolecular and small molecule activators and inhibitors of the protein–protein interactions of Rho GTPases with their regulatory proteins in a multiplexed miniaturized format.

Key words: Suspension bead array, Rho GTPases, Nucleotide analogues, RhoGDIα, Prenyl derivatives, Posttranslational modification, Protein–protein interaction

1. Introduction

Rho GTPases are implicated in many cellular processes including cell proliferation/differentiation, cell motility, apoptosis, and membrane trafficking. Their deregulation severely effects cell homeostasis and is involved in a wide range of diseases, e.g., neurofibromatosis type 2 or different types of cancer like breast cancer, non-Hodgkin's lymphoma, and colon cancer (1, 2). Rho GTPases act as molecular switches that are turned "on" and "off" by a variety of extracellular stimuli. While the GDP-bound form represents the "off" state, GTP binding leads to the activation of the protein. RhoGDIα is only bound in the "off" state and inhibits the dissociation

Francisco Rivero (ed.), *Rho GTPases: Methods and Protocols*, Methods in Molecular Biology, vol. 827,
DOI 10.1007/978-1-61779-442-1_18, © Springer Science+Business Media, LLC 2012

of GDP from the Rho GTPase, which is necessary to become an active protein (3, 4). The binding of Rho GTPases to RhoGDIα is facilitated by the insertion of a prenyl moiety of the GTPase into a hydrophobic cavity of the RhoGDI molecule. This interaction has been shown to be essential for the formation of a tight complex between the two proteins and compounds that inhibit this interaction are thought to have therapeutic effects (5–8).

Inactivation of Rho GTPases occurs after infection with pathogenic *Yersinia* strains such as *Y. enterocolitica*, *Y. pseudotuberculosis*, and *Y. pestis*. These Gram-negative bacteria inject effector proteins into their eukaryotic target cells (9, 10). One of these effectors, YopT, a cysteine protease, cleaves various Rho GTPases at the C-terminal prenylated cysteine, thus in effect removing the isoprenyl moiety from the proteins. This in turn leads to the release of the Rho GTPases from the membrane, and this removal from their primary site of action disrupts the actin cytoskeleton of the target cells and leads to cell rounding (11–16).

The Luminex-bead-based protein microarrays presented here consist of sets of addressable color-coded beads (∅ 5.6 μm), each coated with a defined capture molecule. The amount of captured target proteins is visualized with appropriate reporter systems. Sensitivity, reliability, and accuracy are similar to those observed with standard microtiter plate-based ELISA assays. Bead-based flow cytometric approaches – suspension arrays – enable the simultaneous determination of multiple parameters from a minute amount of sample material. Thus, they are perfectly suited for the analysis of complex cellular signaling networks as well as the detailed analysis of protein–protein interactions within these networks.

Here, we describe a miniaturized, bead-based functional assay for Rho GTPase activation and inhibition. Individual bead types are coated with a rat anti-HA antibody to allow the immobilization of HA-tagged Rho GTPases. The corresponding binding partner GST–RhoGDIα can be visualized using the appropriate tag-specific detection antibody. The Rho GTPase–RhoGDIα complex can be assembled during the assay, but only when GDP is bound to the Rho protein, whereas exchange of bound GDP for GTPγS prevents complex formation. In addition, the methodology allows demonstrating that prenyl derivatives and YopT can interfere with the Rho GTPase–RhoGDIα interaction.

2. Materials

2.1. Proteins and Conjugates

1. Recombinant HA-tagged Rho GTPases are expressed as isoprenylated proteins in Sf9 cells, extracted from the membrane, and purified according to Cerione et al. (17) (see Note 1).

2. GST–RhoGDIα is expressed in *E. coli* strain BL21(DE3). A crude soluble bacterial extract is used.

3. Recombinant GST-tagged YopT is expressed in *E. coli* strain BL21(DE3) and purified on glutathione-Sepharose, followed by enzymatic cleavage of the GST-tag according to standard procedures as described in ref. 18. The plasmid was kindly provided by Prof. Jürgen Heesemann (LMU, Munich).

4. PHYCOLINK goat anti-GST-R-Phycoerythrin (Prozyme).

2.2. Immobilization of Capture Antibody

1. Activation Buffer: 100 mM sodium phosphate (Na_2HPO_4), pH 6.2.

2. Coupling buffer: 50 mM 2-(*N*-morpholino)ethanesulfonic acid (MES), pH 5.0.

3. Wash buffer: sterile PBS (137 mM NaCl, 1.47 mM KH_2PO_4, 2.68 mM KCl, pH 7.4) with 0.05% (v/v) Tween 20.

4. 1-Ethyl-3-[3-dimethylaminopropyl]carbodiimide hydrochloride (EDC): 50 mg/mL in DMSO.

5. *N*-Hydroxysulfosuccinimide sodium salt (sulfo-NHS): 50 mg/mL in DMSO.

6. Capture antibody solution: 100 μg/mL anti-HA antibody (clone 3 F10 without proteinous stabilizers) in coupling buffer.

7. Different bead sets of Luminex xMap™ carboxylated microspheres, i.e., with different fluorescence dyes (Luminex Corp.).

8. 1.5-mL Microcentrifuge tubes (Starlab); for bead handling disposables follow the suggestions given on http://www.luminexcorp.com/support/recommendedmaterials/index.html.

9. Ultrafree-MC filter units, with a 0.65-μm pore size (Millipore).

10. 96-well assay plate, non-binding surface (Corning).

11. ELISA blocking reagent (Roche).

2.3. Protein–Protein Interaction Assay

1. 96-Well filter plate MultiScreen (Millipore).

2. 96-Well assay plate, non-binding surface (Corning).

3. LiquiChip™ System fluid 10× concentrate (e.g., Qiagen).

4. Detergents: CHAPS (Carl Roth GmbH & Co.), sodium cholate (Sigma-Aldrich).

5. Nucleotides: GDP (Sigma-Aldrich), GTPγS (Jena Bioscience GmbH), ATPγS (Jena Bioscience GmbH), GppNHp (Jena Bioscience GmbH).

6. 50 mM $MgCl_2$ in ddH_2O.

7. Wash Buffer (WB): 20 mM Tris–HCl, pH 8.0, 100 mM NaCl, 3.75 mM MgCl$_2$, 1 mM EDTA, 1% (w/v) bovine serum albumin (BSA).

8. Complex buffer (CB): 20 mM Tris–HCl, pH 8.0, 100 mM NaCl, 3.75 mM MgCl$_2$, 1 mM EDTA, 1% (w/v) BSA, 30 μM GDP, 0.1 mM phenylmethanesulfonyl fluoride (PMSF), 1 μg/mL leupeptin, 50 μg/mL aprotinin.

9. Solubilization buffer (SB): 20 mM Tris–HCl, pH 8.0, 100 mM NaCl; 3.75 mM MgCl$_2$, 1 mM EDTA, 3 μM GDP, 0.1 mM PMSF, 1 μg/mL leupeptin, 50 μg/mL aprotinin, 1% (w/v) sodium cholate.

10. Exchange buffer (EB): 20 mM Tris–HCl pH 8.0, 100 mM NaCl, 1 mM MgCl$_2$, 2 mM EDTA, 1% (w/v) BSA, 30 μM GDP, 50 μg/mL aprotinin, 1 μg/mL leupeptin, 0.1 mM PMSF.

11. Prenyl derivatives: farnesylthiosalicylic acid (FTS), geranylthiosalicylic acid (GTS), and geranylgeranylthiosalicylic acid (GGTS) were kindly provided by Prof. Yoel Kloog.

2.4. Instruments

1. Luminex 100/200 or FLEXMAP 3D or MAGPIX instrument with corresponding software (Luminex Corp.).

2. Plate shaker, e.g., Thermomixer comfort (Eppendorf) with a plate module.

3. Vacuum station (Millipore).

4. Sonication bath (Bandelin Electronic).

3. Methods

3.1. Coupling of Capture Antibodies to Microspheres

Prior to starting the functional analysis of the proteins Luminex, xMap™ carboxylated microspheres are coated with the appropriate tag-specific capture antibodies (modified from ref. 19).

1. Activation of carboxyl groups on beads: Vortex the bead stock provided by the vendor thoroughly for at least 10 s. Take 3.75×10^6 beads per coupling reaction. Transfer 300 μL of beads from the bead stock (1.25×10^7 beads/mL) to an Ultrafree-MC filter unit. Centrifuge the filter unit briefly at $3,000 \times g$ and discard the flow-through. Wash beads by adding 300 μL of activation buffer, vortex, and centrifuge again as described above. Discard the flow-through and repeat the wash step.

2. Prepare the EDC and sulfo-NHS solutions (50 mg/mL) during the last wash step and add 15 μL of EDC solution and 15 μL of sulfo-NHS solution to 120 μL of activation buffer

(activation mix). Incubate the beads with 150 μL of activation mix for 20 min at room temperature on a shaker (800 rpm). Protect the beads from light during incubation.

3. Wash beads in an Ultrafree-MC filter with 400 μL of coupling buffer, vortex, centrifuge briefly, and discard the flow-through. Repeat the wash step twice.

4. Couple antibodies to activated beads by adding 250 μL of the prepared anti-HA antibody solution to the filter unit (200 μg/mL, see Notes 2 and 3). Mix the beads thoroughly and incubate for 150 min at room temperature in the dark on a shaker (800 rpm).

5. Wash beads with 400 μL of washing buffer, vortex, centrifuge, and discard the flow-through. Repeat the wash step four times in total.

6. Add 100 μL of ELISA blocking reagent containing 0.05% (w/v) sodium azide to the filter unit (see Note 4) and vortex. Resuspend the beads by pipetting up and down, and transfer the bead suspension to a 1.5-mL microcentrifuge tube. Repeat this step and pool the bead suspensions in one microcentrifuge tube. A bead suspension volume of 200 μL remains. For long-term storage, keep the beads at 4°C in the dark.

7. Determine recovery of antibody-coated beads: Dilute the antibody-coated bead stock 1:500 in a 1.5-mL microcentrifuge tube. Vortex for at least 10 s. Take 100 μL of the dilution and transfer to a microtiter plate (two replicates). Place the microtiter plate for 30 min at 650 rpm on a shaker at room temperature. In order to determine the bead concentration in the stock solution, with the Luminex 100 system, the settings are as follows: sample size 50 μL, time-out 80 s, total beads 10,000. Calculate the bead concentration as follows: beads per μL = number of beads divided by 30 and multiplied by a dilution factor of 500. The recovery is usually approximately 90% (see Note 5).

3.2. Competitive Nucleotide Exchange: Complex Formation

A singleplex setup can be used to analyze the formation of the Rho GTPase–RhoGDIα heterodimeric complex, including the effects when the GDP bound intrinsically to the GTPase is replaced by nonhydrolyzable nucleotide analogues such as GTPγS, GppNHp, or ATPγS. The assay setup is schematically depicted in Fig. 1a, b. All procedures can be carried out at room temperature unless otherwise stated. Technical triplicates are recommended.

1. Block a 96-well filter plate with 100 μL/well of WB.

2. Incubate a suitable number of Luminex xMap™ capture antibody-coated microspheres (see Subheading 3.1) with 60 μL (50 μg/mL) of a distinct HA-tagged Rho GTPase in CB containing 30 μM GDP for 1 h on a 96-well plate with shaking

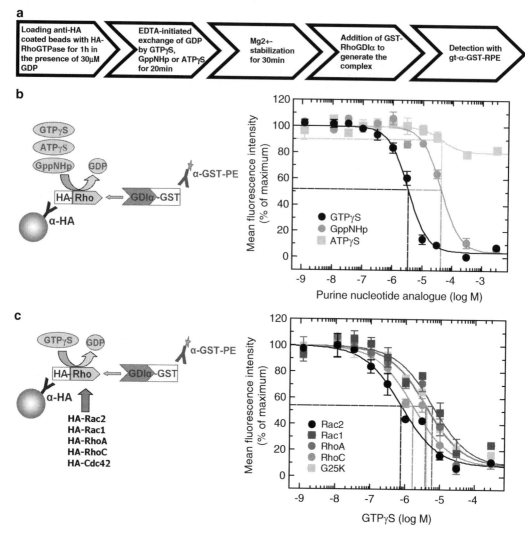

Fig. 1. Effects of nucleotide analogues on complex formation. (a) Flow chart of the assay procedure. (b) In a singleplex approach, HA-Rac2 was captured onto rat anti-HA-coated beads and competitive GDP exchange was done at increasing concentrations of GTPγS, GppNHp, and ATPγS. GST–RhoGDIα was added and the formed heterodimeric complex was detected with goat-anti-GST-RPE. The figure shows the results of a single representative experiment done in triplicates. (c) For multiplexing, HA-Rac2, HA-Rac1, HA-RhoA, HA-RhoC, and HA-G25K were captured onto rat-anti-HA-coated beads of different bead sets and the guanine nucleotide exchange with GTPγS was carried out as described in (a). The graph shows the mean dose–response curves averaged from three experiments, each done in triplicates. Shown is the percentage of the control response, which was determined in the presence of 30 μM GDP, but absence of competing other nucleotides. Reproduced from ref. 5 with permission from Wiley-VCH Verlag GmbH & Co. KGaA.

at 650 rpm. Predilution of the prenylated Rho GTPase in detergent containing SB is recommended to maintain the proteins in solution.

3. Inject the HA-Rho GTPase-loaded beads into the previously blocked filter plate containing WB (step 1).

4. Remove liquid with a vacuum station and wash the beads that remain on the filter membrane three times with WB containing 0.1% (w/v) sodium cholate to eliminate unspecifically bound protein (see Note 6).

5. Add 35 µL of EB containing 3,000, 300, 30, 10, 3, 1, 0.3, 0.1, 0.03, 0.01, and 0.001 µM GTPγS, GppNHp, or ATPγS and 30 µM GDP to each well and incubate for 20 min at 30°C with shaking.

6. Add 4.8 µL of 50 mM MgCl$_2$ to each well to a final concentration of 6 mM to stabilize nucleotide binding and incubate for 30 min at 15°C with shaking (see Note 7).

7. Dilute GST–RhoGDIα in CB and add 5 µL to each well of the filter plate to yield a final concentration of 4 µg/mL.

8. Following incubation for 1 h at 650 rpm, perform two wash steps with 100 µL/well of WB containing 0.5% (w/v) CHAPS.

9. Add 30 µL/well of a 2.5-µg/mL goat-anti-GST-RPE solution in WB to each well and incubate for 2 h with shaking at 650 rpm.

10. After two wash steps with 100 µL/well of WB, resuspend beads in 100 µL/well of WB. Read out in Luminex 100 (settings: sample size 80 µL, time out 60 s, 100 events, bead gate 7,000–15,000).

3.3. Effect of Prenylmimetic Compounds

Compounds that specifically interfere with the Rho GTPase–RhoGDIα interaction are likely to affect cell functions regulated by Rho GTPases (6, 20, 21). This assay provides a useful screening tool for the detailed analysis of prenylcysteine analogues in a multiplexed setup. The assay setup and representative results are shown in Fig. 2. All steps can be carried out at room temperature. Technical triplicates are recommended.

1. Dissolve the prenyl derivatives in DMSO to 20× stock solutions and dilute to 80, 40, 20, 10, 5, 2.5, 1.25, and 0.65 µM with DMSO (see Note 7).

2. Prepare a 4 µg/mL GST–RhoGDIα solution in CB and add 38 µL/well to a 96-well assay plate.

3. Add 2 µL of the appropriately prediluted compound solutions to each well of the assay plate already containing GST–RhoGDIα and incubate for 90 min with shaking at 650 rpm. Final concentrations of prenyl derivatives should be 4,000, 2,000, 1,000, 500, 250, 125, 62.5, and 31.25 µM with a DMSO concentration of 5% (v/v).

4. Transfer 35 µL of the preincubated GST–RhoGDI solution onto a filter plate already containing 2,000 HA-Rho GTPase

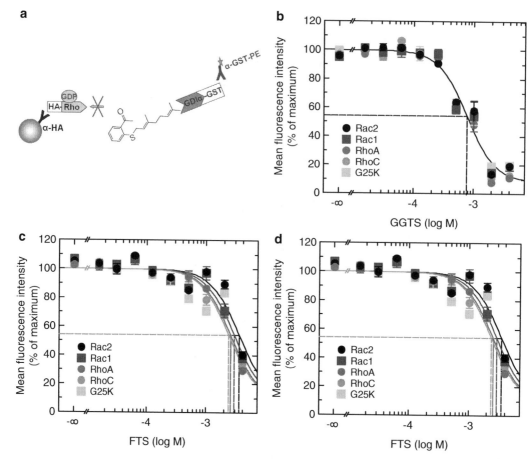

Fig. 2. Multiplexed screening of prenyl derivatives in a 5-plex assay. Different rat-anti-HA-coated bead sets were loaded with HA-Rac2, HA-Rac1, HA-RhoA, HA-RhoC, and HA-G25K, while FTS, GTS, and GGTS were diluted and incubated with GST–RhoGDIα. The bead mixture was then transferred to the assay plate and the preincubated RhoGDIα was added. For detection, goat anti-GST-RPE was added to each well. The assay setup is schematically shown in (a). The graphs show the mean dose–response curves for GGTS (b), GTS (c) and FTS (d) averaged from two experiments, each done in duplicates. The percentage of the control response is shown, which was determined in the absence of prenyl derivatives. Reproduced from ref. 5 with permission from Wiley-VCH Verlag GmbH & Co. KGaA.

loaded beads per well (see Subheading 3.2), in the presence of 30 μM GDP.

5. Incubate for 1 h with shaking at 650 rpm.

6. Proceed as described in Subheading 3.2, steps 8–10.

3.4. Optional Multiplex Setup

Due to the large variety of RhoGTPases – many of which may interact with RhoGDIs – appropriate assays are required to study these interactions under various conditions in parallel. The possibility to combine different beads sets with capture antibody-coated beads which are loaded with different Rho GTPases allows the simultaneous analysis of different Rho GTPase – regulatory protein interactions in a single experiment.

1. Dilute HA-tagged Rho GTPases to 50 μg/mL in SB.

2. Incubate rat anti-HA-coated beads of different bead sets (for coating the beads with capture antibody see Subheading 3.1) with individual HA-tagged Rho GTPases in the presence of 30 μM GDP in separate wells of a 96-well assay plate for 1 h.

3. Wash twice with WB containing 0.1% (w/v) sodium cholate, resuspend in 100 μL of WB, and pool the Rho GTPase-loaded beads.

4. Adjust the volume of the bead mix by centrifugation at $4,000 \times g$ for 5 min and subsequent removal of excess buffer. Use the bead mix in the analysis as described above (see Subheading 3.2, step 3–10, and Subheading 3.3, steps 4–6).

5. Prior to the readout, the different bead sets used in the experiment have to be specified in the Luminex Software.

As shown in Fig. 1c, the formation of the complexes between the five Rho GTPases and GST–RhoGDIα was inhibited in a concentration-dependent manner by replacing bound GDP with GTPγS, with small differences in the IC_{50} values observed. In a comparable manner, the multiplexed setup was used to analyze the effect of FTS, GTS, and GGTS on the formation of the heterodimeric complex (Fig. 2b–d). GGTS compromised the complex formation of all tested HA-Rho GTPases and GST–RhoGDIα with similar IC_{50} values, whereas FTS and GTS were only effective at high concentrations.

3.5. YopT Cleavage Assay

C-terminal posttranslationally modified Rho GTPases have been shown to be differentially sensitive to proteolytic cleavage upstream of the isoprenylated, terminal cysteine residue by the cysteine protease YopT (11). Therefore, Rho GTPases are treated with recombinant YopT and used to perform the Rho GTPase–RhoGDIα interaction assay (see Subheading 3.2). All steps are performed at room temperature unless otherwise stated.

1. Incubate 60 μL (25 μg/mL) of the prenylated Rho GTPases with 5 μL of purified YopT enzyme on a 96-well half-area plate in CB for 30 min at 37°C with shaking at 650 rpm.

2. Add 2,000 beads per well of rat anti-HA-coated beads to a blocked filter plate, as described in Subheading 3.2, steps 1–3.

3. Add the preincubated Rho protein to the bead mixture and incubate for 1 h with shaking at 650 rpm.

4. Remove liquid with a vacuum station and wash the beads, which remain on the filter membrane three times with WB (see Note 7).

5. Proceed as described in Subheading 3.2, steps 5–11.

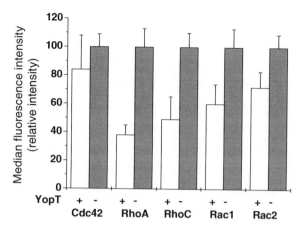

Fig. 3. YopT cleavage assay. The experiment was performed as described in Fig. 1c, except that the Rho GTPases were incubated with YopT prior to RhoGDIα binding. The signal decrease of YopT-treated compared to nontreated GTPases was about 60% for RhoA, 50% for RhoC, 40% for Rac1, 30% for Rac2, and only about 15% for Cdc42. The graph shows the means ± SEM of four independent experiments done in duplicates. The percentage of the control response determined in the presence of 30 μM GDP and absence of YopT is shown. Reproduced from ref. 18 with permission from Wiley-VCH Verlag GmbH & Co. KGaA.

Representative results are shown in Fig. 3. Treatment of Rho GTPases with YopT led to a loss of median fluorescence intensity of about 60% in the case of RhoA, followed by 50% for RhoC, 40% for Rac1, 30% for Rac2, and only about 15% for Cdc42 (isoform 2) compared to nontreated GTPases. The loss of the prenyl moiety of the RhoGTPases leads to a decrease in relative binding affinity and is in accordance to previously reported results (11).

4. Notes

1. Expression of Rho GTPases in Sf9 cells rather than bacteria is crucial to ensure the posttranslational modification of the C-terminus and thus interaction with regulator and effector proteins.

2. It is important that the antibody solutions used for the immobilization and biotinylation procedures are free of primary-amine-containing compounds, e.g. glycine or Tris–HCl. Primary amines would interfere with the EDC/sulfo-NHS coupling procedure.

3. If novel immunoassays are to be developed, different antibody concentrations for the coupling procedure need to be tested in order to define the best assay conditions.

4. Handle with care, sodium azide is toxic.

5. Coupling efficiency can be controlled by the detection of the immobilized IgG using phycoerythrin-conjugated antispecies-specific antibodies, e.g., R-Phycoerythrin-conjugated donkey anti-rat IgG (Jackson ImmunoResearch Laboratories).

6. Blot excess wash buffer from the bottom of the filter plate with a paper towel to avoid leakage of the wells due to capillary forces.

7. Chelation of Mg^{2+} with EDTA can be used to destabilize guanine nucleotide binding to Rho GTPases, e.g., ref. 22. This allows the replacement of bound GDP by other nucleotides. By adding Mg^{2+} ions in molar excess of EDTA, the nucleotide exchange can be stopped.

8. Ensure proper solubilization of prenyl derivatives in DMSO.

Acknowledgments

This work was supported by the Deutsche Forschungsgemeinschaft (DFG), grants JO 687/2-1 and GI138/5-1. We are grateful to Prof. Yoel Kloog, Department of Neurobiochemistry, Tel Aviv University, Israel, for providing the prenyl derivatives.

References

1. Ellenbroek, S.I., and Collard, J.G. (2007) Rho GTPases: functions and association with cancer. *Clin Exp Metastasis* 24, 657–672.

2. Okada, T., Lopez-Lago, M., and Giancotti, F.G. (2005) Merlin/NF-2 mediates contact inhibition of growth by suppressing recruitment of Rac to the plasma membrane. *J Cell Biol* 171, 361–371.

3. Van Aelst, L., and D'Souza-Schorey, C. (1997) Rho GTPases and signaling networks. *Genes Dev* 11, 2295–2322.

4. Wennerberg, K., and Der, C.J. (2004) Rho-family GTPases: it's not only Rac and Rho (and I like it). *J Cell Sci* 117, 1301–1312.

5. Schmohl, M., Rimmele, S., Pötz, O., Kloog, Y., Gierschik, P., Joos, T.O., Schneiderhan-Marra, N. (2010) Protein-protein-interactions in a multiplexed, miniaturized format a functional analysis of Rho GTPase activation and inhibition. *Proteomics* 10, 1716–1720.

6. Kloog, Y., and Cox, A.D. (2004) Prenyl-binding domains: potential targets for Ras inhibitors and anti-cancer drugs. *Semin Cancer Biol* 14, 253–261.

7. Hoffman, G.R., Nassar, N., and Cerione, R.A. (2000) Structure of the Rho family GTP-binding protein Cdc42 in complex with the multifunctional regulator RhoGDI. *Cell* 100, 345–356.

8. Rimmele, S., Gierschik, P., Joos, T.O., and Schneiderhan-Marra, N. (2010) Bead-based protein-protein interaction assays for the analysis of Rho GTPase signaling. *J Mol Recognit* 23, 543–550.

9. Cornelis, G.R. (2002) *Yersinia* type III secretion: send in the effectors. *J Cell Biol* 158, 401–408.

10. Cornelis, G.R., (2002) The *Yersinia* Ysc-Yop 'type III' weaponry. *Nat Rev Mol Cell Biol* 3, 742–752.

11. Fueller, F., and Schmidt, G. (2008) The poly-basic region of Rho GTPases defines the cleavage by *Yersinia enterocolitica* outer protein T (YopT). *Protein Sci* 17, 1456–1462.

12. Shao, F. (2008) Biochemical functions of *Yersinia* type III effectors. *Curr Opin Microbiol* 11, 21–29.

13. Shao, F., Vacratsis, P.O., Bao, Z., Bowers, K.E., Fierke, C.A., and Dixon, J.E. (2003) Biochemical characterization of the *Yersinia* YopT protease: cleavage site and recognition elements in Rho GTPases. *Proc Natl Acad Sci USA* 100, 904–909.

14. Shao, F., Merritt, P.M., Bao, Z., Innes, R.W., and Dixon, J.E. (2002) A *Yersinia* effector and a *Pseudomonas* avirulence protein define a family of cysteine proteases functioning in bacterial pathogenesis. *Cell* 109, 575–588.

15. Aepfelbacher, M., Trasak, C., Wilharm, G., Wiedemann, A., Trulzsch, K., Krauss, K., Gierschik, P., and Heesemann, J. (2003) Characterization of YopT effects on Rho GTPases in *Yersinia enterocolitica*-infected cells. *J Biol Chem* 278, 33217–33223.

16. Fueller, F., Bergo, M.O., Young, S.G., Aktories, K., and Schmidt, G. (2006) Endoproteolytic processing of RhoA by Rce1 is required for the cleavage of RhoA by *Yersinia enterocolitica* outer protein T. *Infect Immun* 74, 1712–1717.

17. Cerione, R.A., Leonard, D., and Zheng, Y. (1995) Purification of baculovirus-expressed Cdc42Hs. *Methods Enzymol* 256, 11–15.

18. Rimmele, S., Gierschik, P., Joos, T.O., and Schneiderhan-Marra, N. (2010) Bead-based protein-protein interaction assays for the analysis of Rho GTPase signaling. *J Mol Recognit* 23, 543–550.

19. Poetz, O., Schneiderhan-Marra, N., Henzler, T., Herget, T., and Joos, T. O. (2011) Receptor tyrosine kinase inhibitor profiling using bead-based multiplex sandwich immunoassays. In Kuester, B. (ed) Kinase Inhibitors. Springer, New York.

20. Chuang, T.H., Bohl, B.P., and Bokoch, G.M. (1993) Biologically active lipids are regulators of Rac.GDI complexation. *J Biol Chem* 268, 26206–26211.

21. Mondal, M.S., Wang, Z., Seeds, A.M., and Rando, R.R. (2000) The specific binding of small molecule isoprenoids to rho GDP dissociation inhibitor (rhoGDI). *Biochemistry* 39, 406–412.

22. Caruso, M.E., Jenna, S., Beaulne, S., Lee, E.H., Bergeron, A., Chauve, C., Roby, P., Rual, J.F., Hill, D.E., Vidal, M., Bossé, R., and Chevet, E. (2005) Biochemical clustering of monomeric GTPases of the Ras superfamily. *Mol Cell Proteomics* 4, 936–944.

Chapter 19

Use of Phage Display for the Identification of Molecular Sensors Specific for Activated Rho

Patrick Chinestra, Isabelle Lajoie-Mazenc, Jean-Charles Faye, and Gilles Favre

Abstract

We describe a phage display approach to select active Rho-specific scFv sensors. This in vitro technique allows preserving the antigen conformation stability all along the selection process. We used the GTP locked RhoBQ63L mutant as antigen against the Griffin.1 library composed of a human synthetic $V_H + V_L$ scFv cloned in the pHEN2 phagemid vector. The method described here has permitted to identify an scFv that discriminates between the activated and the inactivated form of the Rho subfamily.

Key words: Phage display, scFv, Molecular sensors, Activated Rho

1. Introduction

Rho proteins are intracellular switches that cycle between an inactive state bound to GDP and an active state bound to GTP. Once activated, Rho proteins bind to effectors that trigger different intracellular signalling pathways controlling a wide range of functions (1), the deregulation of which participates in pathological processes including cancer (2). By contrast to *ras* genes, no mutated constitutively active form of *rho* has been found so far (3). Moreover, several human tumors feature aberrant expression and activation of Rho (4), emphasizing the role of Rho activation in oncogenesis. The determination of activated Rho in cells is challenging and would represent a significant progress in the understanding of their biological role.

Several methods to detect active Rho have been described. A widely used method consists of precipitating active Rho with the

Francisco Rivero (ed.), *Rho GTPases: Methods and Protocols*, Methods in Molecular Biology, vol. 827,
DOI 10.1007/978-1-61779-442-1_19, © Springer Science+Business Media, LLC 2012

Rho-binding domain (RBD) of Rhotekin fused to glutathione-*S*-transferase (GST) from protein lysates (5), whereas the Cdc42/Rac interactive binding (CRIB) domain of PAK1 is favoured for Rac1 and Cdc42 activation assays (6). While this is an effective method to determine endogenous global Rho activities, it does not provide any information about the subcellular distribution of active Rho. One means to address this problem is to visualize active Rho using an RBD-GFP fusion construct (7–9) or using GST-RBD with an anti-GST dye conjugated antibody (10). We propose here an alternative method to characterize the activated Rho level by selecting scFv antibodies through phage display technology. Phage display is appropriate to maintain the antigen in its active state as the selection proceeds entirely in vitro. Indeed, this technique has been successful in generating precise sensors of molecular conformation of GTPases (11, 12).

To select a conformation-specific scFv against the active form of Rho we used the GTP locked Q63L RhoB mutant (13) against the Griffin.1 library (14), generously provided by the Medical Research Council (Cambridge, England). This library, containing at least 10^9 independent clones, is composed of a human synthetic $V_H + V_L$ scFv cloned in the pHEN2 phagemid vector. It has been designed to introduce a 6xHistidine tag and a *c*-myc tag fused to the C terminus of the antibody fragment in order to purify and to detect the soluble selected scFvs, respectively.

One clone (C1) has been selected that discriminates between the activated versus the inactivated form of Rho, without recognizing Rac or Cdc42. The scFvC1 has been expressed in fusion with a fragment of phage coat protein pIII to improve the antigen recognition allowing *in cellula* immunofluorescence studies (15).

2. Materials

2.1. Rho GTPases Purification

1. *Escherichia coli* strain BL21 containing the GST-Rho construct or GST alone (expression pGST parallel vector is available from the authors' laboratory). Transformed bacteria are stored at −80°C in LB or 2× YT medium containing 20% (v/v) glycerol.

2. LB broth medium: 10 g of tryptone, 5 g of yeast extract, 10 g of NaCl. Add water to a volume of 1 L and sterilize by autoclaving for 15 min at 121°C.

3. 100 mg/mL Ampicillin stock solution in water, store small aliquots at −20°C.

4. LB-A: LB broth medium, 100 µg/mL ampicillin.

5. Isopropyl-β-ᴅ-thiogalactopyranoside (IPTG): 1 M Stock solution in water, store small aliquots at −20°C.

6. Glutathione-Sepharose 4B beads (GE Healthcare).

7. 10× Phosphate buffer saline (PBS): 10.5 mM KH_2PO_4, 30.0 mM Na_2HPO_4, 1.54 M NaCl, pH 7.4.

8. Lysis buffer A: 50 mM Tris–HCl, pH 7.5, 150 mM NaCl, 10 mM $MgCl_2$, 1% (v/v) Triton X-100, 1 mM DTT, 10 μM GDP, proteases inhibitors (Sigma-Aldrich).

9. Ni–NTA Agarose beads (Qiagen).

10. AcTEV™ protease (Invitrogen).

11. SDS-PAGE sample buffer: 0.0625 M Tris–HCl, pH 6.8, 2% sodium dodecylsulfate, 10% glycerol, 20 mM dithiothreitol, 0.04% bromophenol blue.

12. 12% SDS-PAGE precast gels (Bio-Rad).

13. Bovine serum albumin (BSA).

14. Coomassie blue staining solution: SimplyBlue SafeStain (Invitrogen).

2.2. Phage Display

1. *E. coli* TG1 and HB2151 strains were generously provided by the Medical Research Council (Cambridge, England).

2. M13KO7 helper phage (New England Biolabs Inc.).

3. 2× YT broth medium: 16 g of Tryptone-B, 10 g of yeast extract-B, 5 g of NaCl. Add water to a volume of 1 L and sterilize by autoclaving for 15 min at 121°C.

4. 20% Glucose solution, filter sterilized (0.22 μm).

5. M9 salts solution: Weigh 64 g of $NaH_2PO_4 \cdot 7H_2O$, 15 g of KH_2PO_4, 2.5 g of NaCl, and 5 g of NH_4Cl. Add water to a volume of 1 L and sterilize by autoclaving for 15 min at 121°C.

6. M9 minimal agar medium: Add 7.5 g of bacto-agar to 384 mL of deionized water. Sterilize by autoclaving for 15 min at 121°C and cool to 50°C. Add 100 mL of M9 salts solution, 1 mL of 1 M $MgSO_4$ solution, 10 mL of 20% glucose solution, and 5 mL of 20% bacto casamino acids (Difco) solution.

7. Top agar: weigh 7.5 g bacto-agar and add water to a volume of 1 L. Sterilize by autoclaving for 15 min at 121°C.

8. Antibiotics: 100 mg/mL ampicillin at –20°C; 50 mg/mL kanamycin (Sigma-Aldrich).

9. 2× YT-AK: 2× YT broth, 100 μg/mL ampicillin, 25 μg/mL kanamycin.

10. 2× YT-GA: 2× YT broth, 1% glucose, ampicillin 100 μg/mL.

11. 2× YT-GA agar plates: 16 g of Tryptone-B, 10 g of yeast extract-B, 5 g of NaCl, 15 g of Agar-B. Add water to a volume of 1 L and sterilize by autoclaving for 15 min at 121°C. Cool to 55°C and add glucose to a final concentration of 1% and ampicillin to 100 μg/mL. Pour in 9-cm Petri dishes.

12. TE: 10 mM Tris–HCl, pH 7.5, 0.1 mM EDTA.

13. PEG/NaCl: 20% Polyethylene glycol 8000, 2.5 M NaCl.

14. MPBS: PBS containing 3% nonfat dried milk powder.

15. PBST: PBS containing 0.1% Tween 20.

16. Triethylamine (TEA): 700 μL of TEA in 50 mL of water, diluted on the day of use.

17. Large bioassay dish, TC Dish 245×245×25 (Nunc).

18. Oligonucleotides:LMB3(5′-ACAGGAAACAGCTATGACC-3′) and pHENseq (5′-CTATGCGGCCCCATTCAG-3′).

2.3. ELISA

1. PBST: PBS containing 0.1% Tween 20.

2. 96-Well Maxisorp plates (Nunc).

3. 3,3′,5,5′-Tetramethylbenzidine (TMB).

4. Sterile 96-well microwell plates (Nunc).

5. Antibodies: Anti-M13 (HRP) monoclonal antibody (GE Healthcare), anti-*c*-myc antibody (HRP) (Novus Biologicals), anti-GST Antibody (GE Healthcare), anti-RhoA monoclonal Antibody (26C4) (Santa Cruz Biotechnology), anti-RhoB antibody (119) (Santa Cruz Biotechnology), goat anti-rabbit IgG (H + L)-HRP conjugate (Bio-Rad Life Science Research), goat anti-mouse IgG (H + L)-HRP conjugate (Bio-Rad Life Science Research), donkey anti-goat IgG-HRP (Santa Cruz Biotechnology, Inc.).

6. 2YT-A: 2× YT, 100 μg/mL ampicillin.

7. Lysis buffer B: 50 mM Tris–HCl, pH 7.5, 150 mM NaCl, 5 mM $MgCl_2$, 1% (v/v) Triton X-100, 1 mM DTT.

8. Glutathione coated 96-well plate (Pierce, Thermo Fisher Scientific Inc.).

9. Guanosine 5′-diphosphate sodium salt (GDP): 100-mM Stock solution in water, store small aliquots at –80°C.

10. Guanosine 5′-[γ-thio]triphosphate tetralithium salt (GTPγS): 100-mM Stock solution in water, store small aliquots at –80°C.

11. Ethylenediaminetetraacetic acid disodium salt (EDTA): 0.5 M, pH 8 stock solution.

12. Magnesium chloride, 2 M stock solution.

13. 1 M sulphuric acid.

2.4. Production and Purification of Soluble Monoclonal scFv

1. 2× YT-A: 2× YT, 100 μg/mL ampicillin.

2. 1 M Isopropyl-β-ᴅ-thiogalactopyranoside (IPTG) stock solution in water, store small aliquots at –20°C.

3. TSE buffer: 50 mM Tris–HCl, pH 8, 0.5 M sucrose, 1 mM EDTA.

4. 1,3-Diaza-2,4-cyclopentadiene, Glyoxaline (Imidazole).

5. Ni–NTA Agarose beads (Qiagen).

6. W buffer: 50 mM Tris–HCl, pH 8, 250 mM NaCl, 10 mM imidazole.

7. E buffer: 50 mM Tris–HCl, pH 8, 500 mM NaCl, 250 mM imidazole.

8. Tris/NaCl buffer: 50 mM Tris–HCl, pH 8, 150 mM NaCl.

9. Econo-pac column (Bio-Rad).

10. PD10 desalting columns (GE Healthcare).

2.5. Equipment

1. Under-and-over turn table.

2. 50-mL Centrifuge tubes, capable of withstanding $13,000 \times g$ for 30 min (e.g., BD Biosciences).

3. Centrifuge capable of supporting a rotor for 96-well microplates, (e.g., Eppendorf 5810R).

4. Branson sonifier (Branson Ultrasonics Corporation).

5. Microplate reader (Labsystem Multiskan Multisoft; AIE).

6. Gel electrophoresis system: Mini Protean III (Bio-Rad).

3. Methods

3.1. Rho GTPase Purification

For purification of Rho proteins (RhoA, RhoB, or RhoC), the construct used is the human cDNA sequence of wild-type (wt) Rho or activated (Q63L) mutant of Rho cloned into the expression vector pGST parallel (16). In this vector, GST fusion proteins are expressed from a *tac* promoter after induction with IPTG as these vector carry the *lac^{Iq}* gene. The GTPases are synthesized as GST fusion proteins in *E. coli*. They are loaded onto glutathione-coupled beads. Beads can be used directly or the GTPase can be eluted with glutathione. The GST-Tag can be removed by action of the AcTEV™ protease as the vector contains a specific cleavage site for this protease.

3.1.1. Growth of Plasmid-Bearing Bacteria

1. Inoculate 100 mL of LB-A with 20–50 μL of –80°C stock glycerol. For GST production, inoculate 10 mL.

2. Incubate overnight at 37°C, with shaking.

3. Inoculate 1 L of LB-A with the 100 mL overnight culture in a 3-L Erlenmeyer flask. For GST inoculate 100 mL with the 10 mL overnight culture.

4. Grow to $OD_{600} \sim 0.8$ (this usually takes 1 h).

5. Add IPTG to a final concentration of 0.1 mM. Incubate for 4 h at 30°C with shaking.

6. Centrifuge for 15 min at 2,500×*g* and 4°C.

7. Carefully remove the supernatant.

8. Wash the bacterial pellet by suspending it in 50 mL of PBS followed by a centrifugation for 15 min at 2,500×*g* and 4°C. The pellet can be stored at –20°C.

3.1.2. Preparation of Cell Extract and Beads for GST Binding

1. Keep the bacterial pellet on ice, and suspend cells completely in 20 mL of ice-cold lysis buffer A. The suspension should be homogeneous.

2. Transfer the suspension into a 50-mL centrifuge tube.

3. Sonicate on ice five to ten times for 10 s each. Cool for 1 min on ice between sonication bursts to prevent the lysate from overheating (see Note 1).

4. Pellet the bacterial debris by centrifugation for 30 min at 17,000×*g* at 4°C.

5. In the meantime, 500 µL of glutathione-sepharose 4B bead slurry are washed three times in lysis buffer A to remove the 20% ethanol storage buffer. For this add 500 µL of lysis buffer A, suspend the beads by inverting the tube gently, collect the beads by brief centrifugation and remove the supernatant without disturbing the bead pellet. After wash, resuspend beads with 500 µL of lysis buffer A.

3.1.3. Binding of GST-Protein to Beads

1. Carefully transfer the clarified bacterial lysate (corresponding to the supernatant from Subheading 3.1.2, step 4) to a 50-mL centrifuge tube.

2. Add the washed glutathione beads to the lysate.

3. Incubate rotating continuously on an under-and-over turn table for 2 h at 4°C.

4. Centrifuge for 1 min at 500×*g* at 4°C.

5. Carefully discard the supernatant without disturbing the bead pellet and add 5 mL of ice-cold lysis buffer A.

6. Wash the bead pellet three times with 5 mL of ice-cold lysis buffer A by gentle inversion of the tube and centrifugation for 1 min at 500×*g* at 4°C.

7. Suspend the beads in 300 µL of lysis buffer A plus 10% glycerol.

8. Remove a 25-µL aliquot to check the protein preparation by SDS-PAGE.

9. Divide the suspended beads into small aliquots (200 µL) and store at –80°C.

3.1.4. Elution of GST Fusion Protein

1. Add four bead volumes of lysis buffer A containing 10 mM glutathione.

2. Mix gently to resuspend the beads.

3. Incubate rotating continuously on an under-and-over turn table for 15 min at 4°C to elute protein from the beads.

4. Centrifuge for 5 min at $500 \times g$ at 4°C.

5. Carefully decant the supernatant (containing eluted proteins) into a fresh microcentrifuge tube.

6. Repeat elution three times and check the three eluates separately for purified protein by SDS-PAGE and Coomassie blue staining. The yield of fusion protein can be increased by repeating the elution step and pooling the eluates.

7. Adjust the purified protein to 10% glycerol and store aliquots at −20°C.

8. Remove a 25-µL aliquot to check the protein preparation by SDS-PAGE.

3.1.5. Removal of the GST-Tag

1. Incubate GST-beads pellet (from Subheading 3.1.3, step 6) overnight at 4°C with a mix of 75 µL of TEV buffer 20×, 15 µL of 0.1 M DTT, and 100 U of AcTEV™ protease in 1.5-mL final volume (according to the manufacturer's protocol).

2. After cleavage of the fusion protein, remove AcTEV™ protease by affinity chromatography on Ni–NTA resin with 150 µL of Ni–NTA beads (see Note 2).

3. Centrifuge for 5 min at $500 \times g$ at 4°C to clarify the supernatant containing the untagged protein.

4. Add 10% glycerol and store at −80°C.

5. Remove a 25µL aliquot to check the protein preparation by SDS-PAGE.

3.1.6. GST Fusion Protein Quantification

1. Add to the 25-µL aliquot of beads (from Subheading 3.1.3, step 8) or supernatant (from Subheading 3.1.4, step 8 or Subheading 3.1.5, step 5) 10 µL of 2.5× SDS-PAGE sample buffer. Heat at 95°C for 5 min.

2. On a 12% SDS-PAGE gel, load the samples and 1, 5, and 10 µg of denatured BSA.

3. Perform electrophoresis followed by a Coomassie blue staining.

4. Estimate the amount of GST–protein bound to the beads by comparing band intensities to the known amounts of BSA.

3.2. Antibody Fragments Phage Display and Selection

3.2.1. Helper Phage Production

This protocol is detailed according to the Griffin.1 library protocol that is no longer available on line (see Note 3).

1. Inoculate 5 mL of 2× YT medium with a TG1 (see Note 4) colony from a minimal medium plate (see Note 5) and grow overnight at 37°C with shaking.

2. Next day subculture by diluting 1:100 the overnight culture into fresh 2× YT medium and grow shaking until $OD_{600} \sim 0.2$.

Infect 200 µL of bacterial culture with 10 µL of serial dilutions of M13KO7 helper phage (in order to obtain well separated plaques) in a water bath at 37°C without shaking for 30 min.

3. Add 3 mL of molten H-top agar (42°C) and poor onto warm 2× YT agar plates. Allow setting and then incubate overnight at 37°C.

4. Pick a small isolated plaque into 3–4 mL of an exponentially growing culture of TG1 (OD_{600} ~ 0.5) and grow for about 2 h at 37°C with shaking.

5. Inoculate into 500 mL of 2× YT in a 3-L Erlenmeyer flask and grow for 1 h at 37°C with shaking.

6. Add kanamycin to a final concentration of 50 µg/mL and grow shaking at 37°C for a further 8–16 h.

7. Pellet the bacteria by centrifugation in 500-mL bottles for 15 min at $10,800 \times g$ at 4°C.

8. Transfer the phage supernatant into 500-mL bottles and add 1/5 volume of PEG/NaCl solution, mix well, and incubate for a minimum of 30 min on ice.

9. Pellet the phages by centrifugation for 15 min at $10,800 \times g$ at 4°C (see Note 6).

10. Redissolve the phage pellet in 20 mL TE and filter sterilize the stock through a 0.45-µm filter.

11. Titre the helper phage stock by infecting 200 µL of bacterial culture with 10 µL of serial dilutions of M13KO7 helper phage and plating 100 µL on 2× YT agar plates containing 25 µg/mL kanamycin (see Note 7).

12. Store the phage at a dilution of about 1×10^{12} colony-forming units (cfu)/mL at 4°C for short-term storage (several weeks) and at −20°C for long-term storage (see Note 8).

3.2.2. Library Packaging

1. Gently thaw a library glycerol stock on ice (see Note 9).

2. Inoculate 500 mL of 2× YT-GA in a 3-L Erlenmeyer flask with the library glycerol stock.

3. Grow at 37°C with shaking until OD_{600} ~ 0.5.

4. Transfer 25 mL (10^{10} bacteria cells) from this culture to a 50-mL centrifuge tube and add 2×10^{11} cfu of M13KO7 helper phage for infection (see Note 10), mix gently.

5. Incubate without shaking in a 37°C water bath for 30 min.

6. Spin the infected cells at $3,300 \times g$ for 10 min. Discard the supernatant to remove the glucose and the excess of helper phage and suspend the pellet in 30 mL of fresh 2× YT-AK medium (see Note 11).

7. Transfer the suspension to a 3-L Erlenmeyer flask, add 470 mL of 2× YT-AK and incubate at 30°C overnight with shaking.

8. Spin the culture at $10,800 \times g$ for 10 min (or $3,300 \times g$ for 30 min) in 500-mL bottles (see Note 12).

9. Transfer the phage supernatant to 500-mL bottles and add one-fifth volume of PEG/NaCl solution, mix well, and incubate for a minimum of 1 h on ice.

10. Pellet the phages by centrifugation for 30 min at $10,800 \times g$ and 4°C. Suspend the pellet in 40 mL of water, transfer into a 50-mL centrifuge tube, and add 8 mL of PEG/NaCl solution. Mix and leave for at least 20 min at 4°C.

11. Spin at $10,800 \times g$ for 10 min (or $3,300 \times g$ for 30 min) at 4°C and aspirate off the supernatant.

12. Spin briefly and aspirate off the remaining drops of PEG/NaCl solution.

13. Suspend the pellet in 5 mL of PBS and spin for 10 min at $11,600 \times g$ at 4°C in a microcentrifuge to remove most of the remaining bacterial debris.

14. Titrate the phage stock by making two serial dilutions of 1:100 (10 μL into 990 μL PBS) and use 10 μL of the second one to infect 1 mL of TG1 at an $OD_{600} \sim 0.5$.

15. Incubate for 30 min at 37°C in a water bath.

16. Spread 100 μL of three 1:100 serial dilutions on 2× YT-GA agar plates and grow overnight at 37°C.

17. Count the colonies and calculate the cfu titre according to the dilution. The phage stock should be about $10^{12}-10^{13}$ cfu/mL.

3.2.3. Selection by Panning

The first round of selection is the most important. It may be desirable to perform it under low stringency conditions (high concentration of antigen, short washes, and no competitive selection) to avoid loosing rare binders. More stringent conditions will be employed in later rounds. We chose to select scFv's to mutant Q63L of RhoB (locked in GTP binding structure) (13) expressed as a GST-fusion protein captured on glutathione-beads to prevent immobilization-induced conformational changes that could happen by coating on immunotubes (12). During rounds 2–5, phage repertoires are depleted of RhoB unspecific phage by preincubation with GST captured on glutathione-beads prior to selecting on GST-RhoBQ63L. During the fifth round of selection, the repertoire was also depleted of conformational unspecific phage by adding free wt RhoB (mainly in the GDP bound form) to the mixture during the selection. The first round of selection was performed as follows:

1. Block a 5-mL polypropylene tube with 5 mL of MPBS for 1 h at room temperature.

2. Wash three times with PBS.

3. Incubate 10^{13} cfu of scFv's phage library in 5 mL of MPBS with 10 μg of GST-RhoBQ63L captured on glutathione beads

(see Note 13), rotating continuously on an under-and-over turn table for 90 min at room temperature.

4. During this time, block a 1.5-mL microcentrifuge tube with 1.5 mL of MPBS.

5. Pellet the beads by centrifugation for 20 s at $500 \times g$.

6. Discard 4 mL of supernatant and transfer the remaining 1 mL into the blocked microcentrifuge tube.

7. Pellet the beads by centrifugation for 20 s at $500 \times g$ and discard the supernatant.

8. Wash the beads with PBST by gently mixing followed by centrifugation for 20 s at $500 \times g$ for bead separation. Following this procedure, perform consecutive washing steps ten times with PBST then five times with PBS.

9. Elute captured phage by adding 1 mL of TEA and rotating continuously for 10 min on an under-and-over turn table.

10. During the incubation, prepare a tube with 0.5 mL of 1 M Tris, pH 7.4 ready to add the eluted 1 mL phage for quick neutralization.

11. After elution pellet the beads by centrifugation for 20 s at $500 \times g$ and transfer the eluted phage to the tube containing 0.5 mL of 1 M Tris, pH 7.4. Phages can be stored at 4°C or used to infect TG1 as in step 14.

12. Add another 0.2 mL of 1 M Tris, pH 7.4 to the beads to neutralize the remaining captured phage.

13. Take 9.25 mL of an exponentially growing culture of TG1 and add 0.75 mL of the eluted phage. Also add 1 mL of the TG1 culture to the beads (see Note 14). Incubate both cultures for 30 min at 37°C in a water bath without shaking to allow for infection.

14. Pool the 10 mL and 1 mL of the infected bacteria and take 100 μL to make three 1:100 serial dilutions to be spread on 2× YT-GA agar plates for titration and grow overnight at 37°C.

15. Take the remaining infected bacteria and centrifuge for 10 min at $3,300 \times g$. Discard the supernatant, suspend the pellet in 1 mL of 2× YT medium, and plate on a large bio-assay dish of 2× YT-GA agar.

16. Grow at 30°C overnight, or until colonies are visible.

17. Next day add 5–6 mL of 2× YT with 15% glycerol to the bioassay dish and harvest the colonies with a rubber policeman.

18. Inoculate 50–100 μL of the scraped bacteria to 100 mL of 2× YT-GA medium and store the remaining bacteria at –70°C (see Note 15).

19. Grow the bacteria at 37°C with shaking until $OD_{600} \sim 0.5$ (it should take about 2 h).

20. Infect 10 mL of this culture with approximately 2×10^{11} cfu of M13KO7 helper phage (see Note 10) and incubate in a 37°C water bath without shaking for 30 min.

21. Pellet the infected bacteria by centrifugation for 10 min at $3,300 \times g$. Suspend the pellet gently in 50 mL of 2× YT-AK medium and incubate at 30°C overnight with shaking.

22. Next day take 40 mL of the overnight culture and pellet the bacteria for 10 min at $10,800 \times g$ or 30 min at $3,300 \times g$.

23. Precipitate the phage by adding 8 mL of PEG/NaCl solution to the supernatant. Mix and leave for at least 1 h at 4°C.

24. Pellet the phage by centrifugation for 10 min at $10,800 \times g$ (or for 30 min at $3,300 \times g$) at 4°C, then discard the supernatant.

25. Spin briefly and aspirate off the remaining drops of PEG/NaCl solution.

26. Suspend the pellet of phage in 2 mL of PBS, transfer the solution into two microcentrifuge tubes, and centrifuge for 10 min at $11,600 \times g$ in a microcentrifuge to remove most of the remaining bacterial debris.

27. Titrate this phage solution as described in Subheading 3.2.2, steps 14–17 and use it for the next round of selection.

    The following second, third, and fourth round of selection should be performed as follows.

28. Block 2 microcentrifuge tubes with 1.5 mL of MPBS for 1 h at room temperature.

29. Wash three times with PBS.

30. Incubate 10^{12} cfu of phage from the precedent round of selection in 1 mL of MPBS with 20 μg of GST captured on glutathione-beads (to remove unspecific phage), rotating continuously on an under-and-over turn table for 90 min at room temperature.

31. Pellet the beads by centrifugation for 20 s at $500 \times g$.

32. Transfer the unbound phage supernatant in the second blocked microcentrifuge tube and incubate with 10 μg of GST-RhoBQ63L captured on glutathione-beads rotating continuously on an under-and-over turn table for 90 min at room temperature.

33. Continue the procedure as described in steps 8–27.

34. The fifth round of selection should be performed as the fourth one except that the phage incubation with 10 μg glutathione-beads captured GST-RhoBQ63L is made in the presence of 80 μg of free RhoB wt in order to deplete the library of conformationally unspecific phage.

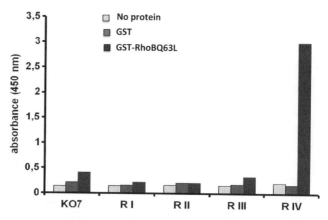

Fig. 1. Polyclonal phage ELISA showing an enrichment of phage population specific for RhoBQ63L: 10^{10} cfu of phage population from the different rounds of selection (RI–RIV) were analyzed for binding to GST (*grey columns*), to GST-RhoBQ63L (*black columns*) or in the absence of antigen (*white columns*) immobilized on glutathione beads. Bound phages were detected with HRP-labelled anti-M13 using TMB as substrate. Results are expressed as absorbance at 450 nm. Helper phage (KO7) was used as a negative control.

3.3. Screening of Active Conformational Antibody Fragments

3.3.1. Polyclonal Phage ELISA

A successful panning results in an enrichment of a phage population specific for the antigen of interest. This can be monitored by screening the populations of phage produced at each round of selection for binding to the selected antigen by ELISA. It can be carried out with glutathione-beads with captured GST-RhoBQ63L instead of glutathione-coated 96-well plates in order to obtain a large amplification-specific signal (Fig. 1) (see Note 16).

1. Block 10 μg of glutathione-beads with captured GST-RhoBQ63L or GST into microcentrifuge tubes with 1.5 mL of MPBS. Include a negative control consisting of the corresponding volume of glutathione beads without protein.

2. Wash the beads three times with PBS.

3. Add 1 mL of MPBS containing 10^{11} cfu of phage from the four rounds of selection (see Subheading 3.2.3) and M13KO7 helper phage as a negative control.

4. Incubate rotating continuously on an under-and-over turntable for 90 min at room temperature.

5. Wash three times with 1 mL of PBST and once with PBS.

6. Add 1 mL of a 1:5,000 dilution of anti-M13 (HRP) antibody in MPBS and incubate rotating continuously on an under-and-over turntable for at least 30 min at room temperature.

7. Wash three times with 1 mL of PBST and once with PBS.

8. Add 200 μL of TMB reagent and incubate for a few minutes at room temperature until a blue colour develops. Do not allow the beads to aggregate at the bottom by gently flicking the tubes.

9. Stop the reaction by adding 50 μL of 1 M sulphuric acid. The colour should turn yellow.

10. Transfer 100 μL to a 96-well microplate and read the OD at 450 nm with a microplate reader.

3.3.2. Identification of Clones with Active Conformational Specificity

Once the enrichment of phage population specific for the antigen of interest has been checked, identification of individual positive clones can be carried out by ELISA. The procedure consists first of identifying positive clones for the active mutant of RhoB by comparing the signal against GST-RhoBQ63L and GST (Fig. 2). The negative clones are eliminated and the positive ones are sequenced in order to further characterize only one specimen in case of redundant clones. These experiments can be achieved either by making phages from single infected bacterial colonies or by making soluble monoclonal antibody fragments (see Note 17). To our knowledge it is better to first screen for individual clones by phage ELISA (see Note 18).

1. Inoculate individual colonies from the titre plates of the last round of selection (see Subheading 3.2.3) into 100 μL of 2× YT-GA medium in a sterile 96-well plate and grow overnight at 37°C with shaking (see Note 19).

2. Transfer a small inoculum from this plate (about 5–10 μL) to a second sterile 96-well plate containing 200 μL of 2× YT-GA medium per well (see Note 20). Grow for 1 h at 37°C with shaking.

3. Add to each well of the second plate 25 μL of 2× YT-GA medium containing 10^9 cfu of M13KO7 helper phage and incubate for 30 min at 37°C without shaking. Then incubate for 1 h at 37°C with shaking.

4. Pellet the bacteria by centrifugation for 10 min at $1,800 \times g$ with a rotor for 96-well plates.

5. Suspend the bacterial pellet in 200 μL of 2× YT-AK medium and grow overnight at 30°C with shaking.

6. Pellet the bacteria by centrifugation for 10 min at $1,800 \times g$ and use 100 μL of the phage supernatant to screen for individual clones by monoclonal phage ELISA (see Subheading 3.3.3).

3.3.3. Monoclonal Phage ELISA

1. Rinse three times each well of a glutathione coated 96-well plate with 200 μL of PBST.

2. Coat wells with 100 μL of PBS with 5 mM $MgCl_2$ containing 1 μg of purified GST-RhoBQ63L, GST, or no recombinant proteins. Plan coating several wells to check for the coating of antigens with commercially available antibodies.

3. Incubate for 1 h at room temperature.

4. Wash three times with PBST.

5. Apply 100 μL of phage supernatant from Subheading 3.3.2 , or of MPBS containing 1010 cfu from Subheading 3.3.4. or commercially available antibodies at the appropriate dilutions. Incubate for 90 min at room temperature.

6. Discard the solutions containing unbound ligands and wash three times with 200 μL of PBST.

7. Apply 100 μL of MPBS containing HRP-secondary antibodies (i.e., HRP-anti-M13, HRP-anti-goat antibody, or HRP-anti-rabbit antibody) at the appropriate dilutions. Incubate at least 30 min at room temperature.

8. Discard the solutions containing unbound secondary antibodies and wash three times with 200 μL of PBST.

9. Apply 100 μL of TMB reagent and incubate for a few min at room temperature until a blue colour develops.

10. Stop the reaction by adding 50 μL of 1 M sulphuric acid. The colour should turn yellow.

11. Read the OD at 450 nm with a microplate reader (see Fig. 2).

3.3.4. Production of Monoclonal Phage in Large Quantities

The selected scFv conformational specificity is then assessed by comparing first their interaction with the wild type form of RhoA, RhoB, and RhoC (mainly in the GDP form) and their corresponding active form (Q63L) and second their specificity for the wild-type Rho GTPases loaded with GTPγS in order to avoid any mutation-dependent recognition. It is then worth checking the absence of binding to the active form of other Rho family members such as Rac and Cdc42, for example. This can be done with a large preparation of monoclonal phage as described below:

1. Inoculate 15 mL of 2× YT-GA medium with positive clones from the stock 96-well plate (see Subheading 3.3.2).

2. Grow at 37°C with shaking until $OD_{600} \sim 0.5$.

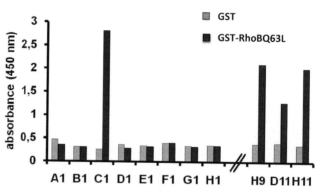

Fig. 2. Identification of positive clones by monoclonal phage ELISA: 48 monoclonal phages from *E. coli* supernatant (only 11 are shown) were analyzed for binding to GST (*grey columns*) and GST-RhoBQ63L (*black columns*) protein immobilized on an ELISA plate. Bound phages were detected with horseradish peroxidase-labelled anti-M13 using TMB as substrate. Results are expressed as absorbance at 450 nm. Four positive clones were identified (C1, H9, D11, and H11) in this set.

3. Infect 10 mL of this culture with around 2×10^{11} cfu of M13KO7 helper phage and incubate in a 37°C water bath for 30 min without shaking. Continue as described in Subheading 3.2.3, steps 21–27.

4. Titrate this phage solution and use it to further characterize the conformational specificity of the selected clones.

3.3.5. Nucleotide Loading of Rho GTPases from Bacterial Crude Extracts

GTPγS loading of purified wild-type GST-RhoB is not easily reproducible, as this protein appears to be very instable. We carried out ELISA by loading Rho GTPases with GDP or GTPγS directly on crude bacterial extract and by coating nucleotide loaded GST-RhoB onto glutathione-coated 96-well plates.

1. Inoculate BL21 *E. coli* harbouring the pGST-RhoA, RhoB, or RhoC plasmids into 10 mL of 2× YT-GA medium and incubate overnight at 37°C with shaking.

2. Dilute 3 mL of this overnight culture into 300 mL of 2× YT-GA medium and grow at 37°C with shaking until $OD_{600} \sim 0.8$.

3. Harvest the bacteria by spinning at $3{,}300 \times g$ for 10 min at RT and suspend the pellet in 300 mL of 2× YT-A containing 100 μM IPTG.

4. Incubate overnight at 20°C with shaking.

5. Harvest the bacteria by spinning at $3{,}300 \times g$ for 10 min at 4°C in six 50-mL centrifuge tubes and discard the supernatant. Freeze the dry bacterial pellets at –20°C (see Note 21).

6. Thaw one pellet and resuspend it into 5 mL of lysis buffer B by vortexing.

7. Sonicate 30 times 5 s on ice (see Note 1).

8. Pellet the bacterial debris by centrifugation at $12{,}500 \times g$ for 15 min at 4°C.

9. Add GDP or GTPγS to a final concentration of 1 mM and 0.1 mM, respectively, to 2.5 mL of supernatant. Adjust to 10 mM with EDTA, mix well, and incubate 1 h at 37°C.

10. Stop the loading reaction by adding $MgCl_2$ to a final concentration of 40 mM.

11. Add 100 μL of nucleotide loaded crude extract per well of the glutathione-coated 96-well plate and incubate 1 h at room temperature.

12. Continue the phage ELISA as described in Subheading 3.3.3, step 4.

3.4. Production and Purification of Soluble Monoclonal scFv from Selected Clones

Before proceeding to the production and purification of soluble monoclonal scFvs in large amounts, we recommend checking their binding capability as soluble fragments. This can be achieved by overexpressing the fusion in TG1 or XL1 blue in the absence of helper phage by induction of the lac promoter with IPTG and by testing the scFv containing supernatant by ELISA (see Note 17). If this is not the case, the inactive antibody fragments can be functionally rescued by fusion to the N-terminal domain of the original phage display fusion partner, filamentous phage p3 protein (15).

In all the cases, soluble monoclonal scFvs can be purified in large amounts by immobilized metal affinity chromatography (17) as the pHEN2 phagemid introduces a poly-histidine tag in C-terminal fusion with the antibody fragment (see http://vbase. mrc-cpe.cam.ac.uk/). The conditions of induction (i.e., IPTG concentration, temperature, and duration) described below should be optimized for each individual antibody fragment depending on its solubility and toxicity.

1. Inoculate *E. coli* carrying the plasmid of interest into 10 mL of 2× YT-GA medium and incubate overnight at 37°C with shaking.

2. Inoculate 6 mL of this culture in a 3-L Erlenmeyer flask containing 600 mL of 2× YT-GA medium and grow at 37°C with shaking until $OD_{600} \sim 0.8$ (it takes approximately 3 h).

3. Harvest the bacteria by spinning at $3,300 \times g$ for 10 min at room temperature and suspend the pellet in 600 mL of 2× YT-A containing 100 μM IPTG.

4. Incubate overnight (no more than 16 h) at 30°C with shaking.

5. Harvest the bacteria by spinning at $6,200 \times g$ for 10 min at 4°C and discard the supernatant.

6. Resuspend the bacterial pellet in 40 mL of TSE and incubate 20–30 min on ice.

7. Harvest the bacteria by spinning at $6,200 \times g$ for 10 min at 4°C. Collect the supernatant corresponding to the periplasmic enriched fraction.

8. Sonicate this supernatant two times 10 s to shear residual bacterial chromosome, which can clog the column.

9. Centrifuge at $12,500 \times g$ for 60 min at 4°C to remove bacterial debris.

10. Adjust the scFv containing supernatant to 250 mM NaCl and 10 mM imidazole and add 1 mL of 50% Ni–NTA agarose slurry equilibrated with Tris/NaCl buffer. Incubate at 4°C for 60 min rotating continuously on an under-and-over turn table.

11. Load the supernatant–Ni–NTA agarose mixture into an Econo-pac column and wash with 20 mL of cold W buffer.

12. Capp the column bottom outlet and elute by adding 2.5 mL of cold buffer E. Mix well and let it stand for 15 min at 4°C.

13. Remove the bottom cap and collect the eluate.

14. In order to remove the excess of imidazole load the eluate into a PD10 desalting column equilibrated with Tris/NaCl buffer and elute with 3.5 mL of Tris/NaCl buffer.

15. For long-term storage, adjust the purified scFv to 10% glycerol and store aliquots at −20°C.

4. Notes

1. The frequency and intensity of sonication should be adjusted such that complete lysis occurs without overheating. The settings for sonication will vary from sonicator to sonicator, and should be adjusted for each laboratory. The lysate becomes viscous as the cells release their DNA. The viscosity should disappear with further sonication as the DNA is sheared.

2. The AcTEV™ protease contains a poly-histidine tag.

3. Other scFv libraries have been described which could be used according to this protocol (18–20). Moreover, biosensors against the active conformation of Ras have been selected by using the intracellular antibody capture (IAC) in scFv and VH formats (21, 22).

4. Some bacterial strains are suitable for helper phage production and phage display as they express the F pilus (i.e. TG1, DH5aF′, XL1 blue). In our hands, XL1 blue are more convenient since they harbour an F′ episome selectable with tetracycline. Moreover, the use of this strain in phage display may decrease recombinant events due to their recombination deficiency (*recA1*).

5. TG1 cells should be grown first on M9 minimal medium plates to select for the F pilus, otherwise their infectability will be reduced.

6. The phage pellet has a white aspect.

7. M13KO7 helper phage encodes a kanamycin-resistance gene that makes the phage titration easier by counting the drug-resistant colonies according to the dilution (corresponding to the infected bacteria) on 2× YT agar plates containing 25 μg/mL kanamycin.

8. We have noticed that phage stored at −20°C becomes less infective. We therefore recommend checking the infectivity of the defrosted phage before using it.

9. An aliquot of the library must represent at least tenfold the library diversity; that is, 10^{10} bacteria cells per ml in the case of the Griffin.1 library. The volume of culture medium in which the library aliquot is diluted should be large enough to obtain a starting $OD_{600} \sim 0.05$.

10. One generally infects bacteria with helper phage at the ratio of 1/20 (i.e., number of bacterial cells/number of helper phage particles) taking into account that 1 OD_{600} is roughly equivalent to 8×10^8 bacteria per mL.

11. Adding ampicillin and kanamycin to the medium allows selecting for bacteria harbouring, respectively the phagemid and the helper phage genome in the case of M13KO7.

12. As suppression of the amber stop codon encoded at the junction of the antibody gene and gIII is never complete, this step is achieved in order not only to concentrate the phage but also to remove any soluble antibody fragments, which can compete with phage during the selection procedure.

13. Alternatively, one can use magnetic glutathione beads instead of glutathione sepharose beads, as they are easier to handle.

14. This step is carried out in order to recover remaining phage still bound to the antigen.

15. It is important to keep a glycerol stock of selected bacteria at each round of selection in order to restart from any step of the selection procedure in case of problems or in case the selection conditions need to be altered.

16. If there is no specific signal against the protein of interest, it is worth checking the antigen coating using an existing monoclonal or polyclonal antibody if available. Also checking for the presence of scFv inserts in the selected phagemids by PCR provides a convenient insight into the success of selection (Fig. 3). It can be carried out without doing phagemid mini-

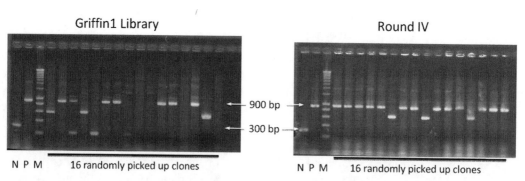

Fig. 3. PCR amplification of scFv from randomly picked colonies: LMB3 and pHENseq primers were used to amplify scFv inserts containing pHEN2 phagemid. The comparison of 16 randomly picked colonies from the Griffin.1 library and round IV highlights the enrichment of scFv during the process of selection. Colonies harbouring the pHEN2 without scFv (N) and with an antithyroglobuline scFv (P) were used as negative and positive controls, respectively. The size of an scFv insert is about 900 bp. The lanes M were loaded with a DNA ladder.

preparations, simply by picking isolated clones on the titre plates and suspending them directly into the PCR mix. Use a pair of oligonucleotides hybridizing on either side of the scFv gene (i.e., LMB3 and pHENseq in the case of the pHEN2 phagemid).

17. The pHEN2 phagemid in which the Griffin.1 library has been cloned carries an amber stop codon (TAG) between the *scFv* gene and the phage protein gene 3. *E. coli* strains used to produce scFv displaying phage such as TG1, XL1 blue, or DH5aF′, harbour the *supE* genotype which allows them to read this codon as glutamic acid thus producing a fusion between the scFv and the p3 which is incorporated onto the phage particles. Nevertheless, it is possible to overexpress this fusion as soluble fragment by induction of the *lac* promoter in the absence of helper phage. Soluble fusion proteins can be monitored by ELISA using a secondary antibody against the *c*-myc tag located at the junction of the scFv and the p3. Alternatively, it is possible to use nonsuppressive strains such as HB2151 that interpret the amber codon as a translation stop signal, thus the scFvs are expressed in soluble form.

18. Nevertheless phage ELISA may present an advantage over the soluble antibody fragment as the anti-p8 antibody can bind to a phage particle at several sites since there are approximately 2,800 copies per phage particles. This can be exploited as a convenient signal amplification system. Antibody fragment ELISA has the disadvantage of eliminating scFvs that have lost their binding capabilities when expressed without the p3 fusion (which happened in the case of the scFvC1 (15) and in the cases described in others studies (23)).

19. To avoid contaminations that can happen between two adjacent wells, we recommend employing 2-mL deepwell sterile microplates containing 500 μL of 2× YT-GA medium.

20. Make stocks of the original 96-well microplates, by adjusting each well to a final concentration of 15% glycerol and storing them at −80°C.

21. Dry bacterial pellets can be stored at −20°C until use for about 1 month.

Acknowledgments

We are grateful to Fiona Sait for generously providing the "Griffin.1 Library. We also want to acknowledge Franck Perez for his helpfull advice.

References

1. Etienne-Manneville, S., and Hall, A. (2002) Rho GTPases in cell biology. *Nature* 420, 629–635.

2. Vega, F.M., and Ridley, A.J. (2008) Rho GTPases in cancer cell biology. *FEBS Lett* 582, 2093–2101.

3. Aznar, S., Fernandez-Valeron, P., Espina, C., and Lacal, J.C. (2004) Rho GTPases: potential candidates for anticancer therapy. *Cancer Lett* 206, 181–191.

4. Gomez del Pulgar, T., Benitah, S.A., Valeròn, P.F., Espina, C., and Lacal, J.C. (2005) Rho GTPase expression in tumourigenesis: evidence for a significant link. *Bioessays* 27, 602–613.

5. Ren, X.D., Kiosses ,W.B., and Schwartz, M.A. (1999) Regulation of the small GTP binding protein Rho by cell adhesion and the cytoskeleton. *EMBO J* 18, 578–585.

6. Sander, E.E., van Delft, S., ten Klooster, J.P., Reid, T., van der Kammen, R.A., Michiels, F., and Collard, J.G. (1998) Matrix-dependent Tiam1/Rac signaling in epithelial cells promotes either cell-cell adhesion or cell migration and is regulated by phosphatidylinositol 3-kinase. *J Cell Biol* 143, 1385–1398.

7. Goulimari, P., Kitzing, T.M., Knieling, H., Brandt, D.T., Offermanns, S., and Grosse, R. (2005) $G_{\alpha 12/13}$ is essential for directed cell migration and localized Rho-Dia1 function. *J Biol Chem* 280, 42242–42251.

8. Bement, W.M., Benink, H.A., and von Dassow, G. (2005) A microtubule-dependent zone of active RhoA during cleavage plane specification. *J Cell Biol* 170, 91–101.

9. Berger, C.D., Marz, M., Kitzing, T. M., Grosse, R., Steinbeisser, H. (2009) Detection of activated Rho in fixed *Xenopus* tissue. *Dev Dyn* 238, 1407–1411.

10. Cascone, I., Audero, E., Giraudo, E., Napione, L., Maniero, F., Philips, M.R., Collard, J.G., Serini, G., and Bussolino, F. (2003) Tie-2-dependent activation of RhoA and Rac1 participates in endothelial cell motility triggered by angiopoietin-1. *Blood* 102, 2482–2490.

11. Nizak, C., Monier, S., del Nery, E., Moutel, S., Goud, B., and Perez, F. (2003) Recombinant antibodies to the small GTPase Rab6 as conformation sensors. *Science* 300, 984–987.

12. Horn, I.R., Wittinghofer, A., de Bruïne, A.P., Hoogenboom, H.R. (1999) Selection of phage-displayed fab antibodies on the active conformation of ras yields a high affinity conformation-specific antibody preventing the binding of c-Raf kinase to Ras. *FEBS Lett* 463, 115–120.

13. Longenecker, K., Read, P., Lin, S.K., Somlyo, A.P., Nakamoto, R.K., Derewenda, Z.S. (2003) Structure of a constitutively activated RhoA mutant (Q63L) at 1.55 Å resolution. *Acta Crystallogr D Biol Crystallogr* 59, 876–880.

14. Griffiths, A.D., Williams, S.C., Hartley, O., Tomlinson, I.M., Waterhouse, P., Crosby, W.L., Kontermann, R.E., Jones, P.T., Low, N.M., Allison, T.J., Prospero, T.D., Hoogenboom, H.R., Nissim, A., Cox, J.P.L., Harrison, J.L., Zaccolo, M., Gherardi, E., and Winter, G. (1994) Isolation of high affinity human antibodies directly from large synthetic repertoires. *EMBO J* 13, 3245–3260.

15. Goffinet, M., Chinestra, P., Lajoie-Mazenc, I., Medale-Giamarchi, C., Favre, G., and Faye, J.C. (2008) Identification of a GTP-bound Rho specific scFv molecular sensor by phage display selection. *BMC Biotechnol* 8, 34–47.

16. Sheffield, P., Garrard, S., and Derewenda, Z. (1999) Overcoming expression and purification problems of RhoGDI using a family of "parallel" expression vectors. *Protein Expr Purif* 15, 34–39.

17. Dubel, S., Breitling, F., Kontermann, R., Schmidt, T., Skerra, A., and Little, M. (1995) Bifunctional and multimeric complexes of streptavidin fused to single chain antibodies (scFv). *J Immunol Methods* 178, 201–209.

18. Knappik, A., Ge, L., Honegger, A., Pack, P., Fischer, M., Wellnhofer, G., Hoess, A., Wolle, J., Pluckthun, A., and Virnekas, B. (2000) Fully synthetic human combinatorial antibody libraries (HuCAL) based on modular consensus frameworks and CDRs randomized with trinucleotides. *J Mol Biol* 296, 57–86.

19. Silacci, M., Brack, S., Schirru, G., Marlind, J., Ettorre, A., Merlo, A., Viti, F., and Neri, D. (2005) Design, construction, and characterization of a large synthetic human antibody phage display library. *Proteomics* 5, 2340–2350.

20. Philibert, P., Stoessel, A., Wang, W., Sibler, A.P., Bec, N., Larroque, C., Saven, J.G., Courtete, J., Weiss, E., and Martineau, P. (2007) A focused antibody library for selecting scFvs expressed at high levels in the cytoplasm. *BMC Biotechnol* 7, 81–97.

21. Tanaka, T., and Rabbitts, T.H. (2003) Intrabodies based on intracellular capture frameworks that bind the RAS protein with

high affinity and impair oncogenic transformation. *EMBO J* 22, 1025–1035.

22. Tanaka, T., Williams, R.L., and Rabbitts, T.H. (2007) Tumour prevention by a single antibody domain targeting the interaction of signal transduction proteins with RAS. *EMBO J* 26, 3250–3259.

23. Jensen, K.B., Larsen, M., Pedersen, J.S., Christensen, P.A., Alvarez-Vallina, L., Goletz, S., Clark, B.F., and Kristensen, P. (2002) Functional improvement of antibody fragments using a novel phage coat protein III fusion system. *Biochem Biophys Res Commun* 298, 566–573.

Chapter 20

Identification of New Interacting Partners for Atypical Rho GTPases: A SILAC-Based Approach

Laura Montani, Damaris Bausch-Fluck, Ana Filipa Domingues, Bernd Wollscheid, and João Bettencourt Relvas

Abstract

In contrast to typical Rho GTPases the regulation of atypical Rho GTPases, such as the members of the RhoBTB subfamily, rarely depends on GEFs and/or GAPs. Instead, they are regulated at the level of their expression, by post-translational modifications, by their rate of degradation as well as through binding of diverse cell-specific interactors. Stable Isotope Labeling by Amino acids in Cell culture (SILAC) is a powerful cutting-edge mass-spectrometry-based technology allowing for protein-interaction studies in vitro with removal of false-positive identifications. In this chapter, we describe how the SILAC technology can be applied to the identification of new interacting partners for atypical – constitutively active – Rho GTPases, i.e. RhoBTB3.

Key words: Atypical Rho GTPases, Protein overexpression, Proteomics, SILAC, Interactome, Immunoprecipitation, Silver staining, Nu-PAGE gradient gel, Mass spectrometry

1. Introduction

Based on sequence homology, the Rho-family of small GTPases can be divided into eight subfamilies (1). Typical small Rho GTPases subfamilies (Rho, Rac, and Cdc42) are binary molecular switches that cycle between an active GTP-bound and an inactive GDP-bound state. The GDP/GTP cycling is subject to tight control by three different classes of regulatory proteins: GTPase activating proteins (GAPs), guanine nucleotide exchange factors (GEFs), and guanine nucleotide dissociation inhibitors (GDIs). Typical Rho GTPase signaling has been the focus of several studies (2–6). On the contrary, the regulation and the signaling cascades modulated by atypical small Rho GTPases subfamilies (Miro, Rnd, RhoBTB, RhoD/Rif, RhoH, and Wrch-1/Chp) are still not well

Francisco Rivero (ed.), *Rho GTPases: Methods and Protocols*, Methods in Molecular Biology, vol. 827,
DOI 10.1007/978-1-61779-442-1_20, © Springer Science+Business Media, LLC 2012

understood (1). Atypical Rho GTPases lack the ability to hydrolyze GTP, with the exception of Miro and Wrch subfamilies whose cluster position in the Rho family is currently under discussion (1, 7). Due to their inability to cycle between GDP and GTP bound states, atypical Rho GTPases are thought to be regulated at the level of their expression (positive regulation) (1, 6), by targeted proteasomal degradation (negative regulation) (8–10), and by post-translational modifications, including palmitoylation (11), prenylation (12), farnesylation (13, 14), and more rarely phosphorylation (10, 14, 15), all of which can determine specific intracellular localization and functionality of these proteins.

Atypical Rho GTPase functions can be modulated in diverse cell populations by different binding interactors, involving domains not found in typical Rho proteins (1). Up to now, the use of classical biochemical approaches, i.e., yeast-2-hybrid assay and co-immunoprecipitation, resulted in the identification of a small number of interacting partners (see Table 1), whose role in atypical Rho GTPases signaling remains mostly unexplored.

It is believed that protein–protein interaction mapping, coupled with experimental validation in orthogonal systems, can provide high-confidence data sets inherently useful for interrogating signaling mechanisms and elucidating individual protein function (16). Stable Isotope Labeling by Amino acids in cell Culture (SILAC) is a practical in vitro labeling strategy for mass spectrometry-based quantitative proteomics (17–19), which has been adapted to identify interacting proteins that co-immunopurify with tagged proteins of interest (20–22) (see Fig. 1). As a result, in the last few years SILAC has become a powerful tool for interactome mapping (21), successfully applied to uncover new insights into different signaling pathways, e.g. integrin-linked kinase (23) and protein-phosphatase-1 (24). SILAC technology can be used to label cellular proteomes through normal metabolic processes, incorporating non-radioactive, stable isotope-containing amino acids into newly synthesized proteins. For interactome studies, cells are first transfected with an epitope tagging vector in which the inserted gene of interest is expressed as a His, Myc, or FLAG tagged fusion protein, or with a control empty tag vector. Transfected cells are then cultured in the presence of "light" (C12) or "heavy" (C13) isotopic amino acids added at the same concentration (see Fig. 1). Incorporation of the amino acids occurs through cell growth, protein synthesis, and turnover. SILAC amino acids have no direct effect on cell morphology or growth rates, and cells cultured for 5–6 doublings ensure close to 100% isotope incorporation (23). Cell protein lysates are then co-immunoprecipitated with an anti-tag antibody and interacting proteins resolved on a SDS-PAGE gradient gel, into protein bands of different molecular weights. Proteins bands are in-gel digested with trypsin to obtain peptides that can be analyzed by tandem mass-spectrometry. Due

Table 1
Previously described atypical Rho GTPases interactors

Atypical Rho GTPase	Interacting protein (reference)	Methodology
Miro	Milton (Gri11, OIP106) (27)	co-IP
	PINK1 (28)	co-IP, MS/MS
RhoBTB1/2/3	Cul3 (12, 29)	Y2H, co-IP
RhoBTB3	Rab9, TIP47 (30)	Y2H, co-IP
RhoD	hDia2C (31)	Y2H, co-IP
	Plexin-A1 (32)	co-IP
RhoH	LCK (33)	co-IP, KA
	Syc (34)	co-IP, KA
	Zap70 (35)	co-IP
RhoU/Wrch-1	FAK (36)	co-IP
	Mfn1/Mfn2 (37)	co-IP
	PAR6(38)	co-IP
	PYK2 (39)	Y2H, co-IP
Rnd1	FRS2β (40)	Y2H, co-IP NMR, X-rayC
	Grb7 (41)	Y2H, co-IP
	Plexin-A1 (42)	co-IP
	Plexin-B1 (43, 44)	Y2H, co-IP
	SCG10 (45)	Y2H, co-IP
	Unc5B (46)	CSBS
Rnd1/2/3	p190RhoGAP (47)	Y2H, co-IP
Rnd2	MgcRacGAP (48)	co-IP
	Plexin-D1 (49)	co-IP
	Pragmin (50)	Y2H
	Rapostlin (51)	Y2H
	Vps4-A (52)	Y2H, co-IP
Rnd3	PKCα (14)	KA
	ROCK1 (15, 53)	Y2H, co-IP
	Socius (13)	Y2H
	Syx (54)	MS/MS

Published identified atypical Rho GTPases interacting proteins and experimental methods used for their detection: *Y2H* yeast-2-hybrid assay, *co-IP* co-immunoprecipitation, *KA* in vitro kinase assay, *CSBS* cell surface binding screen, *MS/MS* tandem mass spectrometry-based assay, *NMR* nuclear magnetic resonance spectroscopy, *X-rayC* X-ray crystallography

to their different mass to charge ration, peptides from light and heavy cell populations remain distinguishable by mass spectrometry, even when mixed, and protein abundances are determined from the relative peptide signal intensities. The incorporation of a negative control permits the relative quantitative comparison of

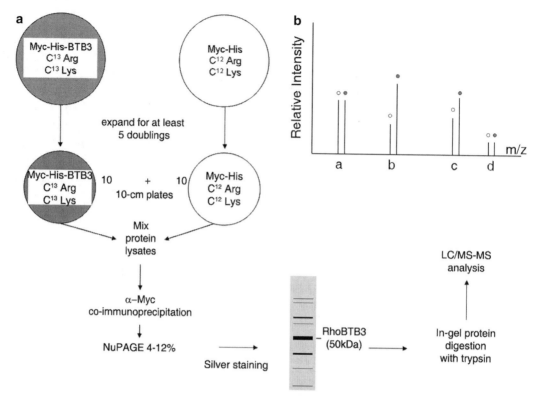

Fig. 1. SILAC interactome study experimental design with example of LC-MS/MS analysis of unspecific and specific identified interacting proteins. (**a**) Schematic experimental design. See text for details. (**b**) Exemplified results for unspecific Myc-His-bound (**a, d**) and specific RhoBTB3-Myc-His-bound (**b, c**) interactors.

two samples and therefore the rapid identification of real hits above a background of contaminants during mass-spectrometry analysis (see Fig. 1). MSQuant and STATQuant, open source software (http://www.msquant.sourceforge.net, https://www.gforge.nbic.nl/projects/statquant/), can then be used to extract SILAC ratios and for follow-up data analysis (25, 26).

2. Materials

All solutions are prepared in filtered deionized water unless otherwise stated.

2.1. Transfection

1. Cell line: adherent Human Embryonic Kidney (HEK) 293 cells (see Note 1).

2. Medium: Dulbecco's Modified Eagle's Medium (DMEM) with 2 mM L-glutamine.

3. Supplements: heat-inactivated fetal bovine serum (FBS), sterile penicillin/streptomycin (P/S; 10,000 U/mL penicillin, 10,000 μg/mL streptomycin stock solution).

4. Additional materials: 6-cm diameter sterile plastic cell culture dishes, sterile water, sterile poly-D-lysine (1% w/v stock solution).

5. Vectors: pcDNA 3.1 (+) Myc-His empty vector (Invitrogen), pcDNA 3.1 (+) RhoBTB3-Myc-His vector (see Notes 2–4).

6. Transfection reagents: PolyJet in vitro DNA Transfection Reagent (Signagen Laboratories).

7. Selection antibiotic: sterile gentamicin (50 mg/mL stock solution).

2.2. SILAC

1. Medium: DMEM-low glucose, without arginine, leucine, lysine (Sigma-Aldrich, D0443), later called SILAC-DMEM (see Notes 5 and 6).

2. Sterile stable isotope-labeled "light" and "heavy" amino acid solutions (see Note 6). Light: L-lysine monohydrochloride (Sigma-Aldrich L8662) 100 mg/mL stock solution in 500 μL aliquots; L-leucine sol: H_2O (Sigma-Aldrich L8912) 50 mg/mL stock solution in 1-mL aliquots; L-arginine sol: H_2O (Sigma-Aldrich A8094) 100 mg/mL stock solution in 500 μL aliquots. Heavy: 13C6 L-arginine sol: HCl (Cambridge Isotope laboratories, CLM-2265) 100 mg/mL stock solution in 500 μL aliquots, 13C6 L-lysine sol: 2HCl (Cambridge Isotope laboratories, CLM-2247) 100 mg/mL stock solution in 500 μL aliquots. Store amino acid solutions at −80°C.

3. Supplements: sterile dialyzed serum (Invitrogen) (see Notes 5 and 7), sterile P/S (see Subheading 2.1, item 3).

4. Additional materials: 10-cm diameter sterile plastic cell culture dishes, sterile enzyme-free cell dissociation buffer PBS-based (Invitrogen) (see Note 8).

2.3. Cell Lysis, Co-immunoprecipitation and Gel Electrophoresis

1. Lysis and co-immunoprecipitation reagents: anti-Myc antibody, Dynabeads Co-Immunoprecipitation Kit (Invitrogen), Dyna Magnet (Invitrogen).

2. Additional: cell scrapers, 1 M NaCl, 1 M $MgCl_2$, Tween-20, EDTA-free proteinase inhibitor (see Note 8).

3. Gel-electrophoresis: Nu-PAGE Novex 4–12% Bis–Tris gradient gels, Nu-PAGE loading buffer, Nu-PAGE reducing solution, MOPS running buffer, gel electrophoresis apparatus.

4. Vacuum concentrator (e.g. SpeedVac, Thermo Savant).

2.4. Silver Staining (see Note 9)

1. Solution A: 40% ethanol, 10% acetic acid.

2. Solution B: 30% ethanol.

3. Solution C: 0.02% $Na_2S_2O_3$.

4. Solution D: 0.1% $AgNO_3$.

5. Solution E: 3% Na_2CO_3, 0.05% formalin.

6. Solution F: 1% glycine.

2.5. Mass Spectrometry Sample Preparation

1. Solution G: 5 mM tris(2-carboxyethyl)phosphine (TCEP), 50 mM NH_4HCO_3.

2. Solution H: 12 mM iodoacetamide, 50 mM NH_4HCO_3.

3. Solution I: 0.1% formic acid, 50% methanol.

4. Solution J: 0.1% formic acid.

5. Solution K: 2.5% formic acid.

6. Additional materials: sequence-grade modified trypsin, acetonitrile, methanol, ZipTipC18 pipette tips with a bed of chromatography media (Millipore).

3. Methods

All cell culture procedures are carried out under sterile conditions.

3.1. Transfection

1. Prepare "normal" medium for cell growth and transfection by adding 10% heat-inactivated FBS and 1% P/S to DMEM with 2 mM L-glutamine. Prepared medium can be stored at 4°C.

2. Coat three 6-cm cell culture dishes (two for follow-up transfection and one for control) with 2 μg/mL poly-D-lysine (1:500 dilution of stock solution in water) for 30 min. Wash cell culture dishes twice with sterile water and allow them to air dry under the tissue culture hood.

3. Grow mycoplasma-free HEK 293 cells (see Note 1) on the above prepared cell culture dishes at 37°C, 5% CO_2, to 80% confluence.

4. One hour before transfection, remove medium from two of the plates and add 2.8 mL of fresh "normal" medium.

5. Just before transfection, prepare two tubes with 100 μL of serum-free medium. Add 5 μg of plasmid DNA (either RhoBTB3-overexpressing or control) (see Notes 2–4). Vortex gently and spin down briefly. For each of the two plates, dilute 7.5 μL of PolyJet Reagent into 100 μL of serum-free medium. Vortex gently and spin down briefly. Add the diluted PolyJet immediately to the DNA solution. Vortex gently and spin down briefly. Incubate 15 min at room temperature.

6. Add the transfection medium dropwise to the corresponding culture dish, and incubate with the cells overnight at 37°C, 5% CO_2.

7. The following day, remove the transfection medium and replace with fresh "normal medium" to which gentamicin (1:1,000) has been added, in order to select transfected cells for antibiotic resistance (see Note 4). Add selection medium also to the third (nontransfected) plate, serving as a control for antibiotic selection (see Note 10).

3.2. SILAC

1. Prepare SILAC-modified medium in two different formulations:

   Heavy SILAC medium. Add one aliquot each of 13C6 L-lysine, 12C6 L-leucine, and13C6 L-arginine (see Note 6), 50 mL of sterile heat-inactivated dialyzed FBS (see Note 7), and 5 mL of penicillin/streptomycin stock solution to 500 mL of SILAC-DMEM.

   Light SILAC medium. Add one aliquot each of 12C6 L-lysine, 12C6 L-leucine, 12C6 L-arginine (see Note 6), 50 mL of sterile heat-inactivated dialyzed FBS (see Note 7), 5 mL of penicillin/streptomycin stock solution, to 500 mL of SILAC-DMEM.

2. Following selection (see Subheading 3.1, step 7), remove "normal" medium from the two dishes of transfected and selected HEK 293 cells. Add 1 mL of enzyme-free cell dissociation buffer to each cell culture dish and swirl them gently until cells begin to detach (maximum 1 min) (see Note 8). Add 4 mL of SILAC-DMEM to stop dissociation. Collect the cells and the medium into sterile 15-mL tubes. Pellet the cells by centrifugation at $300 \times g$ for 5 min.

3. Remove the medium. Add 10 mL of "Light SILAC medium" to the pellet of HEK 293 cells previously transfected with control vector and 10 mL of "Heavy SILAC medium" to the pellet of HEK 293 cells previously transfected with RhoBTB3-overexpressing vector.

4. Carefully resuspend the cells by gently pipetting up and down and plate them onto freshly prepared poly-D-lysine coated 10-cm dishes.

5. Maintain cells in culture for five doublings (see Notes 11 and 12). Upon reaching 70% confluence, dissociate and pellet them by centrifugation as described at $300 \times g$ for 5 min. Resuspend each pellet in 2 mL of light (control) or heavy (RhoBTB3-overexpressing) SILAC medium. Transfer 1 mL of cell suspension to a freshly prepared poly-D-lysine -coated 10-cm dish containing 9 mL of light or heavy SILAC medium (see Note 13 and Fig. 1).

3.3. Cell Lysis, Co-immunoprecipitation, and Gel Electrophoresis

1. Prepare the cell lysis buffer by adding to the diluted co-immunoprecipitation extraction buffer (provided in the Dynabeads co-immunoprecipitation kit) an EDTA-free (see Note 8) proteinase inhibitor and 1 M NaCl to reach a final concentration of 100 mM (see Note 14).

2. Couple 7.5 mg of washed Dynabeads to 32.5 mg of anti-Myc antibody overnight at 37°C.

3. On the following day, collect the antibody-coupled Dynabeads on the magnet and wash according to the manufacturer's protocol. Resuspend at a concentration of 10 mg/mL.

4. Weight an empty 15-mL tube. Collect cells by scraping with a cell scraper from the 20 replicate plates and pool all in the tube (see Note 13). Pellet harvested cells by centrifugation at $500 \times g$ for 5 min at 4°C. Discard the medium.

5. Weight the tube with the cell pellet and subtract the weight of the empty tube to calculate the exact weight of the pellet.

6. Resuspend cells in a 1:9 ratio (cell mass : lysis buffer volume) in lysis buffer. Incubate on ice for 15 min. Centrifuge at $2,600 \times g$ for 5 min at 4°C to remove debris. Transfer the supernatant into a new tube and proceed immediately with co-immunoprecipitation.

7. Resuspend washed antibody-coupled Dynabeads in cell lysate on an orbital rotator for 45 min at 4°C (see Note 15). Recollect beads on the magnet, wash them according to the manufacturer's protocol and finally dry proteins using a vacuum concentrator.

8. Resuspend the dehydrated proteins in Nu-PAGE sample loading buffer, adding NuPAGE reducing solution, to reach a final volume of 20 μL. Boil the sample for 5 min at 95°C. Cool on ice for 5 min and resolve the entire protein sample onto a single lane of a Nu-PAGE Novex 4–12% Bis–Tris gradient gel.

3.4. Silver Staining (see Note 9)

1. Fix the gel in solution A for 1 h under gentle agitation. Wash twice 15 min in solution B and once in water.

2. Sensitize the gel 1 min in solution C under gentle agitation. Wash 3× 3 s in water.

3. Incubate the gel in solution D for 20 min at 4°C under gentle agitation. Wash 3× 3 s in water.

4. Develop the gel in freshly prepared solution E (see Note 16). Keep the gel in solution until bands become visible.

5. Stop the developing process with solution F before strong background staining appears. Wash 3× 10 min in water.

3.5. Mass Spectrometry Sample Preparation

1. Excise all visible bands, cut each of them into small pieces and transfer each band into a separate clean tube. Dehydrate 3× 10 min in 200 μL of acetonitrile. Wash once in 50 μL of water and dehydrate again once. Dry samples in a vacuum concentrator.

2. Incubate the samples in 50 μL of solution G for 15 min at 60°C. Remove the solution and incubate the samples in solution H for 1 h at room temperature protected from the light. Remove the solution. Dehydrate and dry as in step 1.

3. Rehydrate the sample in 50 μL of 12.5 μg/mL trypsin, 50 mM NH$_4$HCO$_3$, pH 8.3 for about 30 min at 4°C and digest overnight at 37°C.

4. Centrifuge samples at 13,000×g for 5 min at room temperature. Transfer supernatant to clean tubes (first supernatant). Repeat centrifugation after incubating the remaining gel pieces for 15 min with 50 μL of water (second supernatant), and 3× with 50 μL of 5% formic acid, 50% acetonitrile (third, fourth, and fifth supernatants). Combine all supernatants and centrifuge at 13,000×g for 5 min at room temperature. Transfer the supernatant into a clean tube.

5. Purify and concentrate peptides through ZipTipC18 pipette tips for subsequent tandem mass spectrometry analysis. ZipTips are prewetted with solution I and equilibrated 3× with solution J. Samples are then pipetted 10× through the respective tips for binding. Wash tips 5× with solution J. Elute peptides with 10 μL of solution K into a clean tube for subsequent tandem mass spectrometry analysis.

4. Notes

1. Once mycoplasma-free HEK 293 cells in culture have been established, we recommend preparing a frozen stock from an early passage to ensure a renewable source of cells in case of future mycoplasma contamination.

2. In order to express a functional tag-fusion protein able to maintain normal cellular localization and to shuttle between cellular compartments, when generating constructs for Rho GTPases, it is essential to conserve a free C terminus for post-translational modification. The tag of choice has therefore to be fused with the protein N terminus.

3. For interactome studies the use of a small tag, i.e., myc (1.2 kDa), His (~1 kDa), or FLAG (~1 kDa), is highly recommended as high molecular weight tags, such as GFP (26.9 kDa), could mask binding sites.

4. In order to ensure that only overexpressing cells are kept in culture, as to maximize the co-immunoprecipitation yield, the overexpression construct must contain an antibiotic resistance gene for mammalian cells, e.g., neomycin or puromycin resistance genes.

5. Modulation of atypical Rho GTPases function in different cell types might depend on binding to different interacting partners. Therefore, the study of atypical Rho GTPases interactomes in different cell lines and/or mammalian primary cells could be of

interest. It is important to note that different cell types might require different media and additional growth factors, and could show decreased growth/proliferation in the presence of dialyzed medium. Any growth medium is suitable for SILAC-based interactome studies providing that it is depleted of the correspondent heavy and light labeled amino acids.

6. By using substituted arginine and lysine amino acids, proteins are labeled specifically at sites of trypsin cleavage, which is convenient for subsequent analysis of tryptic peptides by mass spectrometry. A combination of labeled arginine and lysine ensures that every tryptic peptide contains a quantifiable aminoacid.

7. Use of dialyzed FBS is essential to ensure removal of "normal" amino acids from the serum.

8. EDTA is incompatible with several protein complexes. Therefore, no trypsin–EDTA should be used to detach cells and no EDTA-containing proteinase inhibitor cocktail should be used in the protein lysis buffer.

9. It is possible to visualize protein bands by standard Coomassie-blue staining. However, silver staining provides higher sensitivity of detection.

10. Antibiotic selection is normally considered to be completed when all the cells in the control plate are dead.

11. In order to perform a SILAC-based experiment, cells should be maintained in the modified medium for at least five passages to ensure that the incorporation of labeled amino acids is close to 100% (23). We have observed that overexpression of atypical Rho GTPases might influence cytoskeleton dynamics, leading to changes in cell cycle progression, reduced cell proliferation, loss of cell adhesion, and/or increased rate of apoptosis. Such changes may ultimately preclude the maintenance of cultures for the required time. Therefore, we highly recommend to always quantify the overexpression of the protein of interest by western blot, as well as checking the effect of the overexpression on cell behavior in normal medium over 2–3 doublings.

12. As the present protocol is for transient transfection, we strongly advise to check for protein overexpression by western blot using protein lysates of transfected cells maintained in normal medium over five doublings.

13. To ensure incorporation of labeled amino acids is close to 100%, cells have to be maintained in culture for at least five doublings. We suggest splitting cells 1:2 in poly-d-lysine-coated 10-cm dishes every time they reach about 70–80% confluence. This will lead to a final amount of 16 plates for each condition. We recommend the use of (at least) ten 10-cm dishes per condition (C12 vs. C13, for a total of 20 dishes) for the follow-up

co-immunoprecipitation. The remaining cells/plates can be used for western blots to confirm the maintenance of protein overexpression and/or be fixed in 4% PFA for follow-up immunocytochemistry and co-localization studies.

14. In order to further maximize the number of binding partners while maintaining binding specificity, different co-immunoprecipitation buffer formulations can be tested, as according to the Dynabeads manufacturer's protocol (Invitrogen). If necessary, we advice testing different conditions using transfected cells maintained in "normal" medium, to reduce costs. In this case, to allow comparison, the two samples (control and overexpressor) are immunoprecipitated separately and run onto two separate gel lanes following co-immunoprecipitation. The SILAC experiment should then be executed with the buffer of choice.

15. Longer incubation times will increase the risk of unspecific binding. However, they might be necessary in case of poor binding affinity antibodies.

16. It is important to add fresh formalin, just immediately before use. Once solution turns turbid, replace immediately with the same volume of new fresh solution.

Acknowledgments

We thank Dr. Francisco Rivero for kindly providing us a RhoBTB3 expressing vector. This work was supported by the FP7 program (FP7-PEOPLE-IEF-2008/236777/AXOGLIA), COMPETE (FCOMP-01-0124-FEDER-011182), and the Fundação para a Ciência e a Tecnologia (PTDC/SAU-NEU/69831/2006; PTDC/SAU-NEU/099007/2008).

References

1. Aspenstrom, P., Ruusala, A., and Pacholsky, D. (2007) Taking Rho GTPases to the next level: the cellular functions of atypical Rho GTPases. *Exp Cell Res* 313, 3673–3679.

2. Hall, A., and Lalli, G. (2010) Rho and Ras GTPases in axon growth, guidance, and branching. *Cold Spring Harb Perspect Biol* 2, a001818.

3. Heasman, S.J., and Ridley, A.J. (2008) Mammalian Rho GTPases: new insights into their functions from in vivo studies. *Nat Rev Mol Cell Biol* 9, 690–701.

4. Ridley, A.J. (2006) Rho GTPases and actin dynamics in membrane protrusions and vesicle trafficking. *Trends Cell Biol* 16, 522–529.

5. Tybulewicz, V.L., and Henderson, R.B. (2009) Rho family GTPases and their regulators in lymphocytes. *Nat Rev Immunol* 9, 630–644.

6. Vega, F.M., and Ridley, A.J. (2008) Rho GTPases in cancer cell biology. *FEBS Lett* 582, 2093–2101.

7. Saras, J., Wollberg, P., and Aspenstrom, P. (2004) Wrch1 is a GTPase-deficient Cdc42-like protein with unusual binding characteristics and cellular effects. *Exp Cell Res* 299, 356–369.

8. Aspenström, P., Fransson, A., and Saras, J. (2004) Rho GTPases have diverse effects on the organization of the actin filament system. *Biochem J* 377, 327–337.

9. Wennerberg, K., and Der, C.J. (2004) Rho-family GTPases: it's not only Rac and Rho (and I like it). *J Cell Sci* 117, 1301–1312.

10. Chardin, P. (2006) Function and regulation of Rnd proteins. *Nat Rev Mol Cell Biol* 7, 54–62.

11. Chenette, E.J., Abo, A., and Der, C.J. (2005) Critical and distinct roles of amino- and carboxyl-terminal sequences in regulation of the biological activity of the Chp atypical Rho GTPase. *J Biol Chem* 280, 13784–13792.

12. Berthold, J., Schenkova, K., Ramos, S., Miura, Y., Furukawa, M., Aspenström, P., and Rivero, F. (2008) Characterization of RhoBTB-dependent Cul3 ubiquitin ligase complexes--evidence for an autoregulatory mechanism. *Exp Cell Res* 314, 3453–3465.

13. Katoh, H., Harada, A., Mori, K., and Negishi, M. (2002) Socius is a novel Rnd GTPase-interacting protein involved in disassembly of actin stress fibers. *Mol Cell Biol* 22, 2952–2964.

14. Madigan, J.P., Bodemann, B.O., Brady, D.C., Dewar, B.J., Keller, P.J., Leitges, M., Philips, M.R., Ridley, A.J., Der, C.J., and Cox, A.D. (2009) Regulation of Rnd3 localization and function by protein kinase C alpha-mediated phosphorylation. *Biochem J* 424, 153–161.

15. Riento, K., Totty, N., Villalonga, P., Garg, R., Guasch, R., and Ridley, A.J. (2005) RhoE function is regulated by ROCK I-mediated phosphorylation. *EMBO J* 24, 1170–1180.

16. Cusick, M.E., Klitgord, N., Vidal, M., and Hill, D.E. (2005) Interactome: gateway into systems biology. *Hum Mol Genet* 14 Spec No. 2, R171–181.

17. Ong, S.E., Foster, L.J., and Mann, M. (2003) Mass spectrometric-based approaches in quantitative proteomics. *Methods* 29, 124–130.

18. Ong, S.E., and Mann, M. (2006) A practical recipe for stable isotope labeling by amino acids in cell culture (SILAC). *Nat Protoc* 1, 2650–2660.

19. Ong, S. E., and Mann, M. (2007) Stable isotope labeling by amino acids in cell culture for quantitative proteomics. *Methods Mol Biol* 359, 37–52.

20. Foster, L.J., Rudich, A., Talior, I., Patel, N., Huang, X., Furtado, L.M., Bilan, P.J., Mann, M., and Klip, A. (2006) Insulin-dependent interactions of proteins with GLUT4 revealed through stable isotope labeling by amino acids in cell culture (SILAC). *J Proteome Res* 5, 64–75.

21. Mann, M. (2006) Functional and quantitative proteomics using SILAC. *Nat Rev Mol Cell Biol* 7, 952–958.

22. Selbach, M., and Mann, M. (2006) Protein interaction screening by quantitative immuno-precipitation combined with knockdown (QUICK). *Nat Methods* 3, 981–983.

23. Dobreva, I., Fielding, A., Foster, L.J., and Dedhar, S. (2008) Mapping the integrin-linked kinase interactome using SILAC. *J Proteome Res* 7, 1740–1749.

24. Trinkle-Mulcahy, L., Andersen, J., Lam, Y.W., Moorhead, G., Mann, M., and Lamond, A.I. (2006) Repo-Man recruits PP1 gamma to chromatin and is essential for cell viability. *J Cell Biol* 172, 679–692.

25. Mortensen, P., Gouw, J.W., Olsen, J.V., Ong, S.E., Rigbolt, K.T., Bunkenborg, J., Cox, J., Foster, L.J., Heck, A..J., Blagoev, B., Andersen, J.S., and Mann, M. (2010) MSQuant, an open source platform for mass spectrometry-based quantitative proteomics. *J Proteome Res* 9, 393–403.

26. van Breukelen, B., van den Toorn, H.W., Drugan, M.M., and Heck, A.J. (2009) StatQuant: a post-quantification analysis tool-box for improving quantitative mass spectrometry. *Bioinformatics* 25, 1472–1473.

27. Fransson, S., Ruusala, A., and Aspenström, P. (2006) The atypical Rho GTPases Miro-1 and Miro-2 have essential roles in mitochondrial trafficking. *Biochem Biophys Res Commun* 344, 500–510.

28. Weihofen, A., Thomas, K.J., Ostaszewski, B.L., Cookson, M.R., and Selkoe, D.J. (2009) Pink1 forms a multiprotein complex with Miro and Milton, linking Pink1 function to mitochondrial trafficking. *Biochemistry* 48, 2045–2052.

29. Wilkins, A., Ping, Q., and Carpenter, C.L. (2004) RhoBTB2 is a substrate of the mammalian Cul3 ubiquitin ligase complex. *Genes Dev* 18, 856–861.

30. Espinosa, E.J., Calero, M., Sridevi, K., and Pfeffer, S.R. (2009) RhoBTB3: a Rho GTPase-family ATPase required for endosome to Golgi transport. *Cell* 137, 938–948.

31. Gasman, S., Kalaidzidis, Y., and Zerial, M. (2003) RhoD regulates endosome dynamics through Diaphanous-related Formin and Src tyrosine kinase. *Nat Cell Biol* 5, 195–204.

32. Zanata, S.M., Hovatta, I., Rohm, B., and Puschel, A.W. (2002) Antagonistic effects of Rnd1 and RhoD GTPases regulate receptor activity in Semaphorin 3A-induced cytoskeletal collapse. *J Neurosci* 22, 471–477.

33. Wang, H., Zeng, X., Fan, Z., and Lim, B. (2011) RhoH modulates pre-TCR and TCR signalling by regulating LCK. *Cell Signal* 23, 249–258.

34. Oda, H., Fujimoto, M., Patrick, M.S., Chida, D., Sato, Y., Azuma, Y., Aoki, H., Abe, T., Suzuki, H., and Shirai, M. (2009) RhoH plays critical roles in Fc epsilon RI-dependent signal transduction in mast cells. *J Immunol* 182, 957–962.

35. Gu, Y., Chae, H.D., Siefring, J.E., Jasti, A.C., Hildeman, D.A., and Williams, D.A. (2006) RhoH GTPase recruits and activates Zap70 required for T cell receptor signaling and thymocyte development. *Nat Immunol* 7, 1182–1190.

36. Alan, J.K., Berzat, A.C., Dewar, B.J., Graves, L.M., and Cox, A.D.(2010) Regulation of the Rho family small GTPase Wrch-1/RhoU by C-terminal tyrosine phosphorylation requires Src. *Mol Cell Biol* 30, 4324–4338.

37. Misko, A., Jiang, S., Wegorzewska, I., Milbrandt, J., and Baloh, R.H. (2010) Mitofusin 2 is necessary for transport of axonal mitochondria and interacts with the Miro/Milton complex. *J Neurosci* 30, 4232–4240.

38. Brady, D.C., Alan, J.K., Madigan, J.P., Fanning, A.S., and Cox, A.D. (2009) The transforming Rho family GTPase Wrch-1 disrupts epithelial cell tight junctions and epithelial morphogenesis. *Mol Cell Biol* 29, 1035–1049.

39. Ruusala, A., and Aspenström, P. (2008) The atypical Rho GTPase Wrch1 collaborates with the nonreceptor tyrosine kinases Pyk2 and Src in regulating cytoskeletal dynamics. *Mol Cell Biol* 28, 1802–1814.

40. Harada, A., Katoh, H., and Negishi, M. (2005) Direct interaction of Rnd1 with FRS2 beta regulates Rnd1-induced down-regulation of RhoA activity and is involved in fibroblast growth factor-induced neurite outgrowth in PC12 cells. *J Biol Chem* 280, 18418–18424.

41. Vayssiere, B., Zalcman, G., Mahe, Y., Mirey, G., Ligensa, T., Weidner, K.M., Chardin, P., and Camonis, J. (2000) Interaction of the Grb7 adapter protein with Rnd1, a new member of the Rho family. *FEBS Lett* 467, 91–96.

42. Toyofuku, T., Yoshida, J., Sugimoto, T., Zhang, H., Kumanogoh, A., Hori, M., and Kikutani, H. (2005) FARP2 triggers signals for Sema3A-mediated axonal repulsion. *Nat Neurosci* 8, 1712–1719.

43. Oinuma, I., Katoh, H., Harada, A., and Negishi, M. (2003) Direct interaction of Rnd1 with Plexin-B1 regulates PDZ-RhoGEF-mediated Rho activation by Plexin-B1 and induces cell contraction in COS-7 cells. *J Biol Chem* 278, 25671–25677.

44. Tong, Y., Chugha, P., Hota, P.K., Alviani, R.S., Li, M., Tempel, W., Shen, L., Park, H.W., and Buck, M. (2007) Binding of Rac1, Rnd1, and RhoD to a novel Rho GTPase interaction motif destabilizes dimerization of the plexin-B1 effector domain. *J Biol Chem* 282, 37215–37224.

45. Li, Y.H., Ghavampur, S., Bondallaz, P., Will, L., Grenningloh, G., and Puschel, A.W. (2009) Rnd1 regulates axon extension by enhancing the microtubule destabilizing activity of SCG10. *J Biol Chem* 284, 363–371.

46. Karaulanov, E., Bottcher, R.T., Stannek, P., Wu, W., Rau, M., Ogata, S., Cho, K.W., and Niehrs, C. (2009) Unc5B interacts with FLRT3 and Rnd1 to modulate cell adhesion in *Xenopus* embryos. *PLoS One* 4, e5742.

47. Wennerberg, K., Forget, M.A., Ellerbroek, S.M., Arthur, W. T., Burridge, K., Settleman, J., Der, C.J., and Hansen, S.H. (2003) Rnd proteins function as RhoA antagonists by activating p190 RhoGAP. *Curr Biol* 13, 1106–1115.

48. Naud, N., Toure, A., Liu, J., Pineau, C., Morin, L., Dorseuil, O., Escalier, D., Chardin, P., and Gacon, G. (2003) Rho family GTPase Rnd2 interacts and co-localizes with MgcRacGAP in male germ cells. *Biochem J* 372, 105–112.

49. Uesugi, K., Oinuma, I., Katoh, H., and Negishi, M. (2009) Different requirement for Rnd GTPases of R-Ras GAP activity of Plexin-C1 and Plexin-D1. *J Biol Chem* 284, 6743–6751.

50. Tanaka, H., Katoh, H., and Negishi, M. (2006) Pragmin, a novel effector of Rnd2 GTPase, stimulates RhoA activity. *J Biol Chem* 281, 10355–10364.

51. Fujita, H., Katoh, H., Ishikawa, Y., Mori, K., and Negishi, M. (2002) Rapostlin is a novel effector of Rnd2 GTPase inducing neurite branching. *J Biol Chem* 277, 45428–45434.

52. Tanaka, H., Fujita, H., Katoh, H., Mori, K., and Negishi, M. (2002) Vps4-A (vacuolar protein sorting 4-A) is a binding partner for a novel Rho family GTPase, Rnd2. *Biochem J* 365, 349–353.

53. Riento, K., Guasch, R.M., Garg, R., Jin, B., and Ridley, A.J. (2003) RhoE binds to ROCK I and inhibits downstream signaling. *Mol Cell Biol* 23, 4219–4229.

54. Goh, L. L., and Manser, E. (2010) The RhoA GEF Syx is a target of Rnd3 and regulated via a Raf1-like ubiquitin-related domain. *PLoS One* 5, e12409.

Part V

Rho GTPases in Non-mammalian Model Organisms

Chapter 21

Using Zebrafish for Studying Rho GTPases Signaling In Vivo

Shizhen Zhu and Boon Chuan Low

Abstract

Rho small GTPases play pivotal roles in a variety of dynamic cellular processes including cytoskeleton rearrangement, cell migration, cell proliferation, cell survival, and gene regulation. However, their functions in vivo are much less understood. Recently, the zebrafish, *Danio rerio,* has emerged as a powerful model organism for developmental and genetic studies. Zebrafish embryos have many unique characteristics, such as optical transparency, external fertilization and development, and amenability for various molecular manipulations including morpholino oligo-mediated gene knockdown, mRNA or DNA overexpression-induced gain of function or rescue, in situ hybridization (ISH) with riboprobes for gene expression, western blot for protein analysis, small-molecule inhibition on signaling pathways, and bioimaging for tracking of molecular events. Taking many of such advantages, we have demonstrated the role of *rhoA* small GTPase in the control of gastrulation cell movements and cell survival during early zebrafish embryogenesis, linking RhoA functions to at least the noncanonical Wnt, Mek/Erk, and Bcl2 signaling nodes in vivo. Here, we describe the use of such techniques, including gene knockdown by morpholino oligo, functional rescue by mRNA overexpression, microinjection, ISH, western blot analysis and pharmacological inhibition of signaling pathways by small molecule inhibitors, with special considerations on their merits, potential drawbacks, and adaptation which could pave the way to our better understanding of the roles of various classes of small GTPases in regulating cell dynamics and development in vivo.

Key words: Rho small GTPase, Zebrafish, Mopholino oligos, mRNA rescue, Microinjection, In situ hybridization, Western blot, Small molecule inhibitor

1. Introduction

1.1. The Zebrafish Model

Zebrafish has emerged as an excellent vertebrate model for both developmental and genetic studies. Zebrafish embryos are optically transparent and their embryonic development occurs externally and rapidly, which makes it possible to visualize the development of internal tissues, and to trace cell migration in living embryos (1, 2). A large number of embryos can be produced from one pair of fish

Francisco Rivero (ed.), *Rho GTPases: Methods and Protocols*, Methods in Molecular Biology, vol. 827,
DOI 10.1007/978-1-61779-442-1_21, © Springer Science+Business Media, LLC 2012

each week, which is suitable for large-scale mutagenesis and high-throughput drug screening (3, 4). Furthermore, zebrafish is also amenable to various cellular, molecular, and genetic techniques, allowing us to study the signaling mechanism underlying physiological and pathophysiological processes and to understand networks among different in vivo signaling pathways (2). Most importantly, the zebrafish genome has high homology to that of human, and the majority of key signaling pathways are conserved between zebrafish and human (5). Thus, zebrafish can be used as a powerful in vivo animal model system to determine the role of genes of interest in physiological and/or pathological processes and unravel the molecular mechanism underlying their activities in vivo.

1.2. Rho Small GTPases in Zebrafish

The roles of Rho small GTPases in various cellular processes have been extensively studied in in vitro cell culture systems and their biochemical and molecular bases have been well described (see accompany chapters in this series). With increasing attraction of the zebrafish as a powerful in vivo model system for human physiology and diseases (2, 6–8), more studies have now focused on the roles of small GTPases in regulating cell/tissue dynamics and their roles in embryonic and whole-animal development in vivo. Studies from our laboratory and others have shown that Rho small GTPases including RhoA, Rac, and Rap, play critical roles in regulation of a variety of cell movements during zebrafish development (9–14). For example, RhoA mediates planar cell polarity signaling and cooperates with Rac to control gastrulation convergence and extension movements (9, 10) and neural crest directional migration (11). The mechanism underlying the regulation of a single-cell movement in vivo has been beautifully addressed in migrating germ cells. Using fluorescence resonance energy transfer, Kardash et al. (12) demonstrated that Rac1 and RhoA were localized and activated in the front of migrating germ cells, which led to formation of actin-rich structures and generation of traction forces, subsequently driving germ cells migration. Also in primordial germinal cells, Palamidessi et al. (15) have shown that membrane trafficking of Rac by clathrin- and Rab5-mediated endocytosis led to formation of actin-based migratory protrusions and migration of primordial germinal cells during zebrafish development. Besides such elegant studies on the demonstration of the role of Rho small GTPases in controlling of cell movement in vivo, we have utilized both genetic and biochemical tools and unveiled previously unexpected roles of RhoA in promoting cell survival during zebrafish embryogenesis through Mek/Erk and Bcl2 pathways (16). Taken together, zebrafish has emerged as an excellent in vivo model system to determine the role of Rho GTPases during development. In the following sections, some of the commonly used techniques in zebrafish study will be introduced. Many of the general techniques and resources can be referred to and obtained from ZFIN, an excellent Zebrafish Model Organism Database http://www.zfin.org/cgi-bin/webdriver?MIval=aa-ZDB_home.apg.

1.3. Gene Knockdown by Morpholino Oligos

A morpholino oligo (MO) is a 25-bp long synthetic morpholino-modified antisense oligonucleotide which is complementary to a target RNA (17). MOs can specifically bind to their target sequences leading to gene knockdown through two mechanisms, blocking the translational and splicing events. Translational-blocking oligos can block initiation of translation by targeting the 5' UTR and/or the first 25 bases of the coding sequence in the genes of interest, while splice-blocking oligos can modify pre-mRNA splicing leading to either exon skipping or intron retention by targeting splice junctions or splice regulatory sites of the genes of interest (17). Thus, the former oligos can target both maternal and zygotic mRNA leading to translational block, while the latter can only target zygotic mRNA, eliciting gene knockdown effect at slightly later stage. However, splice-blocking MOs are more popular compared to translational blocking MOs because their knockdown effect can be easily validated by RT-PCR. In contrast, the effect for translational-blocking MOs has to be examined by western blot analyses, which is difficult owing to the lack of effective antibody against the zebrafish proteins. Another potential advantage for splice-blocking MOs is that they can cause exon excision. Thus, they are useful for studying functions of a specific protein domain by selectively removing the exon(s) encoding that particular domain while rendering the remaining sequence of the gene in-frame.

By simple microinjection, MOs can be easily delivered into zebrafish embryos and they will last for a few days in vivo (17, 18). Thus, MOs have become one of the most specific and popular knockdown technologies to study gene functions during zebrafish early embryonic development. However, similar to other sequence-specific knockdown technique, such as short interfering RNAs, MOs may cause off-target effects (19). Thus, to confirm the specificity of MOs, at least two different MOs targeting nonoverlapping sequences of the same gene are required and both MOs should induce a consistent phenotype. To further evaluate the target-specificity of MOs, mRNA should be introduced to rescue the MO-induced phenotype.

1.4. Functional Rescue by mRNA Overexpression

To validate a MO-mediated knockdown effect, we should restore the functional loss by performing rescue experiments. This is usually done by re-examining the normal development in MO-injected embryos after re-introduction and overexpression of that particular gene of interest (18–20). The rescue mRNA is synthesized by in vitro transcription. The template for mRNA synthesis can be zebrafish gene without MO-targeting sequence or its conserved homolog from other species. In principle, the rescue mRNA won't be targeted by MO, thus, the loss of endogenous gene function caused by MO knockdown can be restored by ectopically introduced rescue mRNA (18–20). To perform the rescue experiment, in vitro synthesized mRNA and MO can be co-injected into zebrafish embryos and the phenotype of the injected embryos can be scored to determine the efficiency of the rescue. In addition to the overexpression of the MO-targeted gene for the rescue experiment, mRNA of potential downstream effectors or targets of knockdowned gene can be also applied, which is useful to further delineate genetic interactions of signaling pathways in vivo.

1.5. Microinjection Technique

Microinjection involves injecting very little (nano-liter) volume of sample into the zebrafish embryos. It is easiest and the most efficient delivery technique in zebrafish (21). Depending on the purpose of the study and the type of materials to be delivered, various foreign materials can be microinjected into either the blastomere or the yolk at one-cell or at a later stage (22). Because of the ease and effectiveness of microinjection, the MO knockdown and mRNA rescue techniques mentioned before have become essential and popular techniques to determine gene function during zebrafish development (23, 24). Besides, microinjection is also broadly applied to gain of function studies, generation of stable transgenic lines, analysis of cell autonomous effects, and cell fate determination by delivering RNA, DNA, cells, or small molecules into developing embryos (25, 26).

1.6. In Situ Hybridization with Riboprobes

In situ hybridization (ISH) is used to localize and detect mRNA in whole-mount zebrafish embryos or zebrafish tissue sections using a riboprobe (27). Antisense riboprobes are nucleotide sequences complementary to the specific mRNA of interest, which can be synthesized in vitro using RNA polymerase. Normally the sense riboprobe synthesized from the same sequence is served as a negative control for hybridization. To easily detect and visualize the localization of the hybridized probe to its target, several probe-labeling methods have been developed. Recently, nonradioactive labeling methods, such as digoxigenin labeling and FITC labeling, are more popularly used, because they increase histological resolution, shorten the development time, and can be handled more easily compared to the radioactive labeling method (28). Thus, ISH has been extensively applied to examine endogenous gene expression in vivo. Besides, it is also important for characterizing developmental defects caused by loss or gain of a gene function by detecting the change of expression pattern with different cell lineage-specific markers in zebrafish embryos (29).

1.7. Western Blot Analysis

Western blot analysis is a well-established technique in biochemical and molecular biology studies. While suffering from a general lack of good antibodies in the earlier stage of establishing the zebrafish as an in vivo model, more and more reliable sources of such crucial reagents are now available. The standard process can be well adapted for zebrafish studies except that the yolk of zebrafish embryos needs to be removed before protein extraction. This is because the high proportion of yolk proteins in early embryos can interfere with the detection of cellular proteins (30). Western blot analysis is not only useful to examine endogenous protein expression in embryos or adult tissues; it is also important to validate the effectiveness of MO knockdown or gene overexpression by detecting their expression levels or their downstream targets.

1.8. Pharmacological Inhibition on Signaling Pathway by Small Molecule Inhibitors

One of the amazing advantages of using zebrafish embryos is that they are easily accessible for small molecules or drugs treatment. This makes zebrafish a powerful model system for large-scale drug screening and assessments, and for delineation of signaling pathway interactions in vivo (3, 4, 31). Small molecule inhibitors can be simply added to the egg water and penetrate into embryos through their skin. They can be given or removed from egg water at chosen time-points, which makes them more flexible than MOs in eliciting an inhibitory effect. Similarly to MO knockdown, potential off-target effects and cytotoxicity may occur at high doses of inhibitor. However, one can "titrate" those reagents in a broader or narrower range and still monitor for subtle morphological perturbations. Furthermore, multiple inhibitors for one or more targets can be applied to achieve specific or combinatorial inhibition.

2. Materials

2.1. Zebrafish

Zebrafish are maintained in accordance with The Zebrafish Book (32). Embryos are staged according to Kimmel et al. (33).

2.2. MO Design and Preparation for Gene Knockdown

1. 1× Danieau buffer: 58 mM NaCl, 0.7 mM KCl, 0.4 mM $MgSO_4$, 0.6 mM $Ca(NO_3)_2$, 5.0 mM HEPES pH 7.6. Store at room temperature.

2. Mopholino antisense oligo stock solution: Dissolve the MOs (Gene Tools) in distilled water or 1× Danieau buffer to make a 1 mM stock solution. Aliquot and store at −80°C.

2.3. mRNA In Vitro Synthesis for Functional Rescue

1. PCR purification kit (Qiagen).

2. mMESSAGE mMACHINE Kit (Applied Biosystems).

3. RNase-free DNase I (Applied Biosystems).

4. RNeasy mini kit (Qiagen).

2.4. Microinjection

1. Egg Water stock (60×): weigh 17.53 g of NaCl, 0.76 g of KCl, 2.91 g of $CaCl_2$, 4.88 g of $MgSO_4$, and dissolve in 1 L of reverse osmosis water (see Note 1). To make 1× egg water, dilute the egg water stock 60× with reverse osmosis water and add methylene blue to a final concentration of 0.01%. Store at RT.

2. Agarose.

3. Glass capillaries (World Precision Instruments).

4. Micropipette puller P-97 (Sutter Instruments).

5. Microinjector PL1-100 (Harvard Apparatus).

6. Narishige Micromanipulator (Tritech Research).

7. Microloader tips (Eppendorf).

8. Leica M420 dissecting microscope (Leica).

2.5. Whole-Mount In Situ Hybridization

1. DIG RNA labeling mix (Roche Diagnostics).

2. RNase inhibitor (40 U/µL) (Roche Diagnostics).

3. 5× transcription buffer (Roche Diagnostics).

4. RNA polymerase (T7, Sp6, or T3) (Roche Diagnostics).

5. DEPC water (Applied Biosystems).

6. Phosphate-buffered saline (PBS): 0.8% NaCl, 0.02% KCl, 0.02 M phosphate, pH 7.3.

7. Fix buffer: 4% paraformaldehyde in PBS. Dissolve 40 g of paraformaldehyde in PBS by heating at 60°C. Fill to 1 L with PBS and store at –20°C in 50-mL aliquots (see Note 2).

8. PBST: PBS plus 0.1% Tween-20.

9. Methanol (Fisher Scientific) (see Note 3).

10. Proteinase K stock: 10 mg/mL of proteinase K (Roche Diagnostics) in milliQ water.

11. Prehybridization buffer (Hyb–): 50% formamide, 5× SSC, 0.1% Tween-20. Store at –20°C (see Note 4).

12. Hybridization buffer (Hyb+): Hyb (–) plus 5 mg/mL torula (yeast) RNA type VI (Applied Biosystems) and 50 µg/mL heparin (Sigma Aldrich). Store at –20°C.

13. 20× SSC: dissolve 175.3 g of NaCl and 88.2 g of sodium citrate in 1 L of distilled water, adjust pH to 7.0.

14. 1× SSCT: 1× SSC plus 0.1% Tween-20.

15. 2× SSCT/50% formamide: 2× SSC, 50% formamide, 0.1% Tween-20.

16. 60× Phenylthiourea (PTU) solution: dissolve 1.8 g of PTU powder in 1 L of milliQ water (see Note 5). Store at room temperature.

17. Blocking buffer: 2% goat (or sheep) serum, 2 mg/mL BSA in PBST.

18. Anti-digoxigenin–alkaline phosphatase (AP) or anti-FITC-AP antibody (Roche Diagnostics).

19. Staining buffer: 100 mM Tris–HCl pH 9.5, 50 mM $MgCl_2$, 100 mM NaCl, 0.1% Tween-20.

20. Staining solution: BCIP/NBT AP substrate kit (Vector Laboratories) Add one drop of each solution per 2.5 mL of staining solution.

2.6. Western Blot

1. RIPA lysis buffer: 1% NP-40, 0.5% deoxycholic acid, 0.1% SDS in PBS.

2. 6× loading buffer: 0.2 M Tris–HCl pH 6.8, 25% (v/v) glycerol, 25% (v/v) SDS, 12.5% (v/v) 2-mercaptoethanol, 0.005% (w/v) bromophenol blue.

3. SDS running buffer: 25 mM Tris, 192 mM glycine, and 0.1% (w/v) SDS.

4. Transfer buffer: 20 mM Tris, 150 mM glycine, 20% methanol, and 0.038% SDS.

5. TBS: 50 mM Tris–HCl, pH 7.4, and 150 mM NaCl.

6. TBST: TBS plus 0.1% Tween-20.

7. Blocking buffer: 5% skimmed milk powder in TBST.

8. Amersham ECL chemiluminescent labeling and detection reagents (GE Healthcare).

9. Protease inhibitor cocktail (Roche).

10. Hand-held homogenizer (Sigma).

11. Kontes pellet pestle (Fisher Scientific).

12. Mini-PROTEAN Tetra Systems (Bio-Rad).

13. Nitrocellulose membrane (Thermo Scientific).

3. Methods

3.1. MO Design and Preparation for Gene Knockdown

1. Selection of target sequence. For the translational-blocking MO, the 5′ UTR sequence through the first 25 bases of coding sequence of the gene of interest can be used as target. For the mismatch control MO, around 5-bp along the length of the gene-specific MO should be substituted with mismatched nucleotides with the remaining sequence unchanged. For splice-blocking MO, the nucleotide sequence at exon–intron or intron–exon junctions of the gene of interest can be used for MO design (see Note 6). Check for self-complementarity to avoid dimerization. MOs should have 40–60% GC content and no stretches of four or more contiguous G to maintain a high-target affinity and solubility in water.

2. MOs can be dissolved in distilled water and Danieau buffer to make a 1-mM stock solution and stored at –80°C in aliquots.

3. MOs should be heated up at 65°C for 10 min before application for microinjection and unused MOs can be stored at 4°C temporarily.

3.2. mRNA In Vitro Synthesis for Functional Rescue

1. Linearize 10 μg of plasmid DNA at 3′ end of the gene of interest with a suitable restriction enzyme, and run an agarose gel with 2 μL of the digestion reaction to verify that the plasmid is completely linearized.

2. Purify the linearized DNA with a PCR purification kit.

3. Use the linearized DNA as template for in vitro synthesis with the mMESSAGE mMACHINE kit. For a typical 20-μL reaction, mix 1 μg of linearized DNA, 2 μL of 10× reaction buffer, 10 μL of 2× NTP/Cap, 2 μL of enzyme mix (T7, T6, or T3), top up with nuclease-free water, and incubate the reaction at 37°C for 2 h.

4. Add 1 μL of RNase-free DNase I and incubate for 15 min at 37°C to remove the template DNA.

5. Clear up the RNA using the RNeasy mini kit. Elute RNA from the column with 40 μL of nuclease-free water.

6. Determine the concentration of RNA spectrophotometrically and store in aliquots at –80°C.

3.3. Microinjection

1. Make a microinjection plate by pouring approximately 20 mL of hot 1.5% agarose in egg water into a 100×15 mm Petri dish and laying down a plastic mold with wedge-shaped troughs. The mold is about 40 mm wide×65 mm long and the grooves made from the mold are 50 mm long and 3 mm apart. Remove the mold after the agarose is completely solidified (see Note 7).

2. Pull a microinjection needle using the micropipette puller with the following parameters: heat 575, pull 45, velocity 40, time 50.

3. Break off the tip of the needle with blade or forceps (see Note 8).

4. Backload the needle with MO or/and RNA using a microloader. Shake the needle to eliminate any bubbles (see Note 9).

5. Hook the needle with the microinjector and adjust the micromanipulator to produce a desirable injection volume.

6. Transfer one-cell stage embryos into the troughs of the plate and orient them with their animal pole towards the pipette (see Note 10).

7. Inject the MOs or/and RNA into the cell or yolk, about 1/5 volume of cell diameter or 1/10 volume of yolk (23) (see Note 11).

8. Following injection, transfer embryos into a clean dish, incubate at 28.5°C and monitor their development regularly under the dissecting microscope (see Fig. 1).

3.4. Whole-Mount In Situ Hybridization

Carry out all experiments with 1 mL of total volume of solution at room temperature unless otherwise specified.

1. Gene-specific antisense riboprobe is synthesized using DNA linearized at its 5′ end as template. The reaction is performed at 37°C for 2 h containing 1 μg of linearized DNA, 4 μL of 5× transcription buffer, 2 μL of DIG RNA labeling mix, 1 μL of RNase inhibitor (40 U/μL), 1 μL of RNA polymerase (50 U/μL, T7, Sp6, or T3), and DEPC water to top up to a total volume of 20 μL (see Note 12).

2. Following the reaction, add 2 μL of RNase-free DNase I to digest the template DNA at 37°C for 15 min (see Note 13).

3. Clean up and recover the RNA using the RNeasy mini kit, dilute the RNA probe with hybridization buffer at 1:100 dilution and store at −20°C (see Note 14).

4. Fix embryos at appropriate time points with fresh fix buffer and rock overnight at 4°C (see Note 15). Embryos older than 24 hpf need to be incubated in egg water with 0.003% PTU to prevent pigmentation (see Note 16).

5. Wash embryos in PBST once quickly, followed by twice 5 min.

6. Replace PBST with 100% methanol and incubate for 5 min. Then replace with fresh methanol and leave embryos at −20°C *overnight* (see Note 17).

7. Wash embryos in PBST once quickly, followed by twice 5 min.

8. Digest with proteinase K for embryos older than 30 hpf according to the following chart (see Note 18):

Age of embryos	μL of proteinase K (10 mg/mL)	Length of time (min)
<30 hpf	No treatment	
30–40 hpf	1 μL	20
40–52 hpf	2 μL	20
3–5 dpf	10 μL	20
6–10 dpf	10 μL	30

9. Quickly rinse embryos in PBST once, followed by one wash in PBST for 5 min.

10. Fix embryos with fresh fix buffer for 20 min at room temperature.

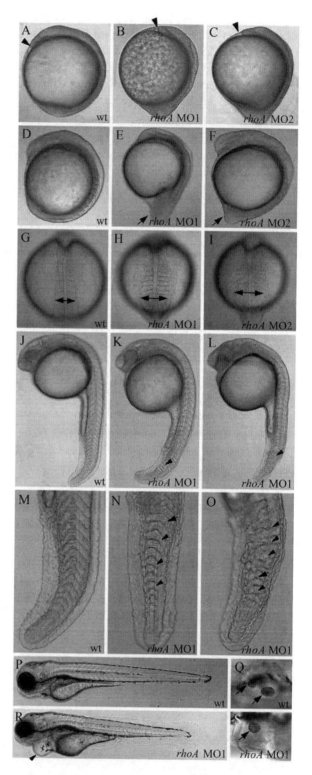

Fig. 1. RhoA is required for zebrafish gastrulation and tail formation. (**a–s**) *RhoA* Morpholino phenotype (morphant) compared to wild type. Two *rhoA* morpholinos have been applied and they caused similar phenotypes. (**a–c**) 1-Somite stage embryos, lateral view, dorsal to the right; *arrowheads* mark the anterior limit of the hypoblast layer. (**d–i**) 6–8 somites

11. Wash embryos in PBST once quickly, followed by 2× 15 min.

12. Replace PBST with 1 mL of Hyb–, rock at 68°C for at least 30 min.

13. Replace Hyb– with 400 µL of preheated Hyb+, and rock at 68°C for at least 1 h.

14. Remove as much of Hyb+ as possible without exposing the embryos to air.

15. Add at least 100µL of preheated Hyb+ containing 20–100 ng of riboprobe and rock overnight at 68°C.

16. Remove the probe and store it at –20°C (see Note 19).

17. Wash embryos in preheated 2× SSCT/50% formamide once quickly, followed by 2× 30 min at 68°C.

18. Wash embryos in preheated 2× SSCT twice 15 min at 68°C.

19. Wash embryos in preheated 0.2× SSCT twice 30 min at 68°C.

20. Wash embryos in PBST once quickly at room temperature, followed by 2× 10 min.

21. Replace PBST with blocking buffer and incubate with gentle shaking for at least 1 h at room temperature.

22. Replace blocking buffer with antibody (anti-digoxygenin-AP or anti-FITC-AP, at 1:5,000 dilution in blocking buffer) and rock overnight at 4°C.

23. Remove antibody and store it at 4°C. The antibody can be reused.

24. Wash embryos in PBST once quickly, 2× 30 min at room temperature.

25. Wash embryos in staining buffer once quickly, 2× 10 min at room temperature.

26. Add staining solution and incubate at room temperature or at 4°C overnight in the dark (see Note 20).

27. Stop staining reaction by washing in PBST 3× 5 min at room temperature with gentle rocking.

28. Replace PBST with 80% glycerol and store embryos at 4°C. The staining stays unchanged for months (see Fig. 2).

Fig. 1. (continued) stage embryos. (**d–f**) Lateral view, dorsal to the right. *Arrows point* at the misprotruded tail (**e, f**). (**g–i**) dorsal view, animal pole to the top; *arrows* in (**h, i**) show the extended somites and notochord. (**j–o**) 24 hpf embryos; (**m–o**) enlarged tail region of 24 hpf embryos; (**k, n**) show mild defect in head and tail region comparing to (**l, o**). *Arrowheads* in (**k, l, n, o**) indicate the malformed somites; *arrows* in (**n, o**) highlight the undulated notochord. (**p–s**) 4–5 dpf embryos, *arrowhead* in (**r**) indicates the heart edema; (**q, s**) enlarged ear region, *arrows* indicates one otolith is missing in (**s**). Overall, *rhoA* morphants display a shorten anterior–posterior body axis, vegetally mispositioned head, detached tail from yolk, and malformed somites. Reprinted from ref. 9 with permission from Elsevier.

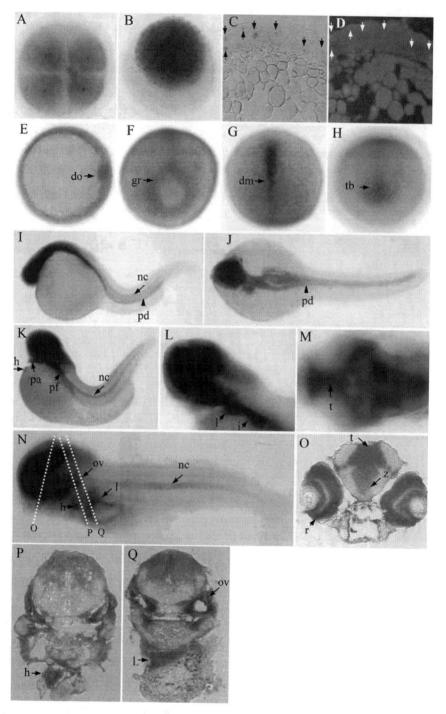

Fig. 2. In situ hybridization analyses for zebrafish *rhoA* expression in different stages of embryonic development. (a) 4-cell-stage embryo, animal view. (b) 128-cell-stage, animal view. (c, d) Longitudinal section of 128-cell-stage embryo, lateral view. (c) In situ hybridization; (d) DAPI staining; *arrows* indicating the overlapping expression of *rhoA* with nucleus staining. (e) Shield stage embryo, animal view with dorsal to the right. (f) 90% epiboly stage embryo, vegetal pole view with dorsal to the top. (g, h) Tail bud stage embryo. (g) Dorsal view, anterior to the top. (h) Vegetal pole view with dorsal to the top. (i) 24 hpf, lateral view with anterior to the left. (J) 48 hpf, dorsal view, without body, anterior to the left. (k) 48 hpf, lateral view

3.5. Western Blot Analysis

1. Remove chorines and yolk from embryos before protein extraction (see Note 21).

2. Transfer embryos into a clean tube and remove all the liquid. From now on, the embryos or the lysate should be on ice.

3. Immediately add RIPA lysis buffer with protease inhibitor cocktail in a ratio of 1–2 μL of lysis buffer/embryo. Homogenize the embryos with a hand-held homogenizer for 30 s until the embryo body is completely invisible (see Note 22).

4. Centrifuge the lysate at 4°C at $9,000 \times g$ for 10 min. The supernatant is collected, denatured directly or stored at –80°C in aliquots.

5. To denature the samples, mix embryo lysate with 1/5 volume of 6× loading buffer and boil for 4 min at 95°C.

6. Samples (around 100 μg) are subjected to standard SDS-PAGE followed by transfer onto nitrocellulose membrane and blotting.

3.6. Pharmacological Inhibition of Signaling Pathways by Small Molecule Inhibitors

1. Place around 20 embryos with 2 mL of egg water into a well of a six-well flat bottom plate (see Note 23).

2. Make 2× concentrated small molecule inhibitor solution in egg water and add 2 mL into the wells with embryos (see Note 24).

3. Place the embryos in an incubator at 28.5°C until they reach the proper developmental stage for further assay.

4. Notes

1. Some salts do not dissolve well. It is better to add one salt at a time instead of trying to dissolve all at once.

2. Paraformaldehyde is toxic. Do all weighing and heating in a chemical hood. Fix buffer is light sensitive and temperature sensitive. Wrap the bottle with aluminum foil and keep the solution at 60–65°C during heating. Thawed fix buffer should be stored at 4°C and usually stays good only for around 2 weeks.

Fig. 2. (continued) with anterior to the left. (**l**) 72 hpf, lateral view with anterior to the left. (**m**) 96 hpf, dorsal view with anterior to the left. (**n**) 96 hpf, lateral view with anterior to the left. White lines indicate the location of section shown in (**o, p, q**) respectively. (**o–q**) Anterior cross sections of 96 hpf embryo with dorsal to the top. *d* diencephalon, *dm* dorsal midline, *do* dorsal organizer, *gr* germ ring, *h* heart, *i* intestine, *l* liver, *nc* notochord, *ov* otic vesicle, *pa* pharyngeal arches, *pd* pronephric duct, *pf*, pectoral fin, *r* retina, *t* tectum, *tb* tail bud, *hpf* hours postfertilization. Reprinted from ref. 9 with permission from Elsevier.

3. Methanol is flammable and very toxic if ingested. It is also an irritant to skin and eyes.

4. Formamide is highly corrosive and may be deadly if ingested.

5. PTU powder is toxic and a carcinogen. Weigh the PTU powder in a chemical hood and heat water to dissolve the powder.

6. MOs can be designed by Gene Tools Company for free, but the sequence of the designed MO needs to be examined by BLAST to check for homology with other zebrafish genes. For the translational blocking MO, designing the oligo targeting only the 5′ UTR sequence can facilitate subsequent rescue experiments. This is because rescue RNA can be synthesized in vitro from the start codon without the MO target sequence.

7. The microinjection plate can be prepared in advance and stored refrigerated for months with a small amount of egg water to avoid drying out.

8. The size and quality of the needle tip is crucial for the ease of injection. Too big an opening may damage the embryos, while too small an opening may cause clogging of the needle with yolk material.

9. In general, the injection volume is in a range of 500 pL to 2 nL. The concentration of MO or mRNA to elicit specific inhibition or rescue of gene function without causing toxic effects is highly variable between MOs or mRNA. Thus, MOs or/and mRNA need to be titrated to determine the optimal dose. 1 mM MO stock concentration is a good starting point for optimization. For mRNA rescue experiments, the concentration of mRNA should be the dose at which itself injection would not cause any phenotype.

10. Right before injection, remove excess egg water without drying out the embryos. Too much water can cause embryos moving during injection, but too little water may damage the embryos.

11. MOs can be introduced into the yolks of embryos no later than eight-cell stage, while mRNA should be injected within 1–2 cell stage for ideal distribution in embryos (20). To ensure the success of injection, the sample can be mixed with 0.25% phenol red for visualization. If cells turn red after injection, this indicates that the microinjection was successful.

12. To make an antisense riboprobe specific for a gene of interest, choose a unique sequence of the gene as template, such as the 3′ UTR. The size of the probe can be varied. We have tested probes from 500 bp to 2.5 kb, they all worked well.

13. This step can be omitted.

14. We generally dilute the probe 100× in hybridization buffer. However, weak probes should be less diluted. The probe is stable in hybridization buffer at –20°C, otherwise the undiluted probe should be stored at –80°C.

15. Embryos younger than 18 hpf can be fixed first and dechorined with a forceps later. For embryos older than 18 hpf, it is better to remove the chorine before fixation to avoid a curved tail.

16. Embryos can be switched and continuously incubated in egg water with 0.003% PTU from gastrulation stage onwards.

17. Embryos can be stored in methanol at –20°C for months.

18. Proteinase K activity varies from batch to batch. For a new batch the digestion time needs to be adjusted first. After proteinase K treatment, embryos become very fragile due to loss of protein. Thus, they should be handled gently beyond this step.

19. The probe can be reused a few times and reused probe gives less background.

20. The staining reaction time is usually in a range of 15 min to few hours. However, to detect genes that are weakly expressed, the staining reaction may take a few days.

21. Embryos chorines can be removed manually by using a forceps or a needle (27G1/2), or by incubating in egg water with 1 mg/mL pronase for 5 min at RT. The yolk can be removed by triturating with a finest plastic pipette tip in cold Ringer's solution with EDTA and PMSF. However, we prefer to remove the yolk manually with needles. This method will take longer time, but the yolk can be removed completely and embryos bodies are still intact.

22. The number of embryos required for Western blot varies depending on the abundance of the protein to be detected and the stage of embryos to be assayed. We have used 15–30 embryos at 24 hpf, which worked fine.

23. To place exactly 2 mL of egg water with embryos, we normally cut the pipette tip to make an opening wide enough to pick up embryos. Stir egg water in the petri dish to gather the embryos in the center. Embryos can be easily picked up while pipetting up the desired amount of egg water.

24. The time frame for inhibitor treatment is based on the temporal expression of the genes of interest. To reduce damage to early stage embryos, the inhibitors should be applied to the embryos after 2.5 h postfertilization. To assess dose-dependent effects, inhibitors are diluted serially.

Acknowledgments

This work was supported in part by the Mechanobiology Institute, National University of Singapore, co-funded by the National Research Foundation and the Ministry of Education, Singapore, and also from a grant from the Biomedical Research Council of Singapore, and a fellowship from the Friends for Life.

References

1. Pyati U.J., Look A.T., and Hammerschmidt, M. (2007) Zebrafish as a powerful vertebrate model system for in vivo studies of cell death. *Semin Cancer Biol* 17, 154–165

2. Lieschke G.J, and Currie, P.D. (2007) Animal models of human disease: zebrafish swim into view. *Nat Rev Genet* 8, 353–367

3. Chakraborty, C., Hsu, C.H., Wen, Z.H., Lin, C.S., and Agoramoorthy, G. (2009) Zebrafish: a complete animal model for in vivo drug discovery and development. *Curr Drug Metab* 10, 116–124

4. Peal, D.S., Peterson, R.T., and Milan, D. (2010) Small molecule screening in zebrafish. *J Cardiovasc Transl Res* 3, 454–460

5. Lam, S.H., Wu, Y.L., Vega, V.B., Miller, L.D., Spitsbergen, J., Tong, Y., Zhan, H., Govindarajan, K.R., Lee, S., Mathavan, S., Murthy, K.R., Buhler, D.R., Liu, E.T., and Gong, Z. (2006) Conservation of gene expression signatures between zebrafish and human liver tumors and tumor progression. *Nat Biotechnol* 24, 73–75

6. Storer, N.Y., and Zon L.I. (2010) Zebrafish models of p53 functions. *Cold Spring Harb Perspect Biol* 2, a001123

7. Kari, G., Rodeck, U., and Dicker, A.P. (2007) Zebrafish: an emerging model system for human disease and drug discovery. *Clin Pharmacol Ther* 82, 70–80

8. Veldman, M.B., and Lin, S. (2008) Zebrafish as a developmental model organism for pediatric research. *Pediatr Res* 64, 470–476

9. Zhu, S., Liu, L., Korzh, V., Gong, Z., and Low, B.C. (2006) RhoA acts downstream of Wnt5 and Wnt11 to regulate convergence and extension movements by involving effectors Rho kinase and Diaphanous: use of zebrafish as an in vivo model for GTPase signaling. *Cell signal* 18, 359–372

10. Bakkers, J., Kramer, C., Pothof, J., Quaedvlieg, N.E., Spaink, H.P., and Hammerschmidt, M. (2004) Has2 is required upstream of Rac1 to govern dorsal migration of lateral cells during zebrafish gastrulation. *Development* 131, 525–537

11. Matthews, H.K., Marchant, L., Carmona-Fontaine, C., Kuriyama, S., Larraín, J., Holt, M.R., Parsons, M., and Mayor, R. (2008) Directional migration of neural crest cells in vivo is regulated by Syndecan-4/Rac1 and non-canonical Wnt signaling/RhoA. *Development* 135, 1771–1780

12. Kardash, E., Reichman-Fried, M., Maitre, J.L., Boldajipour, B., Papusheva, E., Messerschmidt, E.M., Heisenberg, C.P., and Raz, E. (2010) A role for Rho GTPases and cell-cell adhesion in single-cell motility in vivo. *Nat Cell Biol* 12, 47–53

13. Smolen, G.A., Schott, B.J., Stewart, R.A., Diederichs, S., Muir, B., Provencher, H.L., Look, A.T., Sgroi, D.C., Peterson, R.T., Haber, D.A. (2007) A Rap GTPase interactor, RADIL, mediates migration of neural crest precursors. *Genes Dev* 21, 2131–2136

14. Tsai, I.C., Amack, J.D., Gao, Z.H., Band, V., Yost. H.J., and Virshup, D.M. (2007) A Wnt-CKIvarepsilon-Rap1 pathway regulates gastrulation by modulating SIPA1L1, a Rap GTPase activating protein. *Dev Cell* 12, 335–347

15. Palamidessi, A., Frittoli, E., Garre, M., Faretta, M., Mione, M., Testa, I., Diaspro, A., Lanzetti, L., Scita, G., and Di Fiore, P.P. (2008) Endocytic trafficking of Rac is required for the spatial restriction of signaling in cell migration. *Cell* 134, 135–147

16. Zhu, S., Korzh, V., Gong, Z., and Low, B.C. (2008) RhoA prevents apoptosis during zebrafish embryogenesis through activation of Mek/Erk pathway. *Oncogene* 27, 1580–1589

17. Nasevicius, A., and Ekker, S.C. (2000) Effective targeted gene 'knockdown' in zebrafish. *Nat Genet* 26, 216–220

18. Bill, B.R., Petzold, A.M., Clark, K.J., Schimmenti, L.A., and Ekker, S.C. (2009) A primer for morpholino use in zebrafish. *Zebrafish* 6, 69–77

19. Robu, M.E., Larson, J.D., Nasevicius, A., Beiraghi, S., Brenner, C., Farber, S.A., and Ekker, S.C. (2007) p53 activation by knockdown technologies. *PLoS Genet* 3, e78

20. Hyatt, T.M., and Ekker, S.. (1999) Vectors and techniques for ectopic gene expression in zebrafish. *Methods Cell Biol* 59, 117–126

21. Xu, Q. (1999) Microinjection into zebrafish embryos. *Methods Mol Biol* 127, 125–132

22. Stuart, G.W., McMurray, J.V., and Westerfield, M. (1988) Replication, integration and stable germ-line transmission of foreign sequences injected into early zebrafish embryos. *Development* 103, 403–412

23. Rosen, J.N., Sweeney, M..F, and Mably, J.D. (2009) Microinjection of zebrafish embryos to analyze gene function. *J Vis Exp* 25, pii: 1115. doi: 10.3791/1115

24. Yuan, S., and Sun, Z. (2009) Microinjection of mRNA and morpholino antisense oligonucleotides in zebrafish embryos. *J Vis Exp* 27, pii: 1113. doi: 10.3791/1113

25. Wang, W., Liu, X., Gelinas, D., Ciruna, B., and Sun, Y. (2007) A fully automated robotic system for microinjection of zebrafish embryos. *PLoS One* 2, e862

26. Holder, N., and Xu, Q. (1999) Microinjection of DNA, RNA, and protein into the fertilized zebrafish egg for analysis of gene function. *Methods Mol Biol* 97, 487–490

27. Broadbent, J., and Read, E.M. (1999) Wholemount in situ hybridization of Xenopus and zebrafish embryos. *Methods Mol Biol* 127, 57–67

28. Brend, T., and Holley, S.A. (2009) Zebrafish whole mount high-resolution double fluorescent in situ hybridization. *J Vis Exp* 25, pii: 1229. doi: 10.3791/1229

29. Paffett-Lugassy, N.N., and Zon, L.I. (2005) Analysis of hematopoietic development in the zebrafish. *Methods Mol Med* 105, 171–198

30. Link, V., Shevchenko, A., and Heisenberg, C.P. (2006) Proteomics of early zebrafish embryos. *BMC Dev Biol* 6, 1

31. Guo, S. (2009) Using zebrafish to assess the impact of drugs on neural development and function. *Expert Opin Drug Discov* 4, 715–726

32. Westerfield, M. (2000) The zebrafish book. A guide for the laboratory use of zebrafish (*Danio rerio*). Univ. of Oregon Press, Eugene.

33. Kimmel, C.B., Ballard, W.W., Kimmel, S.R. Ullmann, B., and Schilling, T.F. (1995) Stages of embryonic development of the zebrafish. *Dev Dyn* 203, 253–310

Chapter 22

Analysis of Rho GTPase Function in Axon Pathfinding Using *Caenorhabditis elegans*

Jamie K. Alan and Erik A. Lundquist

Abstract

We provide information and protocols for the analysis of Rho GTPase function in axon pathfinding in *Caenorhabditis elegans*. The powerful molecular, genetic, imaging, and transgenic tools available in *C. elegans* make it an excellent system in which to study the *in vivo* roles of Rho GTPases. Methods for imaging of axon morphology in Rho GTPase single and double mutants are provided, as well as methods for the construction of transgenic *C. elegans* strains carrying exogenously introduced transgenes that drive the expression of constitutively active and dominant negative mutants.

Key words: Axon guidance, *C. elegans*, Rho GTPases, Cell-specific promoter, Mutant analysis, Transgene

1. Introduction

Caenorhabditis elegans is an excellent model system to study Rho GTPase involvement in axon guidance, because using this nematode it is possible to combine genetic, molecular, and transgenic analysis (1–11). Because *C. elegans* is optically transparent and easily amenable to anatomical imaging, it constitutes an ideal system to observe neuronal development and morphology *in vivo*. Furthermore, *C. elegans* has a well-defined and consistent developmental pattern, where the fate of every cell is known (12, 13). Information about the biology of *C. elegans* can be found in several published sources (14, 15) and also online (*Caenorhabditis* World Wide Web Server; http://www.elegans.swmed.edu/ and Wormbase; http://www.wormbase.org). Here, we discuss methods of analysis of Rho GTPase function in axon pathfinding, including analysis of loss of function alleles as well as analysis of constitutively active and dominant negative alleles using transgenic techniques.

Francisco Rivero (ed.), *Rho GTPases: Methods and Protocols*, Methods in Molecular Biology, vol. 827,
DOI 10.1007/978-1-61779-442-1_22, © Springer Science+Business Media, LLC 2012

Table 1
***Caenorhabditis elegans* promoters expressed in axons useful in axon guidance analysis**

Neuron(s)	Promoter active in the cell(s)
AQR/PQR	*gcy-32* (40, 41)
ALMs/PLMs AVM/PVM	*mec-4* (42)
CAN	*ceh-23* (2, 43)
PDE	*osm-6* (4, 44)
VD/DD	*unc-25* (7, 8, 45)
HSN	*unc-86* (46)
pan-neuronal	*unc-119* (47)

1.1. Visualizing the C. elegans Nervous System

C. elegans has a well-defined nervous system consisting of 302 neurons in the hermaphrodite and 473 neurons in the male (12, 13, 16). The *C. elegans* nervous system can be visualized with standard light microscopy techniques. Visualization of neurons and/ or proteins in the worm can be achieved by using transgenic expression of fluorescent proteins such as green fluorescent protein (GFP) (17), cyan fluorescent protein (CFP), red fluorescent protein (RFP), and yellow fluorescent protein (YFP) (18). In order to visualize specific proteins, GFP or other such fluor can be fused to a particular protein. To visualize specific neurons, expression of GFP (or other fluor) can also be driven by cell-specific neuronal promoters. There are several well-defined cell-specific neuronal promoters listed in Table 1. Details about promoter sequences can be found in the references cited.

1.2. C. elegans Rho Family GTPases

Rho GTPases are Ras-related small GTPases that regulate cytoskeletal organization and dynamics, cell adhesion, motility, trafficking, proliferation, and survival (19). *C. elegans* has many conserved Rho GTPases (9), including Rho (*rho-1*) (20), Cdc-42 (*cdc-42*) (21), two Rac GTPases similar to vertebrate Rac GTPases (*ced-10* and *rac-2*) (2, 22), and one Mtl Rac (*mig-2*) (1). Mtl Rac GTPases (including *mig-2*) differ from canonical Rac GTPases because they contain a N-terminal myristoylation site. There are no known vertebrate GTPases with a structure similar to Mtl, however, recent evidence suggests that *mig-2* may be functionally related to mammalian RhoG (23). Other Rho GTPases in *C. elegans* include a Chp/Wrch1 like molecule (*chw-1*) (24) and another largely uncharacterized Cdc-42 related protein (*crp-1*) that has some similarity to vertebrate TC10 (25), a GTPase family not obvious in the *C. elegans* genome. A list of the *C. elegans* Rho GTPases and their vertebrate homologs is presented in Table 2.

Several *C. elegans* Rho GTPases have been implicated in the control of neuronal cell migration, neuronal development, and axon guidance (9). RHO-1 controls axon and dendrite morphology of neurons (26). MIG-2 and CED-10 act redundantly to control multiple aspects of axon guidance (1–4).

1.3. Genetics of Rho GTPases in C. elegans

Genetic analysis in *C. elegans* is facilitated by the wide availability of deletion alleles. The use of null alleles has been greatly facilitated by the *C. elegans* gene knockout consortium (http://www.celeganskoconsortium.omrf.org/). The strains identified and characterized by the *C. elegans* gene knockout consortium are available for a nominal fee and can be obtained from the *C. elegans* genetic center (CGC, http://www.cbs.umn.edu/CGC/). Using a strain containing a deletion allele allows loss of function analysis of a variety of genes, including Rho GTPases. There are many Rho GTPase deletion alleles currently available from the CGC (Table 2). More information on these alleles can be found online at Wormbase (http://www.wormbase.org).

Certain Rho GTPase deletion alleles, such as specific alleles of both *cdc-42* and *rho-1*, result in embryonic lethality. In these cases and when double mutants result in lethality, genetic balancers must

Table 2
List of *C. elegans* Rho GTPases, their vertebrate homologues, and available alleles

Rho GTPase	Vertebrate homologue	Allele	Mutation type
rho-1	RhoA	*ok2418*	2091 bp deletion
rac-2[a]	Rac1	*ok326*	469 bp deletion
		gk281	unknown
ced-10	Rac1	*lq20*	Missense P34L mutation
		n3246	Missense G60R mutation
		tm597	612 bp deletion
		n1993	Missense V109G mutation
		n3417	977 bp deletion
mig-2	RhoG (functional homologue)	*lq13*	Missense S75F mutation
		ok2273	1192 bp deletion
		mu28	Premature stop
		rh17/gm103	Missense G16 mutation
cdc-42	Cdc42	*ok825*	632 bp deletion in 3′ UTR
		gk388	478 bp deletion
chw-1	Wrch1/Chp	*ok697*	2,030 bp deletion
crp-1	TC10?	*ok685*	TTCTT insertion/659 bp deletion

[a]*rac-2* might be a nonfunctional pseudogene

Table 3
Balancers useful in working with Rho GTPase mutations

Rho GTPase	Balancers[a]	pharyngeal GFP linked
rho-1	nT1 IV;V	Yes
rac-2	nT1 IV;V	Yes
ced-10	nT1 IV;V	Yes
mig-2	szT1 I;X	No
cdc-42	mIn1 II	Yes
chw-1	nT1 IV;V	Yes
crp-1	nT1 IV;V	Yes

[a]The roman numerals indicate the linkage groups that are balanced by the rearrangement

be used to maintain the mutations in a heterozygous state. Genetic balancers are chromosomal re-arrangements that allow mutations to be stably maintained in the worm as heterozygotes. Genetic balancers either suppress recombination or produce inviable recombinant gametes in the region of interest. A list of balancers useful for Rho GTPase mutants is presented in Table 3. Some genetic balancers (such as mIn1 and hT2) have integrated pharyngeal GFP (27–29), which makes selection of unbalanced (non-GFP) homozygotes easily identifiable. More information on specific genetic balancers can be found online using Wormbook (http://www.wormbook.org/chapters/www_geneticbalancers/geneticbalancers.html).

1.4. Constitutively Active and Dominant Negative Rho GTPases

In addition to using deletions alleles, analysis of constitutively activated and dominant negative Rho GTPase proteins is also possible using *C. elegans*. When small GTPases are GDP bound, they are inactive and do not engage and activate downstream effectors (30). When they are GTP bound, they undergo a conformational change, allowing them to engage and activate downstream effectors (30). Small GTPases are positively regulated by guanine nucleotide exchange factors (GEFs), which facilitate the exchange of GDP for GTP (31) and are negatively regulated by GTPase activating proteins (GAPs), which enhance hydrolysis of the terminal phosphate of GTP, resulting in a GDP bound and inactive protein (see Fig. 1) (32). One main type of activating mutation can be engineered in a variety of Rho GTPases at one of two positions, a glycine (G) to valine (V) mutation at the residue analogous to G12 in Ras, or a glutamine (Q) to leucine (L) mutation, analogous to Q61 in Ras. Both of these mutations render the protein GTPase deficient and GAP-insensitive, therefore locking the GTPase in its GTP-bound and constitutively activated form (33–35). The chromosomal mutations

Positive regulator

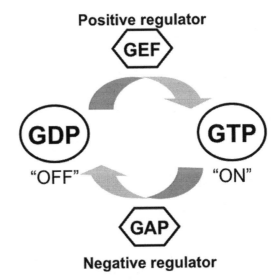

Negative regulator

Fig. 1. Small GTPases are regulated by GDP/GTP binding. GEFs (guanine nucleotide exchange factors) promote GTP binding, resulting in an activated GTPase (31). GAPs (GTPase activating proteins) hydrolyze GTP, leaving the proteins GDP bound and inactive (32).

mig-2(rh17) and *mig-2(gm103)* both cause missense of the G16 residue in MIG-2, analogous to the activating G12 Ras mutation (1, 7). Thus, these mutations can be analyzed as constitutively active forms of the MIG-2/Mtl GTPase. Dominant negative mutations in Rho GTPases can be generated by substituting a threonine (T) to asparagine (N) at the position analogous to T17 in Ras. The T17N mutation locks the GTPase in a nucleotide-free state, trapping GEFs and resulting in a dominant negative GTPase (36). Both constitutively active and dominant negative forms of Rho GTPases can be engineered by using site-directed mutagenesis. Indeed, a transgene expressing a G12V mutant form of CED-10/Rac1-induced ectopic lamellipodia and filopodia in neurons due to constitutive activation of the CED-1/Rac1 GTPase (4). Other dominant chromosomal mutations of *ced-10* and *mig-2* exist (*mig-2(lq13)S75F, ced-10(lq20) P34L,* and *ced-10(n3246)G60R*) (8), although it is not known if these are activating or dominant negative mutations.

1.5. Transgenic Manipulation of Rho GTPases in C. elegans

Expression of most small GTPases, driven by their endogenous promoters, is usually ubiquitous throughout the worm (1, 2). When transgenic constitutively activated CED-10(G12V) is expressed by its own promoter, embryonic lethality results and transgenic lines cannot be established (our unpublished observations). This is likely due to the widespread effects of the activated CED-10/Rac1 on multiple aspects of development. Therefore, it is often useful to express constitutively activated small GTPases in a cell-specific manner, in this case in subsets of neurons that are not required for viability. To do this, expression of these GTPases can be driven by a cell-specific promoter. There are many well-characterized cell-specific neuronal promoters available (Table 1). By first engineering an activating mutation in a small GTPases, and then expressing the protein in a cell-specific manner, two major types of analyses can be conducted. First, the effect of the activated GTPase on the phenotype of the neuron can be observed and quantified. For example, the activated allele, *mig-2(rh17)*, results in axon pathfinding defects in the PDE neuron as well as defects in AQR and PQR migration (7), and transgenic expression of *ced-10(G12V)* results in axon guidance defects as well as ectopic lamellipodia and filopodia and ectopic neurites (4).

If the activated GTPase produces a measurable phenotype in the neuron (e.g., ectopic lamellipodia and filopodia), null alleles in genes encoding putative downstream effectors can be crossed into the activated GTPase background to determine if the gene products are required for the effects of the activated GTPase. By performing this type of analysis, genetic pathways and epistatic relationships can be explored and elucidated. For example, mutations in *unc-115/abLIM*, the Arp2/3 activators *wve-1/WAVE* and *wsp-1/WASP*, and the *rack-1/Receptor for activated C kinase* all suppress the effects of CED-10(G12V) (4, 7, 11), indicating that these molecules act downstream of CED-10/Rac in lamellipodia and filopodia formation.

Once an activating mutation is engineered and a fluorescent tag is attached, the transgene can then be used to generate a transgenic animal. Transgenes in *C. elegans* are maintained as either extrachromosomal arrays or integrated arrays (37). Extrachromosomal arrays are repeated copies of the transgene maintained extrachromosomally in a contiguous array. Extrachromosomal arrays are lost from the germline and mitotically at some frequency and can change their properties over time. Therefore, extrachromosomal arrays may result in difficulties maintaining the transgene and scoring a particular phenotype. These problems can be overcome by maintaining the transgene as an integrated array. Integration involves treating an array-bearing strain with DNA-damaging agents (UV and trimethyl psoralen (TMP)) and selecting for arrays that have been integrated into the genome via nonhomologous end joining of double strand breaks. Integrated arrays offer several advantages including stronger and more reproducible expression patterns and phenotypes. A new method for generating single copy transgenic constructs involves the use of the Mos transposable element. A detailed description of this method is not included here but can be found in ref. 38.

In this chapter, we present methods for (1) generating transgenic animals; (2) obtaining *C. elegans* strains, including strains that harbor *promoter::fluor::gtpase* transgenes and null alleles; (3) introducing *promoter::fluor::gtpase* transgenes into different mutant backgrounds; (4) mounting worms for inspection by fluorescence microscopy; and (5) strategies for scoring neuronal defects in the worm.

2. Materials

2.1. Buffers, Media, and Equipment

1. Nematode growth medium (NGM) plates. To make 1 L of NGM add 2.5 g of tryptone, 3 g of NaCl, and 15 g of agar to 995 mL of water. Add 1 mL of each of the following sterile solutions: 5 mg/mL cholesterol in ethanol, 1 M $CaCl_2$, 1 M $MgSO_4$, 2 M Tris base, and 3.5 M Tris–HCl. Sterilize the mixture by autoclaving. To make NGM plates, pour sterile NGM into a petri dish, which can be done aseptically using a peristaltic pump. Medium size 60-mm diameter plates are typically used for all the procedures outlined in this chapter. Once the plates are poured, allow them to solidify overnight, then store at 4°C in an airtight container.

2. LB broth: combine 10 g of tryptone 5 g of yeast extract and 5 g of NaCl in 1 L of ddH_2O. Autoclave and store at room temperature.

3. *Escherichia coli* strain OP50: This strain is used as a food source for *C. elegans*. To prepare a stock of OP50 to use for seeding NGM plates, aseptically inoculate a 250-mL screw top bottle containing 100 mL of sterile LB broth, and allow growing at room temperature overnight. The resulting OP50 liquid culture should be stored at 4°C and if not contaminated, can be usable for several months.

4. NGM plates seeded with OP50: Drop approximately 0.5 mL of OP50 liquid culture onto each 60-mm NGM plate, using sterile technique. Spread the bacteria drop using a glass rod. Once seeded, allow the plates to dry overnight. Store the plates in an airtight container at room temperature for up to 2 weeks.

5. M9 buffer: combine 3 g of KH_2PO_4, 6 g of Na_2HPO_4, and 5 g of NaCl in 1 L of ddH_2O and autoclave. After the mixture has cooled, add 1 mL of 1 M $MgSO_4$. Store at room temperature.

6. Co-injection markers: *str-1::gfp* (39) and rol-6. The plasmid pRF4 (*rol-6, su1006*) can be obtained from J. Kramer (Chicago, IL).

7. Cosmid DNA. It can be ordered from the Sanger Institute (http://www.sanger.ac.uk/technology/clonerequests/).

8. TMP: dissolve into dimethylformamide or dimethylsulfoxide at a concentration of 1 mg/mL.

9. 2% Agarose (Sigma) in water or in M9, with or without 5 mM NaN_3.

10. A "worm pick" (see Note 1).

11. Glass microscopy slides and 22 × 60 mm coverslips.

12. Mineral oil.

13. Microscopes: a compound microscope with at least a 10× and a 40× objective, a dissecting microscope, a fluorescence dissecting microscope, and an injection scope with a 10× and a 40× objective fitted with a needle arm for microinjection.

14. Microinjector pump (Harvard apparatus PLI-90).

15. Calibrated UV light source (crosslinker).

16. Needle puller P30 (Sutter Instruments) and borosilicate microcapillary pipets (World Precision Instruments).

2.2. Growth and Maintenance of C. elegans

Wild-type (N2) and mutant strains of *C. elegans* can be obtained from the *Caenorhabditis elegans* Genetics center (CGC) (http://www.cbs.umn.edu/CGC) or from individual *C. elegans* laboratories (http://www.elegans.swmed.edu/Worm_labs/). Nematodes are typically grown on NGM plates that have a lawn of the *E. coli* strain OP50. Strains of *C. elegans* are easily maintained by transferring hermaphrodites from old plates to new plates containing a fresh lawn of OP50. Hermaphrodites should be transferred from the old plate when the bacterial lawn has been consumed by the animals. Hermaphrodites will self-fertilize ("selfing") and give rise to approximately 250 progeny. There are two main methods of transferring hermaphrodites: "picking" worms by the use of a platinum wire, and "chunking" worms by using a sterilized spatula. Picking is generally used to plate single worms or to set up matings, while chunking is used for general strain maintenance. Once the animals are plated, they should be maintained at 20°C, unless a specific mutation in the animals dictates maintenance at another temperature (generally 15°C or 25°C).

2.2.1. "Picking" Worms

1. Briefly sterilize the platinum wire of a worm pick in the open flame of an alcohol burner (see Note 1).

2. While looking at the plate through a dissecting microscope, place a worm on the end of the platinum wire of the pick (see Note 2).

3. Quickly transfer the worm to a NGM plate seeded with OP50, using the dissecting microscope to visualize the worm (see Note 3).

4. Sterilize the platinum wire of the worm pick to prevent cross-contamination of strains and contamination with bacteria or mold from the environment. Repeat until the desired number of worms are transferred.

5. When the transfer of worms is completed, store the plate at 20°C.

2.2.2. "Chunking" Worms

1. Sterilize the end of a spatula by dipping it in 95% alcohol and passing it through the flame of an alcohol burner, allowing the alcohol to burn off.

2. After allowing the spatula to cool, use the tip to cut a small square chunk out of the agar.

3. Gently and quickly place the chunk of agar, worm-side down, onto a NGM plate seeded with OP50.

4. Sterilize the spatula to prevent cross-contamination of strains, and to prevent contamination with bacteria or mold from the environment.

5. When the transfer of worms is completed, store the plate at 20°C.

2.3. Promoter ::fluor::gtpase Constructs

C. elegans expression vectors containing GFP or other fluor of interest can be obtained online at Addgene (http://www.addgene.org). The cDNA of the desired GTPase is inserted into the multiple cloning site of the target vector. Once a plasmid is obtained, it is relatively simple to made further modifications, such as changing to a cell-specific promoter, changing a fluor, and/or performing site-directed mutagenesis to introduce activating or null mutations. If necessary, the promoter or fluor from the obtained plasmid can be swapped to a cell-specific promoter or alternate fluor using PCR and conventional cloning. Site-directed mutagenesis can be done using commercially available kits such as the QuickChange® kit (Stratagene).

3. Methods

3.1. Making Transgenic Animals

3.1.1. Preparing the DNA and the Worms for Injection

DNA can be injected into *C. elegans* to create a transgenic animal. Before proceeding, select a co-injection marker to be injected with the DNA of interest. A co-injection marker should be an obvious phenotypic marker such as rol-6 (results in animals that are twisted into a right-handed helix) or gfp-str-1 (results in animals with a single green neuron in the head). Quantitate the plasmids and dilute the DNAs to their proper concentrations in 20 μL of ddH$_2$O. DNA concentrations should be used as follows: experimental plasmid (1–25 ng/μL), rol-6 co-injection marker (12–25 ng/μL), gfp-str-1 co-injection marker (12–25 ng/μL), cosmid DNA (25–50 ng/μL) (see Note 4). A noncontaminated stock of N2 worms should be used for injection purposes (see Note 5). Contamination of the plate may result in darker worms, making the visualization of the gonad more difficult.

3.1.2. Preparing Agarose Pads for Injection

1. Place several 22 × 60 mm coverslips on a benchtop.

2. Prepare a 2% agarose solution in ddH$_2$O (not M9 or other salt buffer) and microwave the mixture until dissolved.

3. Using a dropper, place a drop of liquid agarose onto a coverslip and immediately place another coverslip on top of the drop. The diameter of the agarose pad should be between 15 and 20 mm.

4. After the agarose has solidified, slide the coverslips apart and allow the agarose pad to air dry completely (1 h to overnight).

5. Store the injection pads in a coverslip box at room temperature.

3.1.3. Preparing the Injection Needles and Needle "Breakers"

1. Make fresh needles every time you inject. Prepare a needle using a borosilicate microcapillary pipet, pulled on a needle puller. For a Sutter Instruments P-30 needle puller, the recommended general settings are Heat #1 – 605, Pull – 990. Pull the needles so that the taper is medium (the needle is not too long and not too short), adjusting the heat as necessary to achieve the proper taper. Pull several needles at a time. The needles can be stored and transported in a box, mounted on a strip of modeling clay.

2. The tip of the microinjection needle is very fine and closed at the end. The needle will need to be broken using a thin piece of glass. Grasping both ends of a microcapillary pipet, heat the middle portion in a flame and when hot quickly pull the ends apart, leaving a long string of glass attached to both ends.

3.1.4. Loading the Needle for Injection

1. Load the needle just prior to use. Using a P20 pipetman or a pulled pipet, place a small drop (1 μL or less) onto the open end of the pulled needle.

2. After about 5 min, capillary action will draw the injection solution to the bottom of the needle tip. Be careful not to bump the needle against anything when loading and mounting the needle.

3.1.5. Mounting the Needles on the Needle holder of the Microinjector

1. Place the collar from the needle holder over the open end of the needle, such that the tapered end of the collar faces the point of the needle.

2. Place a small piece of nylon tubing over the open end of the needle.

3. Place the open end of the needle in the needle holder and screw on the needle collar.

3.1.6. Breaking the Needle for Injection

1. Open the main valve on a nitrogen or helium tank that is connected to a microinjection needle arm and holder.

2. Set the "injection" pressure to 10–20 psi and the "out" pressure to about 2 psi.

3. Place a small filament of needle-breaking glass in a drop of mineral oil on a coverslip, and then place the coverslip on the stage of an inverted microscope.

4. Position the glass piece on the coverslip in the center, then focus on it using a low power objective and bright field illumination.

5. Using the micromanipulator, position the needle in the center of the viewing field. Lower the needle so that it just touches the top of the oil.

6. Focus up with the microscope to locate the needle tip. Coordinately bring the tip down while focusing on the tip.

Bring the tip and filament into focus in the same plane. Be careful not to bring the tip down too far, which could result in ramming the tip into the coverslip thus breaking the needle.

7. When the needle tip and filament are in the same plane, depress the foot pedal and check for the ejection of fluid, to see if the needle is already broken. If not, press the foot pedal to apply pressure, and gently ram the tip into the side of the filament by moving the stage. When flow comes out of the needle tip, stop ramming the needle into the filament. Break off only the tip of the needle because a blunt needle will not work well in injections.

8. Remove the coverslip from the stage after raising the needle up, but do not change the X- or Y-axis position.

3.1.7. Mounting the Worms on Injection Pads

1. Place a drop of mineral oil on a prepared agarose pad. Using a worm pick, pick a healthy, nonstarved, young adult worm into the drop of mineral oil. Avoid carrying bacteria with the worm. It is easiest to pick the worm onto the very tip of the worm pick, and then transfer the worm into the oil by gently dragging the animal in the oil.

2. Gently press the worm onto the surface of the agarose pad, making sure its entire body sticks onto the injection pad.

3. After the worm is in the oil, you now have about 5 min to inject the worm and recover it in M9 buffer.

3.1.8. Injecting the Worm

1. Place the agarose pad onto the stage and center the worm using a low power objective.

2. Using techniques similar to needle breaking (see Subheading 3.1.6) lower the needle into the same plane of focus as the worm. Switch to a 40× objective, focus on the worm syncytial gonad and position the needle in the same plane as the syncytial gonad.

3. Using the stage, move the needle into the worm syncytial gonad.

4. When the needle has entered the syncytial gonad, apply injection pressure. If the liquid has entered the gonad, you will observe a small wave of liquid spreading through the gonad (see Note 6). Be careful not to overinject, which will result in a sterile animal.

5. Remove the needle from the worm by moving the stage, changing back to a low power objective, and raising the needle out of the oil.

3.1.9. Recovering the Worm

1. Remove the injection pad and place a drop of M9 directly on top of the worm. The M9 will displace the oil and the worm will move into the M9.

2. Pick the worm out of the M9 and transfer to a seeded NGM plate, directly in another drop of M9.

3. Allow the worm to recover for 30–60 min. After this time, singly plate the worms onto seeded NGM plates.

4. Screen the singly plated worms for expression of the co-injection marker.

5. When the F_1 progeny that express the co-injection marker reach the L3 to L4 stage, singly plate them to seeded NGM plates.

6. Screen for plates that produce F_2 worms that express the co-injection marker.

3.1.10. Integrating Extra-Chromosomal Arrays Using UV Irradiation

1. Pick 15–30 worms that express the array into 50 μL of 30 μg/mL TMP dissolved in M9, on an unseeded NGM plate.

2. Place the plate in the dark for 15 min, then wash 2× with M9 buffer.

3. Place the plates in a UV crosslinker, with the lids removed. UV irradiate with 350 μJ (×100) of 360 nm long wave UV light (at 300 μW/cm^2 for 30 s to 2 min).

4. Singly plate the worms onto seeded NGM plates, then allow them to recover in the dark (5 h to overnight).

5. From the animals that have been mutagenized with UV/TMP, pick 15–30 young adult hermaphrodite P_0 worms to single plates.

6. From the progeny of the P_0 worms, pick 120–150 F_1 worms that express the transgene to single plates.

7. From each F_1 plate, even if there are few progeny, pick 2 F_2 animals that express the transgene.

8. Screen for plates where all the animals express the transgene, indicating that the array has been integrated and has been homozygosed.

9. Outcross the suspected integrants.

3.2. Introducing Promoter::fluor::gtpase Transgenes into Different Mutant Backgrounds

To determine a genetic relationship between Rho GTPases and other molecules, it is often useful to study *promoter::fluor::gtpase* transgenes in different mutant backgrounds. A genetic cross is required to make the combination of the mutant and the transgene, in which a *trans*-heterozygote of the mutant and transgene is created, and the resulting hermaphrodite is allowed to self-fertilize, followed by selection of the double homozygote. Mating of male worms to hermaphrodites is required for these genetic crosses, the details of which are discussed below.

3.2.1. Heat Shocking to Generate Males for C. elegans Crosses

Males arise in populations of *C. elegans* at a low frequency (approximately 0.2%) (15). The frequency of males can be increased to 2–5% by heat-shocking hermaphrodites during germ cell development.

1. Prepare five seeded NGM plates, each containing 5 L4 or young adult hermaphrodites.

2. Incubate the animals at 37°C for 1 h (alternatively 30°C for 6 h).

3. Allow the heat-shocked hermaphrodites to self-fertilize at 20°C.

4. Inspect the resulting progeny for the presence of male worms.

5. Pick the male worms onto seeded NGM plates for crossing (see Note 7).

3.2.2. C. elegans Matings

1. Onto a seeded NGM plate, place 5–10 males (see Subheading 3.2.1). In order to avoid contamination, ensure that no eggs or larvae are accidentally transferred from the male stock. If eggs or larvae are inadvertently transferred, move the males to a seeded NGM plate, or remove the eggs and larvae with a worm pick.

2. Pick 5L4 hermaphrodites to the plate containing the male worms. Using "virgin" L4 hermaphrodites ensures that self-fertilization has not yet occurred, thus promoting maximum mating efficiency.

3. Incubate the plate containing males and hermaphrodites at 20°C (see Note 8).

4. On each of the following 2 days, transfer the males and hermaphrodites to a new seeded NGM plate.

5. On days 3 and 4, inspect the plates for the presence of cross-progeny.

3.2.3. Strategies to Introduce Promoter::gfp::gtpase Transgenes into Different Mutant Backgrounds

Figure 2 shows a schematic for how to cross a *promoter::fluor::gtpase* construct into different mutant backgrounds. In this example, *mig-15(rh148)* is the mutation of interest and *lqIs37* is the *osm-6::cdc-42(G12V)*-integrated transgene (see Note 9). This scheme can be generalized for crossing many combinations of mutant and transgene strains. Animals homozygous for *mig-15(rh148)* are easily identified by phenotype, as the animals are *unc*, *dpy*, and *pvul* (see Note 10). This cross will require three strains, wild-type (N2), *mig-15(rh148)* homozygotes, and *lqIs37* homozygotes.

1. Isolate N2 males as described in Subheading 3.2.1.

2. P_0 cross: mate N2 males to *lqIs37* hermaphrodites as described in Subheading 3.2.2.

3. F1 cross: mate *lqIs37/+* F1 males (they will be GFP+) to *mig-15(rh148)* homozygous hermaphrodites (they will be *unc, dpy, and pvul*). All F_1 males from the P_0 cross will be cross progeny, and any that exhibit GFP fluorescence (as seen by using a fluorescence dissecting microscope) will have inherited the *lqIs37* transgene.

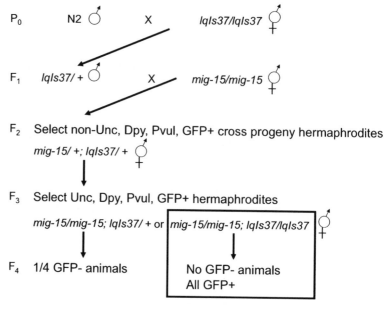

Fig. 2. An example scheme showing how to introduce *promotor::gfp::gtpase* transgenes into different mutant backgrounds. *lqIs37* is the integrated *osm-6::cdc-42(G12V)* transgene and *mig-15* is the *mig-15(rh148)* allele that has an uncoordinated (*Unc*), dumpy body shape (*Dpy*), and protruding vulva (*pVul*) phenotype that is easily observed with a dissecting microscope. *lqIs37* is followed by dominant GFP fluorescence. The generations are noted as parental (P_0) and the first, second, and third filial (F_1–F_3). The P_0 and F_1 generations involve crosses between males and hermaphrodites. The F_2 and F_3 generations involve hermaphrodite self-fertilization. The double homozygote *mig-15/mig-15; lqIs37/lqIs37* is boxed.

4. F_2 generation: Using a fluorescence dissecting microscope, single plate 5–10 GFP + F_2 cross-progeny L4 hermaphrodites to seeded NGM plates. Picking L4 hermaphrodites or "virgin" hermaphrodites ensures that they have not been fertilized by the males on the plate. These F_2 hermaphrodites will be heterozygous for both *lqIs37* and *mig-15(rh148)* (*lqIs37/+; mig-15(rh148)/+*). Allow these F_2 hermaphrodites to self-fertilize.

5. F_3 generation: Approximately one-fourth of the F_3 animals with be homozygous for *mig-15(rh148)* and approximately three-fourths will be GFP+, however, only one-fourth of the animals will be *lqIs37* homozygotes. From one plate of F_2 hermaphrodites singly plate 16 F_3 homozygous *mig-15(rh148)* hermaphrodites (animals will be *unc, dpy,* and *pvul*) that are GFP+ and to seeded NGM plates. The *mig-15(rh148)* GFP + F_3 will be one of two genotypes, *mig-15(rh148)/mig-15(rh148); lqIs37/+* and *lqIs37/lqIs37*.

6. F_4 generation: Inspect F_4 animals from the 16 and F_3 plates for GFP using a fluorescence dissecting microscope. The F_3

animals that are homozygous for *lqIs37* will give rise to all GFP+ and animals and no GFP– animals. Those that were heterozygous for *lqIs37* will give rise to approximately one-fourth GFP- and animals. Save a strain that is homozygous for both *mig-15(rh148)* and *lqIs37* for analysis.

3.3. Mounting C. elegans for Examination by Fluorescence Microscopy

Once the transgene is introduced into mutant backgrounds of interest, the animal is ready to inspect for axon pathfinding defects. First, animals are mounted onto an agarose pad, and then visualized by fluorescence microscopy.

3.3.1. Preparing an Agarose Pad for Mounting C. elegans

1. Melt 2% agarose in M9, then add 5 mM NaN_3 to the mixture (NaN_3 is used to immobilize the worms). This mixture should be maintained at 65°C to maintain it in liquid form. This can easily be done by aliquoting the mixture into glass culture tubes placed in a 65°C heat block.

2. Prepare spacer slides by placing a strip of laboratory tape onto two glass microscope slides. These spacer slides will be used to regulate the thickness of the agarose pad.

3. Align the two spacer slides in parallel and place a third blank glass slide in between the two spacer slides.

4. Using a dropper, place a drop of the melted 2% agarose solution (about 100 μL) onto the middle glass slide. Immediately place another slide on top of the drop of melted agarose solution, perpendicular to and supported by the spacer slides.

5. After waiting approximately 1 min for the agarose to solidify, gently remove the top glass slide, while keeping the agarose pad on the bottom slide. The agarose pad is now ready to use, and can be stored in a humidified chamber for several hours.

3.3.2. Mounting Animals onto the Agarose Pad

1. Place 10 μL of M9 with 5 mM NaN_3 onto the agarose pad.

2. Using a worm pick, select several animals from a NGM plate, while observing them with a dissecting microscope. For many studies young adult worms should be observed, however, experimental conditions might require that earlier larval stages be examined. After placing the animals on a worm pick, transfer them to the drop of M9 containing NaN_3. Be sure to complete the transfer before the drop of M9 solution is evaporated (see Note 11).

3. Slowly and carefully, place a glass coverslip over the agarose pad containing the animals.

3.3.3. Recovering Animals from the Agarose Pad

If required, animals can be recovered from the agarose pad after imaging, even if NaN_3 is used. The animals should be recovered within 30 min to avoid death by NaN_3 or desiccation of the slide.

1. While observing the animal of interest under the compound microscope, establish a landmark by observing the position of the animal of interest, relative to other animals.

2. Transfer the slide to a dissecting microscope. While observing the animal of interest under the dissecting scope, place 5 μL of M9 underneath the edge of the coverslip.

3. Slowly slide the coverslip off to one side and then off the agarose pad, without putting pressure on the pad. Slide the coverslip off while simultaneously observing the animal of interest under the dissecting microscope.

4. Once the coverslip is off, remove the animal of interest from the agarose pad, and transfer it to a seeded NGM plate. Allow the animal to recover on the NGM plate, which can take up to 1 h. To aid in recovery, place 5 μL of M9 over the animal.

3.4. Scoring Defects in C. elegans Neurons

1. Place the prepared slide onto the stage of a compound microscope.

3.4.1. Observing C. elegans Under a Compound Microscope

2. Locate the animals using bright field imaging under low magnification.

3. Turn off the bright field light source and turn on the fluorescent light source to observe the fluor.

4. If necessary, switch to higher magnification to observe the neuron(s) of interest (usually a 40× objective is sufficient to visualize most cells).

5. Once the neuronal phenotype of the neuron(s) has been noted, repeat steps 2–4 until all animals are scored (see Note 12).

3.4.2. Strategies for Axon Pathfinding Scoring Defects in C. elegans Neurons

1. Before observing neurons in a mutant background, the phenotype of the neurons in a wild-type background should be scored first, in order to become familiar with normal variations in the phenotype of the neuron. Since neurons change throughout development, similar stages in both wild-type and mutant backgrounds should be scored (i.e., if one is scoring young adults in a mutant background, young adults should also be scored in a wild-type background).

2. It is important to note that phenotypes will vary depending on the specific neuron being studied. Defects may include (but are not limited to): axon pathfinding defects, extra neurites, branching, ectopic lamellipodia, ectopic filopodia, and reversal of polarity. For PDE neurons, the most common defects are axon-pathfinding defects, extra-neurites, and branching (see Note 13). Several examples of PDE defects are illustrated in Fig. 3. Before scoring, it is important to define the defect to be consistent when scoring animals. For example, an axon pathfinding defect in a PDE neuron would be any PDE axon that does not reach the ventral nerve cord, or takes greater than a 45° turn from the wild-type ventral trajectory.

3. When scoring worms with multiple transgenes or mutant alleles, it is often useful to determine the defects caused by

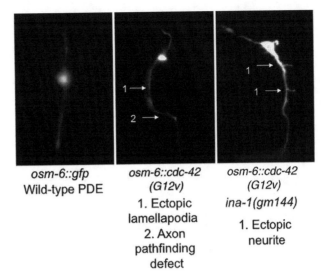

osm-6::gfp
Wild-type PDE

osm-6::cdc-42
(G12v)
1. Ectopic
lamellapodia
2. Axon
pathfinding
defect

osm-6::cdc-42
(G12v)
ina-1(gm144)

1. Ectopic
neurite

Fig. 3. Examples of defects seen in the PDE neuron. (a) A wild-type PDE neuron that extends an axon ventrally from the cell body toward the ventral nerve cord (an *arrow* indicates the axon). (b) A PDE with an ectopic lamellipodium (1) and an axon-pathfinding defect (2). (c) A PDE neuron that displays an ectopic lamellipodium (1) and an ectopic neurite (2). The scale bar in A represents 5 μm.

each individual transgene or mutant allele as well as the double mutant. This type of analysis will help determine if the defects are a result of the transgene or mutant allele alone, or a result of the interaction between the two perturbations.

3.4.3. Statistical Analysis of Defects in C. elegans Neurons

When comparing the neuronal defects of mutant strains to wild-type animals, it is important to determine if there is a statistically significant difference between the strains. The most common statistical tests used in analyzing neuronal defects are the *t*-test and analysis of variance (ANOVA).

4. Notes

1. When making a worm pick, use the flat end of a razor blade to flatten the end of the platinum wire into a shovel shape.

2. When picking up the worm from a plate, it is helpful to coat the end of the worm pick with a globule of OP50, which will serve as "glue" to pick up the worms.

3. To remove the worms from the pick, gently touch the pick to the agar, and allow the animal to crawl off. Alternatively, the worm can be removed by gently dragging the pick across the surface of the agar. Care should be taken to not puncture the agar, because the worms will burrow into any break in the agar surface. When

first learning how to pick worms, transfer one worm at a time. With practice, it is possible to transfer several animals at a time.

4. The ideal concentration of DNA to use for injecting *C. elegans* will vary depending on the construct, and can range from 1 ng/μL to 100 μg/μL. To avoid overexpression artifacts, the smallest amount of DNA possible should be used. To start, a concentration of 15–50 μg/μL is recommended. If expression is seen with these concentrations, it is recommended to titrate down the amount of DNA used until the smallest amount possible is used. If expression is not seen using these concentrations, it is recommended to titrate up as needed.

5. In order to clean up contaminated worm plates, transfer worms with a worm pick to uncontaminated seeded NGM plates several times in 1 day. Repeat this process until the plates are not contaminated.

6. When piercing the gonad with the needle, it is best to enter at approximately a 180° angle, relative to the gonad. This will increase the survival rate of the worms.

7. Once males have been isolated by heat-shocking, it is useful to maintain a male stock of the strain for future use. Using the mating protocol outlined in Subheading 3.2.2, mate the males produced from heat-shocking to hermaphrodites from the same strain. The resulting progeny will be one-half male, and can be used in future crosses.

8. To increase mating efficiency, special mating plates can be used when making crosses. Instead of spreading a lawn of OP50, place a drop of OP50 on a NGM plate, and allow to dry. Place the males and hermaphrodites in this small area of bacteria, which will increase the likelihood of mating.

9. This crossing scheme uses an integrated transgene. If an extrachromosomal array is used, select the animals in each generation that harbor the transgene with a fluorescence dissecting microscope.

10. The phenotype of *mig-15(rh148)* is a visible phenotype where the worm is short and fat (*dpy*), displays uncoordinated motion (*unc*), and has a protruding vulva (*pvul*).

11. Animals can be removed from the worm pick by allowing them to swim off the end into the drop of M9. To fasten this process, slowly drag the pick through the drop of M9.

12. To ensure that all animals are scored only once, start at one end of the agarose pad, and scan in a grid pattern until the whole agarose pad has been scored.

13. Before scoring worms, it is useful to prepare a scoring sheet containing: a place to write the name of the worm strain, the date, and a list of expected phenotypes. This type of sheet can easily be made in a program like Microsoft Excel.

References

1. Zipkin, I.D., Kindt, R.M., and Kenyon, C.J. (1997) Role of a new Rho family member in cell migration and axon guidance in *C. elegans*. *Cell* 90, 883–894.

2. Lundquist, E.A., Reddien, P.W., Hartwieg, E., Horvitz, H.R., and Bargmann, C.I. (2001) Three *C. elegans* Rac proteins and several alternative Rac regulators control axon guidance, cell migration and apoptotic cell phagocytosis. *Development* 128, 4475–4488.

3. Wu, Y.C., Cheng, T.W., Lee, M.C., and Weng, N.Y. (2002) Distinct rac activation pathways control *Caenorhabditis elegans* cell migration and axon outgrowth. *Dev Biol* 250, 145–155.

4. Struckhoff, E.C., and Lundquist, E.A. (2003) The actin-binding protein UNC-115 is an effector of Rac signaling during axon pathfinding in *C. elegans*. *Development* 130, 693–704.

5. Gitai, Z., Yu, T.W., Lundquist, E.A., Tessier-Lavigne, M., and Bargmann, C.I. (2003) The netrin receptor UNC-40/DCC stimulates axon attraction and outgrowth through enabled and, in parallel, Rac and UNC-115/AbLIM. *Neuron* 37, 53–65.

6. Steven, R., Kubiseski, T.J., Zheng, H., Kulkarni, S., Mancillas, J., Ruiz Morales, A., Hogue, C.W., Pawson, T., and Culotti, J. (1998) UNC-73 activates the Rac GTPase and is required for cell and growth cone migrations in *C. elegans*. *Cell* 92, 785–795.

7. Shakir, M.A., Jiang, K., Struckhoff, E.C., Demarco, R.S., Patel, F.B., Soto, M.C., and Lundquist, E.A. (2008) The Arp2/3 activators WAVE and WASP have distinct genetic interactions with Rac GTPases in *Caenorhabditis elegans* axon guidance. *Genetics* 179, 1957–1971.

8. Shakir, M.A., Gill, J.S., and Lundquist, E.A. (2006) Interactions of UNC-34 Enabled with Rac GTPases and the NIK kinase MIG-15 in *Caenorhabditis elegans* axon pathfinding and neuronal migration. *Genetics* 172, 893–913.

9. Lundquist, E.A. (2006) Small GTPases. *WormBook*, 1–18.

10. Quinn, C.C., Pfeil, D.S., and Wadsworth, W.G. (2008) CED-10/Rac1 mediates axon guidance by regulating the asymmetric distribution of MIG-10/lamellipodin. *Curr Biol* 18, 808–813.

11. Demarco, R.S., and Lundquist, E.A. RACK-1 Acts with Rac GTPase Signaling and UNC-115/abLIM in *Caenorhabditis elegans* axon pathfinding and cell migration, *PLoS Genet* 6, e1001215.

12. Sulston, J.E., Schierenberg, E., White, J.G., and Thomson, J.N. (1983) The embryonic cell lineage of the nematode *Caenorhabditis elegans*. *Dev. Biol.* 100, 64–119.

13. Sulston, J.E., and Horvitz, H.R. (1977) Post-embryonic cell lineages of the nematode, *Caenorhabditis elegans*. *Dev Biol* 56, 110–156.

14. Riddle, D.L., Blumenthanl, T., Meyer, B., and Priess J., eds. (1997) *C. elegans* II. Cold Spring Harbor Laboratory Press, Cold Spring Harbor.

15. Wood, W., ed. (1988) The nematode *Caenorhabditis elegans*. Cold Spring Harbor Laboratory Press, Cold Spring Harbor.

16. White, J.G., Southgate, E., Thomson, J.N., and Brenner, S. (1986) The structure of the nervous system of the nematode *Caenorhabditis elegans*. *Philos. Trans. R. Soc. Lond.* 314, 1–340.

17. Chalfie, M., Tu, Y., Euskirchen, G., Ward, W.W., and Prasher, D.C. (1994) Green fluorescent protein as a marker for gene expression. *Science* 263, 802–805.

18. Shaner, N.C., Steinbach, P.A., and Tsien, R.Y. (2005) A guide to choosing fluorescent proteins. *Nat Methods* 2, 905–909.

19. Jaffe, A.B., and Hall, A. (2005) Rho GTPases: biochemistry and biology, *Annu Rev Cell Dev Biol* 21, 247–269.

20. Spencer, A. G., Orita, S., Malone, C. J., and Han, M. (2001) A RHO GTPase-mediated pathway is required during P cell migration in *Caenorhabditis elegans*. *Proc Natl Acad Sci USA* 98, 13132–13137.

21. Kay, A.J., and Hunter, C.P. (2001) CDC-42 regulates PAR protein localization and function to control cellular and embryonic polarity in *C. elegans*. *Curr Biol* 11, 474–481.

22. Reddien, P.W., and Horvitz, H.R. (2000) CED-2/CrkII and CED-10/Rac control phagocytosis and cell migration in *Caenorhabditis elegans*. *Nat Cell Biol* 2, 131–136.

23. deBakker, C.D., Haney, L.B., Kinchen, J.M., Grimsley, C., Lu, M., Klingele, D., Hsu, P. K., Chou, B.K., Cheng, L.C., Blangy, A., Sondek, J., Hengartner, M.O., Wu, Y.C., and Ravichandran, K.S. (2004) Phagocytosis of apoptotic cells is regulated by a UNC-73/TRIO-MIG-2/RhoG signaling module and armadillo repeats of CED-12/ELMO. *Curr Biol* 14, 2208–2216.

24. Tao, W., Pennica, D., Xu, L., Kalejta, R. F., and Levine, A.J. (2001) Wrch-1, a novel member of the Rho gene family that is regulated by Wnt-1. *Genes Dev* 15, 1796–1807.

25. Neudauer, C.L., Joberty, G., Tatsis, N., and Macara, I.G. (1998) Distinct cellular effects and interactions of the Rho-family GTPase TC10. *Curr Biol* 8, 1151–1160.

26. Zallen, J.A., Peckol, E.L., Tobin, D.M., and Bargmann, C.I. (2000) Neuronal cell shape and neurite initiation are regulated by the Ndr kinase SAX-1, a member of the Orb6/COT-1/warts serine/threonine kinase family. *Mol Biol Cell* 11, 3177–3190.

27. Edgley, M.L., Baillie, D.L., Riddle, D.L., and Rose, A.M. (2006) Genetic balancers. *WormBook*, 1–32.

28. Edgley, M.L., and Riddle, D.L. (2001) LG II balancer chromosomes in *Caenorhabditis elegans*: mT1(II;III) and the mIn1 set of dominantly and recessively marked inversions. *Mol Genet Genomics* 266, 385–395.

29. McKim, K.S., Peters, K., and Rose, A.M. (1993) Two types of sites required for meiotic chromosome pairing in *Caenorhabditis elegans*. *Genetics* 134, 749–768.

30. Bourne, H.R., Sanders, D.A., and McCormick, F. (1991) The GTPase superfamily: conserved structure and molecular mechanism. *Nature* 349, 117–127.

31. Rossman, K.L., Der, C.J., and Sondek, J. (2005) GEF means go: turning on RHO GTPases with guanine nucleotide-exchange factors. *Nat Rev Mol Cell Biol* 6, 167–180.

32. Scheffzek, K., and Ahmadian, M.R. (2005) GTPase activating proteins: structural and functional insights 18 years after discovery. *Cell Mol Life Sci* 62, 3014–3038.

33. Gibbs, J.B., Sigal, I.S., Poe, M., and Scolnick, E.M. (1984) Intrinsic GTPase activity distinguishes normal and oncogenic ras p21 molecules. *Proc Natl Acad Sci USA* 81, 5704–5708.

34. Frech, M., Darden, T.A., Pedersen, L.G., Foley, C.K., Charifson, P.S., Anderson, M.W., and Wittinghofer, A. (1994) Role of glutamine-61 in the hydrolysis of GTP by p21H-ras: an experimental and theoretical study. *Biochemistry* 33, 3237–3244.

35. Farnsworth, C.L., Marshall, M.S., Gibbs, J.B., Stacey, D.W., and Feig, L.A. (1991) Preferential inhibition of the oncogenic form of RasH by mutations in the GAP binding/"effector" domain. *Cell* 64, 625–633.

36. Feig, L.A. (1999) Tools of the trade: use of dominant-inhibitory mutants of Ras-family GTPases. *Nat Cell Biol* 1, E25-27.

37. Mello, C., and Fire, A. (1995) DNA transformation. *Methods Cell Biol* 48, 451–482.

38. Frokjaer-Jensen, C., Davis, M.W., Hopkins, C.E., Newman, B.J., Thummel, J.M., Olesen, S.P., Grunnet, M., and Jorgensen, E.M. (2008) Single-copy insertion of transgenes in *Caenorhabditis elegans*. *Nat Genet* 40, 1375–1383.

39. Troemel, E.R., Kimmel, B.E., and Bargmann, C.I. (1997) Reprogramming chemotaxis responses: sensory neurons define olfactory preferences in *C. elegans*. *Cell* 91, 161–169.

40. Yu, S., Avery, L., Baude, E., and Garbers, D.L. (1997) Guanylyl cyclase expression in specific sensory neurons: a new family of chemosensory receptors. *Proc Natl Acad Sci USA* 94, 3384–3387.

41. Chapman, J.O., Li, H., and Lundquist, E.A. (2008) The MIG-15 NIK kinase acts cell-autonomously in neuroblast polarization and migration in *C. elegans*. *Dev Biol* 324, 245–257.

42. Driscoll, M., and Chalfie, M. (1991) The mec-4 gene is a member of a family of *Caenorhabditis elegans* genes that can mutate to induce neuronal degeneration. *Nature* 349, 588–593.

43. Forrester, W.C., Perens, E., Zallen, J.A., and Garriga, G. (1998) Identification of *Caenorhabditis elegans* genes required for neuronal differentiation and migration. *Genetics* 148, 151–165.

44. Collet, J., Spike, C.A., Lundquist, E.A., Shaw, J.E., and Herman, R.K. (1998) Analysis of osm-6, a gene that affects sensory cilium structure and sensory neuron function in *Caenorhabditis elegans*. *Genetics* 148, 187–200.

45. Jin, Y., Jorgensen, E., Hartwieg, E., and Horvitz, H.R. (1999) The *Caenorhabditis elegans* gene unc-25 encodes glutamic acid decarboxylase and is required for synaptic transmission but not synaptic development. *J Neurosci* 19, 539–548.

46. Finney, M., and Ruvkun, G. (1990) The unc-86 gene product couples cell lineage and cell identity in *C. elegans*. *Cell* 63, 895–905.

47. Maduro, M., and Pilgrim, D. (1995) Identification and cloning of unc-119, a gene expressed in the *Caenorhabditis elegans* nervous system. *Genetics* 141, 977–988.

Chapter 23

Protocol for Ex Vivo Incubation of *Drosophila* Primary Post-embryonic Haemocytes for Real-Time Analyses

Christopher J. Sampson and Michael J. Williams

Abstract

The cellular branch of the *Drosophila* larval innate immune system consists of three immunosurveillance (haemocyte) cell types: plasmatocytes, crystal cells, and lamellocytes. In order to examine haemocyte cytoskeletal dynamics or migration, most researchers use embryos or in vitro cell culture systems, but very little is known about the behaviour of post-embryonic haemocytes. The current method employs an ex vivo system, in which post-embryonic haemocytes are isolated for short-term analysis, in order to investigate various aspects of their behaviour during events requiring cytoskeleton dynamics and Rho GTPase signalling.

Key words: Cell migration, Rho GTPases, Innate immunity, Haemocyte, Cell adhesion

1. Introduction

The *Drosophila* embryo model has provided an in vivo platform in which the powerful genetic tools of *Drosophila* can be utilized to manipulate various proteins and pathways, such as Rho-family small GTPases, in order to elucidate the mechanisms controlling cell migration within an organism, rather than in in vitro cell lines (1–4). The ability to visualize embryonic haemocytes in vivo means that later life stages of *Drosophila* have been overshadowed. While the larval stages are used quite heavily to study the humoral and cellular immune responses (5, 6), there is a general lack of understanding of the cellular dynamics of larval haemocytes.

The *Drosophila* larval immunosurveillance cells consist of three haemocyte types: plasmatocytes, crystal cells, and lamellocytes. Plasmatocytes, the most abundant haemocyte in circulation in healthy larvae, are rounded in morphology, ranging from ~20 to

Francisco Rivero (ed.), *Rho GTPases: Methods and Protocols*, Methods in Molecular Biology, vol. 827,
DOI 10.1007/978-1-61779-442-1_23, © Springer Science+Business Media, LLC 2012

35 μm in diameter, and are the *Drosophila* equivalent to professional phagocytes. Plasmatocytes are capable of engulfing bacteria and fungi, and are involved in the primary stages of encapsulation against the endoparasitoid *Leptopilina boulardi* (7–9). Crystal cells are the second most abundant circulating haemocyte type in healthy larvae. They are similar in size to plasmatocytes and contain crystals of the zymogen prophenoloxidase. Crystal cells have been widely associated with coagulation and the innate immune response, in particular with encapsulation of parasitoid wasp eggs. They are responsible for local application of prophenoloxidase resulting in the deposition of melanin (10–14). Lastly, lamellocytes are the largest haemocyte cell type ranging in diameter from ~100 to 250 μm. These cells have an extremely spread morphology. They are very rarely observed in healthy larvae, and are only present in small numbers prior to the pre-pupal stage, but their numbers increase significantly in parasitized larvae. Their main responsibility, known so far, is forming a secondary layer of haemocytes during the encapsulation response (6, 7, 15–17). Their purpose within healthy larvae is unknown, but it is thought that they provide some function during morphogenesis at the pre-pupal stage.

To date, the *Drosophila* larval model has provided a basic characterization of haemocyte actin-cytoskeletal dynamics within an immune context (9, 18, 19), but further information is required to include data on its motile behaviour at the larval instar and pre-pupal stages. Overexpression of wild-type or mutant forms of Rac1 in larval haemocytes demonstrated that Rac1 regulates two distinct pathways to induce F-actin accumulation and lamellipodia formation, involving either the Jun kinase Basket (Bsk) or cofilin (18). In another investigation, it was revealed that Rac1 and Rho1 reciprocally regulate each other's activity in F-actin formation (18, 19). Both of these studies used fixed bled haemocytes to study the effects of either hyperactivating or inhibition of Rho family GTPase function. The novel ex vivo method described here allows researchers to bridge that gap between in vitro and in vivo by recreating a system of incubation, which can support primary *Drosophila* haemocytes isolated from larvae or pre-pupae. Using this technique, one can analyze haemocyte motility, provide an initial characterization of haemocyte behaviour, and study the function of Rho-family GTPases during haemocyte migration in post-embryonic life stages.

2. Materials

2.1. Drosophila Maintenance

All assays are performed on the lab-classified control strain w^{1118}, apart from adhesion and spreading assays, which are detailed below. For GAL4-UAS assays, use the well-described *Hemese-GAL4*

Table 1
Loss-of-function and gain-of-function Rho-family GTPase
D. melanogaster strains available from the Bloomington
Stock Center

Rac1 strains
y¹ w*; Rac1^{J11} P{FRT(w^{hs})}2A/TM6B, Tb¹
P{UAS-Rac1.Myc}2.4, y¹ w*; TM3, Sb¹/TM6B, Tb¹

Rac2 strains
Rac2^Δ ry^{506}
y¹ w^{67c23}; P{SUPor-P}Rac2^{KG05681b} ry^{506}
y¹ w^{67c23}; P{wHy}Rac2^{DG19808}
w^{1118}; P{UAS-Rac2.AR}2-4/CyO, P{GAL4-twi.G}2.2, P{UAS-2xEGFP}AH2.2

Rho1 strains
w*; P{UAS-Rho1.V14}2.1
P{UAS-Rho1.N19}1.3, w*
w*; {UAS-Rho1.N19}2.1
P{UASp-GFP.Rho1}, w*
W*; {UASp-GFP.Rho1}

Cdc42 strains
y¹ w* Cdc42² P{neoFRT}19A
w*; P{UAS-Cdc42.V12}2
w*; P{UAS-Cdc42.N17}3

(*He-GAL4*) (15, 20, 21) or Hemolecting-GAL4 (Hml-Gal4) (21) drivers. Flies are grown and maintained on a standard cornmeal diet: 60% w/v yellow cornmeal, 12% w/v inactive dry yeast, 7.5% v/v inverted sugar syrup (can be substituted by light molasses), 1.4% w/v methyl paraben (Tegosept), and 6% w/v agar at 25°C and under 12/12-h light and dark conditions. GAL4-UAS strains are an exception; they are allowed to mate for 48 h at 25°C before being placed in a 29°C incubator to drive optimal GAL4 expression. The various loss-of-function and gain-of-function Rho-family GTPase strains available from the Bloomington Stock Center are presented in Table 1.

2.2. Reagents and Buffers

All solutions must be prepared with great diligence in order to preserve sterile conditions. Standard reagents are prepared with deionized water. All reagents are made at room temperature underneath a fume hood and then kept at room temperature, except for DHIM and fibronectin, which are kept at 4°C.

1. Phosphate-buffered saline (PBS): 8 g/L NaCl, 0.2 g/L KCl, 1.44 g/L Na_2HPO_4, 0.24 g/L KH_2PO_4, pH 7.4.

2. Laminin-coating buffer: 50 mM Tris–HCl, pH 7.5, 150 mM NaCl.

3. Gelatine (MERCK 1.04080): 0.2% (v/v) solution in PBS. Aliquot into 1.5-mL Eppendorf tubes and store either at room temperature or at 4°C.

4. Laminin (Sigma–Aldrich L2020): 1:40 (v/v) dilution in laminin-coating buffer. Aliquot into 1.5-mL Eppendorf tubes and store at 4°C.

5. Fibronectin (Sigma–Aldrich F0895): 1:40 (v/v) dilution in PBS. Aliquot into 1.5-mL Eppendorf tubes and store at –20°C. Each time fibronectin is needed, a single aliquot is taken from –20°C and kept at 4°C in between use. The solution can be kept for a maximum of 1 month at 4°C.

6. *Drosophila* haemocyte-isolating medium (DHIM): 75% (v/v) filtered Schneiders *Drosophila* medium (VWR, 733-1663) and 25% (v/v) filtered fetal bovine serum. DHIM is always prepared fresh every day (see Note 1).

2.3. Equipment

1. Glass bottom dishes (MatTek corps P35G-1-14-C).

2. Sterile 25-gauge dissecting needles.

3. Sterile level 4 dissecting forceps.

4. Basic stereo light microscope.

5. Bleeding platform made of a transparent cell culture dish lid.

6. Inverted microscope with phase-contrast illumination capability.

7. 20×, 40×, 63× phase-contrast objectives with a minimum numerical aperture of 1.1.

8. CCD camera and imaging software with the ability to acquire time-lapse imaging.

9. Analytical software of choice, e.g. ImageJ (available free at http://rsbweb.nih.gov/ij/).

3. Methods

The extracellular matrix (ECM) is required to complete the ex vivo system so that cellular behaviour is interpreted under conditions representing the in vivo environment. Multiple ECMs can be found within a single system, usually containing collagen, laminin, and fibronectin. Within *Drosophila*, both collagen and laminin are largely abundant, whereas fibronectin does not naturally occur – instead, it is replaced by a fibronectin-like protein. Each ECM consists of specific ligands that interact with various cellular integrins

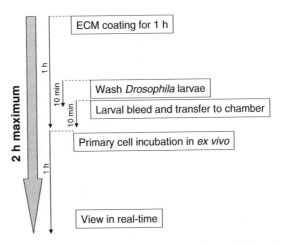

Fig. 1. Timeline indicating approximate time necessary for each stage of the ex vivo protocol.

to create ligand–integrin signalling complexes, which induce both generic and specific cellular behaviours (22–27). The ex vivo system allows for coating with multiple ECM components, individually or mixed, to conform to the study of a specific signalling pathway or cellular response by real-time microscopy.

From start to finish, this protocol should take about 2 h (see Fig. 1). A flow chart outlining the procedure is presented in Fig. 2.

1. Pipette 70–100 µL of ECM solution on to the glass bottom of a sterile MatTek dish. Let the ECM coat the glass for 1 h at 25°C. After 40 min has elapsed into the coating time, with 20 min remaining, collect the desired number of larvae. Six larvae, in total, are collected and washed in distilled water to ensure that the surface of each larva is clean of any debris.

2. Once cleaned, place a 100-µL spot of DHIM on the bleeding platform and then put it under a stereo light microscope. In order to isolate the primary haemocytes, transfer the washed larvae to this spot for bleeding (see Note 2).

3. Bleed each larva using a sterile 25-gauge needle and level 4 dissecting forceps. Place the larva so that their dorsal side is facing up. Using the level 4 forceps, grab the larva at the anterior end, just behind the head, to secure the larva in place. Insert the 25-gauge needle in the posterior end of the larvae, between the tracheae. Apply enough force to the needle to puncture the epithelium, after which pull the needle away gently from the larva. Approximately 0.5 µL of haemolymph, containing the haemocytes, flows out of the puncture wound and into the DHIM. Take great care so that the larval gut remains intact.

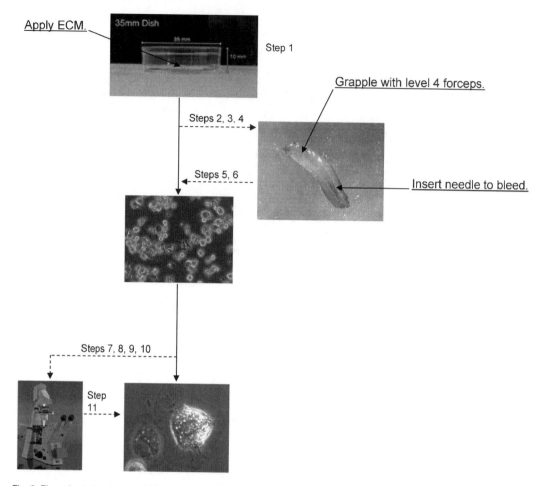

Fig. 2. Flow chart showing in which order to perform each step of the ex vivo protocol.

Using the same 25-gauge needle, repeat the bleeding procedure for the remaining larvae (see Note 3).

4. After bleeding all the larvae, recover the ex vivo MatTek chamber from the 25°C incubator and remove the ECM substrate. Transfer the 100 µL of DHIM, containing the isolated haemocytes, from the bleeding platform to the coated glass bottom of the chamber. Pipette 100 µL of DHIM onto the bleeding platform, where the six larvae were bled in order to uplift any remaining haemocytes (see Note 4). Add this extra 100 µL of DHIM to the MatTek chamber to further supplement the cells present.

5. Transfer the entire chamber to a 25°C incubator for 1 h to allow the cells to adhere and spread across the ECM before viewing.

6. During the final 1 h of incubation, set up the microscope. The basic set-up of the inverted microscope is as follows:

   (a) Stage adapted to allow for visualization of the ex vivo chamber.

   (b) For high-detail analysis, use a 63× Plan-Apochromat phase-contrast objective.

   (c) Set the microscope condenser to the phase-contrast ring matching the objective.

   (d) Fully open the light source.

7. After 1 h at 25°C, transfer the chamber to the microscope stage and add another 300 μL of DHIM. Select the desired objective and switch the microscope to live viewing. Optimize the optical quality of the haemocytes being captured. Perform Köhler illumination in order to optimally focus the image (see Note 5).

8. After focusing the haemocytes into view, reduce the light source aperture to its lowest setting so that the light is focused on one spot. Readjust the condenser stage until the edges of the spot are sharp. If the spot of light is off centre, then readjust the condenser screws to bring the spot back to centre. Once complete, open the light aperture to approximately ¾ open.

9. Stop the live view and select the time-lapse setting. Set the time lapse to have a frame rate of 15 s and capture images for a minimum of 20 min of elapsed time.

10. After completing the time lapse, condense the frames to create an AVI file for viewing and analysis.

The ex vivo technique for incubation of primary *Drosophila* larval haemocytes provides a stage for multiple cellular assays. This technique can be used to analyze basic haemocyte behaviour, determined by general morphology, protrusion activity, and general cytoskeleton dynamic behaviour, across all larval life stages and pre-pupae. Additionally, adhesion and spreading on gelatine and laminin can also be investigated, in relation to the activity of Rho-family GTPases. Finally, this method was used to produce an initial characterization of haemocyte motility at the pre-pupal life stage (Sampson and Williams, manuscript in preparation).

4. Notes

1. Both Schneiders *Drosophila* medium and fetal bovine serum need to be pre-filtered using a 0.2-μm filter tip before mixing. Filtration was required as it was determined that primary cell health had a greater efficacy than in non-filtered plus within both liquids there was a considerable amount of debris which affected the cells' behaviour ex vivo, as well affecting the imagery.

2. It is stated that 20 min should be taken to wash the larvae, but from experience the longer one washes them, the less healthy the cells are. Ensure that most external debris is removed from the larvae, but try to have the larvae washed and in the DHIM mixture within as short a time period as possible.

3. Similar to Note 2, the longer one waits to bleed the larvae, the less healthy the cells are. This requires practice, as it is not easy to bleed these larvae under a stereomicroscope. Ensure that all larvae are aligned with their dorsal side up (facing the user), and anterior end directed towards the left (this varies depending on user's confidence with handling the 25-gauge needle).

4. An excess wash is very important and must not be forgotten. During the bleeding process, a lot of haemolymph is lost when bleeding and removing the larvae. So one must maximize the number of cells available, and simply a wash with 100 μL of DHIM on to the previous spot, where the larvae were bled, achieves this. Haemocytes survive for approximately 8 h in DHIM.

5. The optical quality must be at its best. To begin with, only use the minimum amount of immersion oil, as too much oil affects the path of contrast light from the sample to the detector. As we are using phase-contrast light, it is essential to perform Köhler illumination to any sample in order to gain the optimum focus and resolution. It is easiest to focus the sample first, on the desired cellular population, and then perform Köhler illumination so that one has now focused all phase-contrast light from the condenser on to that single spot. Once this is complete, switch the light from the ocular to the CCD camera and begin time-lapse acquisition.

Acknowledgements

This work was partially supported by funds from the BBSRC, the Royal Society, and the University of Aberdeen.

References

1. Stramer B., Wood, W., Galko, M.J., Redd, M.J., Jacinto, A., Parkhurst, S.M., and Martin, P. (2005) Live imaging of wound inflammation in *Drosophila* embryos reveals key roles for small GTPases during in vivo cell migration. *J Cell Biol* 168, 567–573.

2. Wood, W., Faria, C., and Jacinto, A. (2006) Distinct mechanisms regulate hemocyte chemotaxis during development and wound healing in *Drosophila melanogaster. J Cell Biol* 173, 405–416.

3. Vlisidou, I., Dowling, A.J., Evans, I.R., Waterfield, N., ffrench-Constant, R., and Wood, W. (2009) *Drosophila* embryos as model systems for monitoring bacterial infection in real time. *PLoS Pathog* 5, e1000518.

4. Moreira, S., Stramer, B., Evans, I., Wood, W., and Martin, P. (2010) Prioritization of competing damage and developmental signals by migrating macrophages in the *Drosophila* embryo. *Curr Biol* 20, 464–470.

5. Lemaitre, B., and Hoffmann, J. (2007) The host defense of *Drosophila melanogaster*. *Annu Rev Immunol* 25, 697–743.

6. Williams. M.J. (2007) *Drosophila* hemopoiesis and cellular immunity. *J Immunol* 178, 4711–4716.

7. Russo, J., Dupas, S., Frey, F., Carton, Y., and Brehelin, M. (1996) Insect immunity: Early events in the encapsulation process of parasitoid (*Leptopilina boulardi*) eggs in resistant and susceptible strains of *Drosophila*. *Parasitology* 112, 135–142.

8. Russo, J., Brehelin, M., and Carton, Y. (2001) Haemocyte changes in resistant and susceptible strains of *D. melanogaster* caused by virulent and avirulent strains of the parasitic wasp *Leptopilina boulardi*. *J Insect Physiol* 47, 167–172.

9. Williams, M.J., Ando, I., and Hultmark, D. (2005) *Drosophila melanogaster* Rac2 is necessary for a proper cellular immune response. *Genes Cells* 10, 813–823.

10. Rizki, T.M., Rizki, R.M., and Grell, E.H. (1980) A mutant affecting the crystal cells in *Drosophila melanogaster*. *Wilhelm Roux's Arch* 188, 91–99.

11. Meister, M., and Lagueux, M. (2003) *Drosophila* blood cells. *Cell Microbiol* 5, 573–580.

12. Meister, M. (2004) Blood cells of *Drosophila*: Cell lineages and role in host defence. *Curr Opin Immunol* 16, 10–15.

13. Crozatier, M., and Meister M. (2007) *Drosophila* haematopoiesis. *Cell Microbiol* 9, 1117–1126.

14. Bidla, G., Dushay, M.S., and Theopold, U. (2007) Crystal cell rupture after injury in *Drosophila* requires the JNK pathway, small GTPases and the TNF homolog eiger. *J Cell Sci* 120, 1209–1215.

15. Zettervall, C., Anderl, I., Williams, M.J., Palmer, R., Kurucz, E., Ando, I., and Hultmark, D. (2004) A directed screen for genes involved in *Drosophila* blood cell activation. *Proc Natl Acad Sci USA* 101, 14192–14197.

16. Crozatier, M., Ubeda, J., Vincent, A., and Meister, M. (2004) Cellular immune response to parasitization in *Drosophila* requires the EBF orthologue collier. *PLoS Biol* 2, e196.

17. Markus, R., Laurinyecz, B., Kurucz, E., Honti, V., Bajusz, I., Sipos, B., Somogyi, K., Kronhamn, J., Hultmark, D., and Andó, I. (2009) Sessile hemocytes as a hematopoietic compartment in *Drosophila melanogaster*. *Proc Natl Acad Sci USA* 106, 4805–4809.

18. Williams, M., Wiklund, M., Wikman, S., and Hultmark, D. (2006) Rac1 signalling in the *Drosophila* larval cellular immune response. *J Cell Sci* 119, 2015–2024.

19. Williams, M.J., Habayeb, M.S., and Hultmark, D. (2007) Reciprocal regulation of Rac1 and Rho1 in *Drosophila* circulating immune surveillance cells. *J Cell Sci* 120, 502–511.

20. Kurucz, E., Zettervall, C., Sinka, R., Vilmos, P., Pivarcsi, A., Ekengren, S., Hegedüs, Z., Ando, I., and Hultmark, D. (2003) Hemese, a hemocyte-specific transmembrane protein, affects the cellular immune response in *Drosophila*. *Proc Natl Acad Sci USA* 100, 2622–2627.

21. Goto, A., Kadowaki, T., and Kitagawa, Y. (2003) *Drosophila* hemolectin gene is expressed in embryonic and larval hemocytes and its knock down causes bleeding defects. *Dev Biol* 264, 582–591.

22. D'Souza-Schorey, C., Boettner, B., and Van Aelst, L. (1998) Rac regulates integrin-mediated spreading and increased adhesion of T lymphocytes. *Mol Cell Biol* 18, 3936–3946.

23. Geberhiwot, T., Ingerpuu, S., Pedraza, C., Neira, M., Lehto, U., Virtanen, I., Kortesmaa, J., Tryggvason, K., Engvall, E., and Patarroyo, M. (1999) Blood platelets contain and secrete laminin-8 ($\alpha 4\beta 1\gamma 1$) and adhere to laminin-8 via $\alpha 6\beta 1$ integrin. *Exp Cell Res* 253, 723–732.

24. DeMali, K.A., Wennerberg, K., and Burridge, K. (2003) Integrin signaling to the actin cytoskeleton. *Curr Opin Cell Biol* 15, 572–582.

25. Pellinen, T., and Ivaska, J. (2006) Integrin traffic. *J Cell Sci* 119, 3723–3731.

26. Rose, D.M., Alon, R., and Ginsberg, M.H. (2007) Integrin modulation and signaling in leukocyte adhesion and migration. *Immunol Rev* 218, 126–134.

27. Kim, S.H., Turnbull, J.E., and Guimond, S.E. (2011) Extracellular matrix and cell signalling – the dynamic cooperation of integrin, proteoglycan and growth factor receptor. *J Endocrinol* doi:10.1530/JOE-10-0377.

Chapter 24

Analysis of Rho GTPase Activation in *Saccharomyces cerevisiae*

Gary Eitzen and Michael R. Logan

Abstract

Rho proteins act as molecular switches to control multiple cellular processes. The switch mechanism involves cycling between active and inactive states based on GTP loading and hydrolysis. Assays that quantitatively analyze the GTP loading of Rho proteins have become important molecular tools to decipher upstream signals and mechanisms that regulate activation and de-activation. These assays make use of Rho activation probes constructed from Rho-binding domains of downstream effectors. The utility of these assays comes from effector domains that show selective high affinity interactions with specific subsets of GTP-bound activated GTPases. Here, we describe assays used to analyze yeast Rho GTPase activation.

Key words: Cdc42p, Rho1p, Rho activation, RhoGDI, Rho effector, Yeast

1. Introduction

The Rho family of GTPases act as molecular switches to regulate many cellular processes. Most Rho GTPases can be classified into one of the three most common subtypes Rho, Rac, and Cdc42. The yeast, *Saccharomyces cerevisiae*, has orthologs of Rho and Cdc42, but does not have Rac GTPases.

The yeast genome encodes five Rho subtypes, *RHO1* to *RHO5*, and a single *CDC42* gene. *RHO1* and *CDC42* are essential genes, while *RHO2* to *RHO5* are nonessential and have ill-defined, overlapping functions with *RHO1*. Rho/Cdc42 proteins are involved in cytoskeletal remodeling which is required for several crucial aspects of cellular morphogenesis including the establishment of cell polarity, cell division, organelle biogenesis, and secretion. Rho1p spatially regulates vesicular traffic, driving exocytosis, which facilitates the incorporation of plasma membrane components

Francisco Rivero (ed.), *Rho GTPases: Methods and Protocols*, Methods in Molecular Biology, vol. 827,
DOI 10.1007/978-1-61779-442-1_24, © Springer Science+Business Media, LLC 2012

(1). Vesicles are mobilized along actin cables growing from the formins Bni1p and Bnr1p, which are downstream Rho1p effectors (2, 3). Rho1p is also a regulatory subunit of a cell wall synthesis complex that controls cell integrity through Pkc1p signaling (4). Cdc42p activates the polymerization of branched actin filaments from the Arp2/3 complex (5). In mammalian cells, Cdc42 interacts directly with WASP/WIP to activate polymerization from Arp2/3. However, in yeast, Cdc42p does not directly interact with the WASP/WIP orthologs Las17p/Vrp1p but rather indirectly signals likely through one of three p21-activated kinases (PAK), Cla4p, Ste20p, and Skm1p. These kinases play an essential role in septin organization and cell division (6). Yeast Rho1 and Cdc42 proteins are the subject of several excellent reviews (7–10).

The Rho GTPase switch operates by alternating between an active, GTP-bound state and an inactive, GDP-bound state. The fact that Rho effector proteins form high-affinity complexes specifically with the GTP-bound form of Rho proteins has been exploited experimentally to develop high-affinity purification probes that can be used to monitor the level of activated Rho proteins. The initial assay was developed using the Rho-binding domain (RBD) of the mammalian Rho effector protein, Rhotekin. This RBD motif has been shown to bind specifically to RhoA-GTP (11), and has subsequently also been shown to bind yeast Rho1p-GTP (12). Similarly, the Cdc42 binding domain (CBD) of the mammalian Cdc42 effector protein, PAK1, binds to both mammalian and yeast Cdc42-GTP (13, 14). In yeast, CBDs have been identified in Cla4p and Ste20p, while Rho1p binding domains have been identified in the formins and Pkc1p (12, 15). These domains have been cloned as GST-tagged fusion proteins, which allow affinity isolation ("pull-down" on glutathione-agarose) of GST-CBD/Cdc42p-GTP and GST-RBD/Rho1p-GTP complexes. The amount of activated Rho proteins in pull-down assays is determined by immunoblot using specific antibodies and is compared to control reactions containing maximally activated GTPases via chemically induced nucleotide loading.

Although much progress in the analysis of Rho activation in mammalian cells has been made, relatively few biochemical analyses of Rho activation in *S. cerevisiae* have been reported. Three manuscripts report the detection of activated Rho1p and Cdc42p directly from yeast lysates. Kono et al. show that Rho1p is activated during the cell cycle at the G1/S boundary (15), while Wai et al. show higher levels of Cdc42p-GTP in small budded cells (16). Boulter et al. show levels of activated Cdc42p and Rho1p increase when *RDI1*, the natural inhibitor of Rho proteins, is deleted, even though the overall stability of Rho proteins decrease (17). Using a different approach, we have shown that Cdc42p and Rho1p activation can be stimulated in cell-free extracts containing membrane-bound GTPases and simple reaction buffers (14). Kinetic and pharmacological analyses clearly differentiate the activation mechanisms of Cdc42p and Rho1p (12).

2. Materials

All reagents are purchased from Sigma-Aldrich unless otherwise indicated. Long-term storage of complex buffers is not recommended. Liquids are measured as % based on vol/vol.

2.1. Buffers and Solutions

1. Protease inhibitor cocktail (PIC): PIC is made as a 60× stock and added to buffers just prior to use. 60× PIC: 12 µg/mL leupeptin, 24 µg/mL pepstatin, 24 µg/mL aprotinin, 24 mM *o*-phenanthroline, 6 mM Pefabloc SC.

2. Phosphate-buffered saline (PBS): 137 mM NaCl, 2.7 mM KCl, 4.3 mM Na_2HPO_4, 1.47 mM KH_2PO_4. Adjust to a final pH of 7.4.

3. Bradford reagent and protein standard for protein concentration determination (BioRad).

4. 1.2× SDS-PAGE sample buffer (1.2× SSB): 15 mM Tris–HCl, pH 6.8, 0.5% SDS, 2.4% glycerol, 1.2% β-mercaptoethanol, 0.0025% bromophenol blue.

5. Probe buffer: 25 mM Tris–HCl, pH 7.5, 120 mM NaCl, 0.5 mM DTT, 1 mM $MgCl_2$, 1× PIC, and 5% glycerol.

6. Binding buffer: 20 mM HEPES–HCl, pH 7.5, 60 mM NaCl, 0.1 mM DTT, 5 mM $MgCl_2$, 1× PIC, and 1% Triton X-100.

7. PS buffer: 20 mM PIPES–HCl, pH 6.8, 200 mM sorbitol. This buffer is designed for osmotic stabilization of purified membranes.

8. Fusion Reaction buffer: 120 mM KCl, 5 mM $MgCl_2$, and 1× PIC made in PS buffer.

9. ATP solutions: 5 mM ATP, 5 mM $MgCl_2$, and ATP regenerating system (5 mM ATP, 5 mM $MgCl_2$, 0.5 mg/mL creatine kinase, 200 mM creatine phosphate) made in PS buffer.

10. Yeast lysis buffer: 20 mM HEPES–HCl, pH 7.5, 120 mM NaCl, 1 mM DTT, 1 mM $MgCl_2$, and 2× PIC.

11. Spheroplast buffer: 20 mM potassium phosphate, pH 7.5, 1.8 M sorbitol, 30 mg of yeast lytic enzyme (MP Biomedicals, Cat. No. 8360952). Mix components from high molarity stock solutions in YPD; at least half the volume should be YPD.

12. Ficoll solutions: for 15%, 8%, and 4% ficoll solutions, dissolve, respectively, 15, 8, and 4 g of ficoll 400 (Sigma-Aldrich Cat. No. F4375) in 100 mL of PS buffer. Mix overnight on a rocking platform.

13. 0.5 M Isopropyl β-D-1-thiogalactopyranoside (IPTG): dissolve 12 g of IPTG in 50 mL of water; store as 1-mL aliquots at –20°C.

14. 20% Triton X-100: To 80 mL of double distilled water, add 20 mL of Triton X-100. Make the mixture in a 100-mL cylinder and mix overnight with a small stir bar and stir plate.

15. Yeast cytosol: grind several grams of yeast in PS buffer with 2× PIC by vortexing with 0.5-mm glass beads (5 times 1 min). The mixture should have a paste-like consistency, so add only 1 vol of PS buffer (equal volume to cells) and 0.5 vol of glass beads. Remove glass beads, unbroken cells, and membranes by centrifugation (20,000×g, 4°C, 20 min). Carefully take the upper layer without disturbing the pellet. This cytosol should have a protein concentration >10 mg/mL and should be stored in small aliquots at −80°C until needed.

16. Nucleotide loading buffer: 20 mM HEPES–HCl, pH 7.5, 60 mM NaCl, 0.1 mM DTT, 2.5 mM EDTA, 1× PIC, and 20 μM desired nucleotide (see Note 1).

2.2. Expression of Activation Probes in Escherichia coli

1. *E. coli* strains: use of protease deficient strains is recommended. The *E. coli* BL21strain (Invitrogen) has deletions in outer membrane (*ompT*) and cytosolic (*lon*) proteases. Competent BL21 cells can be purchased from commercial sources and incubation with nanogram quantities of plasmid will result in transformed colonies that are maintained on LB amp agar plates or liquid media.

2. *E. coli* growth media (LB amp): dissolve 5 g of yeast extract, 10 g of tryptone, and 10 g of NaCl in 1 L of water. Sterilize by autoclaving. Add 0.1 g/L ampicillin after autoclaving.

3. Rho1p activation probes are prepared by expressing the RBDs of Rhotekin (amino acids 7–89), Bni1p (amino acids 90–343), or Pkc1p (amino acids 384–632) as GST-tagged fusion proteins (see Note 2). Similarly, Cdc42p activation probes are prepared by expressing the Cdc42 binding fragments of PAK1 (amino acids 67–150), Ste20p (amino acids 331–410), or Cla4p (amino acids 183–243) as GST-tagged fusion proteins (see Note 2). Purified probes are shown in Fig. 1a. All plasmids are available from academic sources and therefore cloning is not necessary (11–15).

4. Grow transformed *E. coli* strains in a shaker incubator at 180 rpm at 37°C. Read the optical density (OD) of the culture at 600 nm using LB medium as a blank. IPTG is added (1 mL of 0.5 M IPTG stock per liter of culture) to induce protein expression.

5. GST-tagged probes are immobilized on glutathione-agarose. Other glutathione resins can also be used (glutathione-Sepharose 4B, GE Healthcare; glutathione-superflow, Clontech). Store resin in 20% ethanol, 0.01% sodium azide at 4°C. Make probes fresh for each experiment.

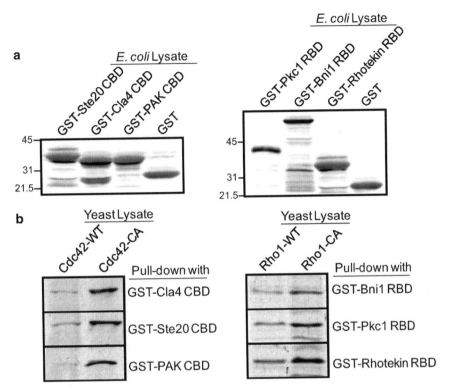

Fig. 1. Efficiency of Rho activation probes. (**a**) Lysates were prepared from *E. coli* expressing the indicated Cdc42p (*left panel*) and Rho1p (*right panel*) activation probes. Probes were bound to glutathione-agarose and eluted by boiling in 1.2× SSB. (**b**) Whole cell lysates were prepared from yeast expressing HA-tagged wild-type (*WT*) or constitutively active (*CA*) Cdc42p (*left panel*) and Rho1p (*right panel*). 0.5 mg of yeast lysates were mixed with 10 µL of bead-bound probes made from effectors with Cdc42/Rho1 binding domains as indicated. Panel (**a**) is reproduced from ref. 12 with permission from Elsevier.

2.3. Yeast Strains and Media

1. Yeast strains: use of protease deficient strains is recommended. Proteinase A (*PEP4*) is required for the maturation of several other vacuolar proteases and therefore the *PEP4*-gene deletion strain, BY4742 (Matα, *his3, leu2, ura3, lys2, pep4*::kanMX4) is often the strain of choice. This strain is available from commercial (Invitrogen) or academic sources (12).

2. Yeast growth medium (YPD): dissolve 10 g of yeast extract, 20 g of peptone, and 20 g of glucose in 1 L of water. Sterilize by autoclaving. Grow yeast strains in a shaker incubator at 180 rpm at 28°C.

2.4. Immunoblot, Antibodies, and Proteins

1. Standard reagents and equipment for performing immunoblot (western blot).

2. Polyclonal rabbit anti-yeast Cdc42p antibodies (y-191) (Santa Cruz Biotechnology).

3. Antibodies against yeast Rho1p can be obtained from academic sources (18, 19).

4. Secondary antibodies that recognize Cdc42p and Rho1p poly-clonal antibodies can be obtained from several commercial sources (see Note 3).

5. Purified recombinant GEF domain, His$_6$-Dbs (Cytoskeleton Inc., Cat. No. GE01). Dbs efficiently activates Rho1p and Cdc42p and can be used for positive reaction controls.

6. Purified recombinant RhoGDI, GST-GDI (Cytoskeleton Inc., Cat. No. GDI01). RhoGDI blocks the activation of Cdc42p and Rho1p and can be used for negative reaction controls.

3. Methods

All buffers, mixtures, and reactions should be kept at 4°C unless indicated otherwise. Protein lysates and fractions potentially containing activated Rho proteins should be rapidly analyzed since intrinsic GTPase activity will result in deactivation and decay of signal (see Note 4). If rapid analysis is not possible, keep samples at –80°C until just prior to analysis.

3.1. Preparation of Cdc42p and Rho1p Activation Probes

Rho/Cdc42 activation kits containing all the buffers, resin, and GST-tagged activation probes necessary to perform these assays can be obtained from several commercial sources (Cytoskeleton, Inc.; Millipore; Thermo Scientific). Alternatively, the activation probes and buffers can be made in-house as outlined in the following steps.

1. Plasmids with cloned GST-tagged activation probes are used to transform *E. coli* BL21 strain. Maintain strains on LB amp agar plates. Inoculate 100 mL of LB amp with a single colony and grow overnight. Dilute the overnight culture one-tenth into fresh LB amp and grow for 1–2 h until an OD$_{600}$ of 0.5 is reached. Induce fusion protein expression by adding 0.5 mM IPTG and grow for 5 h at 30°C.

2. After the 5-h induction period, harvest the bacteria by centrifugation (5,000×g, 5 min, 4°C). Wash cells by resuspending in 100 mL of probe buffer, centrifuge again, and resuspended in 1/100 the original culture volume with probe buffer. Lyse bacteria by sonicating with six 30 s bursts and cooling on ice between bursts. Add Triton X-100 to a final concentration of 1% from a 20% stock. Incubate on ice for 20 min with periodic vortexing. Clear lysates of unbroken cells by centrifugation (10,000×g, 20 min, 4°C). Lysate can be stored at –80°C until required. Protein concentration should be >5 mg/mL.

3. Probes are immobilized for pull-down assays by binding to glutathione-agarose beads. Beads are mix in a ratio of 1 mg of

bacterial lysate (containing GST-tagged probe) per 100 µL of packed glutathione-agarose beads. The mixture is incubated for 30 min at 4°C with constant mixing. Beads are settled by centrifugation ($1,000 \times g$, 30 s, 4°C), and washed three times in binding buffer.

4. Resuspend probe-bound glutathione-agarose beads as a 10% slurry in binding buffer (see Note 5). 10 µL of probe-bound beads is used in pull-downs, which is sufficient to quantitatively pull-down up to 100 nmol of activated GTPase. The efficiency of the different probes is shown in Fig. 1b.

3.2. Preparation of Yeast Whole Cell Lysate for Activation Assays

Basal levels of activated Rho proteins are obtained by directly probing yeast whole cell lysates. However, numerous specialized growth conditions can be applied to suit desired experimental conditions (see Note 6).

1. Grow yeast cultures to an $OD_{600} \sim 1.0$. Harvest cells by centrifugation ($5,000 \times g$, 5 min, 4°C). Wash once in yeast lysis buffer and resuspended in 1/100th the original culture volume of yeast lysis buffer. Samples are immediately frozen by pipetting the slurry dropwise into liquid nitrogen. Frozen cells can be stored at −80°C until all samples are ready for lysis.

2. Yeast whole cell extract is prepared by grinding frozen cell slurries in a coffee grinder containing a few pellets of dry ice. A fine powder is formed after 2–3 min of grinding and the suspension is rapidly thawed by placing the powder into 15-mL conical tubes and briefly incubating at 30°C. Add Triton X-100 to a final concentration of 1% and mix by vortexing. Remove unbroken cells by centrifugation ($20,000 \times g$, 10 min, 4°C). Determine the protein concentration by Bradford assay.

3. Samples should be immediately probed for activated GTPases (see Subheading 3.5) or subjected to additional experimental treatments (see Note 6).

3.3. Preparation of Yeast Vacuoles for Activation Assays

1. Grow a 1-L culture of yeast to $OD_{600} \sim 1.0$. Harvest cells by centrifugation ($5,000 \times g$, 5 min, 4°C). Wash cells once in 50 mM Tris–HCl, pH 9, containing 1 mM DTT. Harvest cells again and incubate 30 min at 30°C in spheroplast buffer.

2. Isolate spheroplasted cells by centrifugation ($2,500 \times g$, 10 min, 4°C). Resuspend the cells in 2.5 mL of 15% ficoll solution. Add 0.5 mg of DEAE–dextran (GE Healthcare Cat. No. 17-0350-01). Gently vortex to dissolve the dextran and incubate on ice for 3 min at 30°C; cool on ice for 3 min.

3. Place the lysate on the bottom of a 12-mL centrifuge tube that fits a Beckman SW41 rotor. Carefully layer 3 mL of 8% ficoll solution followed by 3 mL of 4% ficoll solution. Add PS buffer to within 1 mm of the top of the tube. Solutions should be kept at 4°C.

4. Centrifuge ($120,000 \times g$, 90 min, 4°C) in a Beckman ultracentrifuge. Remove tubes from the rotor. Aspirate lipids that may have floated to the top. Vacuoles will form a white band of membranes at the 0%/4% ficoll interface. Carefully remove vacuoles by pipetting with a 200-µL pipettor.

5. Determine the protein concentration of isolated membrane by Bradford assay. Standard activation assays are done with 30–40 µg of membranes at a concentration of 0.5 mg/mL. Membranes can be stored at –80°C, but are more active when analyzed the day of isolation.

6. Membranes can be directly analyzed or subjected to additional experimental treatments (see Note 6).

3.4. Preparation of In Vitro Reactions for Rho GTPase Activation

Rho GTPase function as membrane proteins, therefore it is of particular interest to examine activation mechanisms associated with membrane fractions. Typical experiments involve incubation of purified membranes in conditions that result in GTPase activation (12, 14). Membranes are purified by differential centrifugation or isopycnic centrifugation as described in Subheading 3.3 (see Note 7). Membrane purification results in at least a tenfold purification over whole cell lysate and therefore represents higher selectivity of analysis.

1. For kinetic analysis of Cdc42p or Rho1p activation, incubate 375 µg of membranes in 750 µL of fusion reaction buffer containing 5 mM ATP, 40 µM GTPγS, and 0.5 mg/mL cytosol at 28°C for 0–90 min (see Notes 8 and 9 and Fig. 2b). We recommend

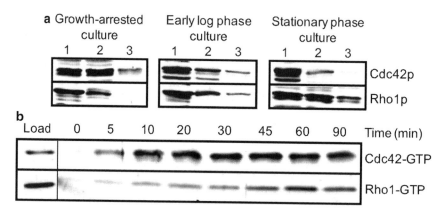

Fig. 2. Analysis of Rho activation in a typical experiment. (a) Analysis of activated Cdc42p and Rho1p from 1 mg of cell lysate. Cells were subjected to growth arrest (10 µg/mL nocodazole for 2 h), early log phase growth ($OD_{600} \sim 0.5$) or stationary phase ($OD_{600} \sim 5$). Lysates were incubated with GST-PAK CBD and GST-Rhotekin RBD, to pull-down activated Cdc42p and Rho1p, respectively. A 5% load control (*lane 1*), a 100% activation control via chemically induced GTPγS exchange (*lane 2*) and the experimental condition (*lane 3*) are shown for each sample. (b) Kinetic analysis of Cdc42p and Rho1p activation on purified membranes. 40 µg of vacuoles were incubated in fusion reaction buffer for each time point. At specific time intervals a sample was removed and analyzed for levels of activated Cdc42p and Rho1p by pull-down with GST-PAK CBD and GST-Rhotekin RBD, respectively. A 10% load control is shown. Panel B is reproduced from ref. 12 with permission from Elsevier.

analyzing samples at 5, 10, 20, 30, 45, 60, and 90 min time intervals to detect early rates of activation and how long activation is sustained.

2. Negative controls can be optionally performed by including 10 μM purified Rdi1p (GST-RhoGDI) the natural inhibitor of Rho1p and Cdc42p activation, in standard reaction mixtures (see Note 10).

3. Positive controls can be optionally performed by including 2 μM purified Rho GEF, (His$_6$-tagged Dbs domain) in standard reaction mixtures.

4. At 5–10 min intervals, briefly vortex the reaction mixture and remove 100 μL. Dilute fivefold in ice-cold PS buffer. Re-isolate membranes by centrifugation (20,000×g, 10 min, 4°C). Aspirate the supernatant and dissolve the membrane pellet in binding buffer. Samples should be immediately probed for activated GTPases (see Subheading 3.5).

5. A 10% load control is taken by removing 10 μL from the reaction mixture. Add 40 μL of 1.2× SSB and heat the sample to 95°C for 5 min. The sample is then ready for SDS-PAGE and immunoblot analysis.

3.5. Affinity Isolation of Activated Cdc42p-GTP and Rho1p-GTP from Lysates and Reactions

Levels of activated (GTP-bound) Rho GTPases from reactions, membrane fractions, or lysates are determined by affinity precipitation with activation probes immobilized on glutathione-agarose. Cdc42p activation is examined by pull-down using GST-CBD beads; Rho1p activation is examined by pull-down using GST-RBD beads (separate tubes).

1. 10 μL of GST-CBD or GST-RBD beads are transferred into a 1.5-mL eppendorf tube by pipetting 100 μL from well-mixed 10% slurry of beads (see Note 5). Samples to be analyzed for activated Rho proteins are added and the volume is adjusted to 1 mL with binding buffer (see Note 8). Typically, 50 μg of purified membrane fraction or 1 mg of whole cell lysate is probed with GST-CBD or GST-RBD beads to pull-down activated Cdc42p-GTP or Rho1p-GTP, respectively.

2. Negative controls for background binding are performed by incubating samples with 10 μL of beads containing GST without CBD and RBD domains. This determines the background binding (0% control).

3. Positive controls for maximum binding are performed by incubating activation probes (GST-CBD or GST-RBD beads) with samples that have undergone chemically induced nucleotide loading of GTPγS using nucleotide loading buffer. Loading of Rho GTPases with a nonhydrolyzable analog of GTP (i.e., GTPγS) determines the maximum binding (100% control).

Incubate samples for 5 min at 30°C in nucleotide loading buffer containing 10 μM GTPγS. Cool on ice for 1 min. Quench the loading reaction by adding MgCl₂ to 10 mM.

4. Samples are incubated at 4°C for 30 min with gentle mixing. Beads are settled by centrifugation ($1,000 \times g$, 30 s, 4°C) and the supernatant is aspirated (see Note 11). The beads are washed three times with 0.5 mL of binding buffer. The binding buffer is fully aspirated after the final wash and 40 μl of 1.2× SSB is added. The samples are heated to 95°C for 5 min and are then ready for SDS-PAGE and immunoblot analysis.

5. Equivalent amounts of sample are subjected to SDS-PAGE and immunoblot analysis using the appropriate primary and secondary antibodies. A typical experiment, including controls is shown in Fig. 2.

6. Band intensities are measured and normalized within an experiment by calculating the ratio as a percentage of the 100% control. This reduces day-to-day experimental variations.

4. Notes

1. Incubation of samples in nucleotide loading buffer results in chemically induced loading of the desired nucleotide. This is normally GTPγS, which allows the determination of maximum activation signal. EDTA strips Mg^{2+} ions that are required for the co-ordination of nucleotides in the binding pocket and free diffusion of nucleotides. Loading reactions are quenched by adding MgCl₂ to 10 mM, which results in "locking" the nucleotide in the binding pocket.

2. All efforts at minimizing processing time to prepare fractions should be explored. It is imperative to process samples rapidly and immediately add activation probes.

3. For immunoblot detection, we prefer fluorescence detection in the far IR range. Typically, we use goat anti-rabbit, guinea pig, or mouse IgG secondary antibodies conjugated to Alexafluor-680 and Alexafluor-750 (Invitrogen) which allow scanning of blots with a far IR digital imager (i.e., Odyssey, Licor Inc.). The Odyssey digital imaging system allows for rapid quantification of band intensity. Alternatively, conventional detection using peroxidase-coupled antibodies can be performed. To ensure accurate band-densitometry we recommend taking an average reading from three different exposures.

4. We cloned and tested several Cdc42/ Rho1 binding domains using the bacterial expression vectors pGEX-2T and pGEX-4T (GE Healthcare). However, we found that Rhotekin and PAK1

derived probes most efficiently detected the activation of yeast Cdc42 and Rho1 (12).

5. Make a 10% slurry of probe-bound beads by adding 900 μL of binding buffer to 100 μL of beads (or similar ratio). Cut the end off a P200 tip with a sharp razor when pipetting the beads.

6. Yeast should be grown at 28°C to early log phase ($OD_{600} \sim 0.8$) for all preparations. However, it is common to prepare lysates from specialized growth conditions. Numerous specialized growth conditions can be applied to suit a desired experiment. These include cell synchronization by treatment with α-factor or nocodazole (15), re-growth of stationary phase cultures, metabolic changes, and temperature shift of mutant strains (16).

7. Specific membrane fractions can be isolated from yeast lysates by isopycnic centrifugation (20). Probe membranes by immunoblot to verify the presence of Rho GTPases. Some membranes have significant levels of both Cdc42p and Rho1p (vacuoles, plasma membrane) (21, 22) while others have only Cdc42p (Golgi) (23) or Rho1p (peroxisomes, secretory vesicles) (19, 24). Rho GTPases localized to the cytosol are complexed with RhoGDI and cannot be activated in the absence of membranes.

8. Yeast lysates or membrane fractions should occupy no more than one-fifth of the volume.

9. Levels of membrane can be adjusted depending on levels of Rho GTPases localized to the membrane. This should be empirically determined by immunoblot. Rho GTPases should be easily detected in a 10% load control by immunoblot.

10. Inhibition of Rho activation can be performed with purified Rdi1p (RhoGDI) as a control. Rdi1p is active in recombinant form. An *E. coli* strain that overproduces GST-Rdi1p is available (12). GST-Rdi1p is bound to glutathione resin in yeast lysis buffer containing 2.5 mM $CaCl_2$. Rdi1p is purified by incubation with 10 U/mL thrombin at 37°C for 2 h. Thrombin is removed by incubation with benzamindine agarose.

11. An unbound sample of the supernatant should be kept for immunoblot analysis. This can be used to show whether quantitative pull-down is achieved.

Acknowledgments

We would like to acknowledge the Canadian Institutes of Health Research and the National Science and Engineering Research Council of Canada for funding projects that study Rho GTPases in yeast.

References

1. Guo, W., Tamanoi, F., and Novick, P. (2001) Spatial regulation of the exocyst complex by Rho1 GTPase. *Nat Cell Biol* 3, 353–360.

2. Dong, Y., Pruyne, D., and Bretscher, A. (2003) Formin-dependent actin assembly is regulated by distinct modes of Rho signaling in yeast. *J Cell Biol* 161, 1081–1092.

3. Imamura, H., Tanaka, K., Hihara, T., Umikawa, M., Kamei, T., Takahashi, K., Sasaki, T., and Takai, Y. (1997) Bni1p and Bnr1p: downstream targets of the Rho family small G-proteins which interact with profilin and regulate actin cytoskeleton in *Saccharomyces cerevisiae*. *EMBO J* 16, 2745–2755.

4. Delley, P.A., and Hall, M.N. (1999) Cell wall stress depolarizes cell growth via hyperactivation of RHO1. *J Cell Biol* 147, 163–174.

5. Lechler, T., Jonsdottir, G.A., Klee, S.K., Pellman, D., and Li, R. (2001) A two-tiered mechanism by which Cdc42 controls the localization and activation of an Arp2/3-activating motor complex in yeast. *J Cell Biol* 155, 261–270.

6. Kadota, J., Yamamoto, T., Yoshiuchi, S., Bi, E., and Tanaka, K. (2004) Septin ring assembly requires concerted action of polarisome components, a PAK kinase Cla4p, and the actin cytoskeleton in *Saccharomyces cerevisiae*. *Mol Biol Cell* 15, 5329–5345.

7. Wu, H., Rossi, G., and Brennwald, P. (2008) The ghost in the machine: small GTPases as spatial regulators of exocytosis. *Trends Cell Biol* 18, 397–404.

8. Perez, P., and Rincón, S.A. (2010) Rho GTPases: regulation of cell polarity and growth in yeasts. *Biochem J* 426, 243–253.

9. Johnson, D.I. (1999) Cdc42: An essential Rho-type GTPase controlling eukaryotic cell polarity. *Microbiol Mol Biol Rev* 63, 54–105.

10. Levin, D.E. (2005) Cell wall integrity signaling in *Saccharomyces cerevisiae*. *Microbiol Mol Biol Rev* 69, 262–291.

11. Ren, X.D., and Schwartz, M.A. (2000) Determination of GTP loading on Rho. *Methods Enzymol* 325, 264–272.

12. Logan, M.R., Jones, L., and Eitzen, G. (2010) Cdc42p and Rho1p are sequentially activated and mechanistically linked to vacuole membrane fusion. *Biochem Biophys Res Commun* 394, 64–69.

13. Benard, V., and Bokoch, G.M. (2002) Assay of Cdc42, Rac, and Rho GTPase activation by affinity methods. *Methods Enzymol* 345, 349–359.

14. Jones, L., Tedrick, K., Baier, A., Logan, M.R., and Eitzen, G. (2010) Cdc42p is activated during vacuole membrane fusion in a sterol-dependent subreaction of priming. *J Biol Chem* 285, 4298–4306.

15. Kono, K., Nogami, S., Abe, M., Nishizawa, M., Morishita, S., Pellman, D., and Ohya, Y. (2008) G1/S cyclin-dependent kinase regulates small GTPase Rho1p through phosphorylation of RhoGEF Tus1p in *Saccharomyces cerevisiae*. *Mol Biol Cell* 19, 1763–1771.

16. Wai, S.C., Gerber, S.A., and Li, R. (2009) Multisite phosphorylation of the guanine nucleotide exchange factor Cdc24 during yeast cell polarization. *PLoS One* 4, e6563.

17. Boulter, E., Garcia-Mata, R., Guilluy, C., Dubash, A., Rossi, G., Brennwald, P.J., and Burridge, K. (2010) Regulation of Rho GTPase crosstalk, degradation and activity by RhoGDI1. *Nat Cell Biol* 12, 477–483.

18. Qadota, H., Python, C.P., Inoue, S.B., Arisawa, M., Anraku, Y., Zheng, Y., Watanabe, T., Levin, D.E., and Ohya, Y. (1996) Identification of yeast Rho1p GTPase as a regulatory subunit of 1,3-β-glucan synthase. *Science* 272, 279–281.

19. Marelli, M., Smith, J.J., Jung, S., Yi, E., Nesvizhskii, A.I., Christmas, R.H., Saleem, R.A., Tam, Y.Y., Fagarasanu, A., Goodlett, D.R., Aebersold, R., Rachubinski, R.A., and Aitchison, J.D. (2004) Quantitative mass spectrometry reveals a role for the GTPase Rho1p in actin organization on the peroxisome membrane. *J Cell Biol* 167, 1099–1112.

20. Chang, J., Ruiz, V., and Vancura, A. (2008) Purification of yeast membranes and organelles by sucrose density gradient centrifugation. *Methods Mol Biol* 457, 141–149.

21. Eitzen, G., Thorngren, N., and Wickner, W. (2001) Rho1p and Cdc42p act after Ypt7p to regulate vacuole docking. *EMBO J* 20, 5650–5656.

22. Wu, H., and Brennwald, P. (2010) The function of two Rho family GTPases is determined by distinct patterns of cell surface localization. *Mol Cell Biol* 30, 5207–5217.

23. Paglini, G., Peris, L., Diez-Guerra, J., Quiroga, S., and Cáceres, A. (2001) The Cdk5-p35 kinase associates with the Golgi apparatus and regulates membrane traffic. *EMBO Rep* 2, 1139–1144.

24. Abe, M., Qadota, H., Hirata, A., and Ohya, Y. (2003) Lack of GTP-bound Rho1p in secretory vesicles of *Saccharomyces cerevisiae*. *J Cell Biol* 162, 85–97.

Chapter 25

Assaying Rho GTPase-Dependent Processes in *Dictyostelium discoideum*

Huajiang Xiong and Francisco Rivero

Abstract

The model organism *D. discoideum* is well suited to investigate basic questions of molecular and cell biology, particularly those related to the structure, regulation, and dynamics of the cytoskeleton, signal transduction, cell–cell adhesion, and development. *D. discoideum* makes use of Rho-regulated signaling pathways to reorganize its cytoskeleton during chemotaxis, endocytosis, and cytokinesis. In this organism the Rho family encompasses 20 members, several belonging to the Rac subfamily, but there are no representatives of the Cdc42 and Rho subfamilies. Here we present protocols suitable for monitoring the actin polymerization response and the activation of Rac upon stimulation of aggregation competent cells with the chemoattractant cAMP.

Key words: *Dictyostelium*, cAMP, Actin polymerization, Pull-down assay, Rac, Chemotaxis

1. Introduction

1.1. Dictyostelium discoideum as a Model Organism

Dictyostelium discoideum is a simple eukaryotic microorganism whose natural habitat is the deciduous forest soil where free-living amoebas feed on bacteria and multiply by equal mitotic division. Exhaustion of the food source triggers a developmental program in which more than 100,000 cells aggregate by chemotaxis toward cAMP to form a multicellular structure. Differentiation and morphogenesis result in the formation of a fruiting body composed of a mass of spores supported by a stalk made of vacuolized dead cells. Numerous advantages make *D. discoideum* a widely accepted and well-suited model organism to investigate basic questions of molecular and cell biology, particularly those related to the structure, regulation, and dynamics of the cytoskeleton, signal transduction, cell–cell adhesion, and development (1). More recently *D. discoideum* is being employed as a model to study the

Francisco Rivero (ed.), *Rho GTPases: Methods and Protocols*, Methods in Molecular Biology, vol. 827,
DOI 10.1007/978-1-61779-442-1_25, © Springer Science+Business Media, LLC 2012

mechanisms of infection by pathogenic bacteria (2). *D. discoideum* has a short life cycle, is easy to cultivate, grows in inexpensive media, can be harvested in large amounts and is amenable to a variety of biochemical and molecular and cell biological techniques (3). Its genome, completely sequenced and largely annotated, can be easily manipulated by means of a growing list of molecular genetics techniques that have allowed considerable advances in the study of biological processes.

1.2. Rho GTPases in D. discoideum

Like other eukaryotic cells, *D. discoideum* makes use of Rho-regulated signal transduction pathways to reorganize its cytoskeleton during chemotaxis and other processes like endocytosis and cytokinesis. In *D. discoideum* the Rho family encompasses 20 members. Based on phylogenetic analyses, several Rho GTPases can be grouped in the Rac subfamily: Rac1a, Rac1b, Rac1c, RacF1, RacF2, and more loosely RacB and the GTPase domain of RacA (the *D. discoideum* RhoBTB ortholog). All other *D. discoideum* Rho GTPases were named Rac for historical reasons but they do not have a clear affiliation, although some are closer to Rac than to members of other subfamilies. There are no representatives of the Cdc42, Rho, or other subfamilies in *D. discoideum* (4).

Like in other organisms, knockout, overexpression and gain-of-function mutants have been used to investigate the physiological roles of *D. discoideum* Rho GTPases. However, the presence of multiple Rac proteins, some of them as very closely related isoforms, has presented a challenge for the elucidation of their function, because of potential functional redundancy. Nevertheless the evidence accumulated so far implicates Rho GTPases in the regulation of chemotaxis and cell motility, endocytosis, and cytokinesis in *D. discoideum*. Studies on knockout mutants have established clear roles for RacB and RacC in the regulation of chemotaxis (5, 6) and studies on strains expressing activated and/or dominant negative Rac1b, RacE, and RacG suggest that these Racs may also be involved (7–9). RacE was the first *D. discoideum* Rho GTPase shown to be essential for cytokinesis (10) and subsequent studies have provided indirect evidence that the three Rac1 isoforms, RacB, RacC, and RacH may also implicated in the regulation of cytokinesis (7, 11–13).

A number of Racs have been implicated in regulating the uptake phase of macropinocytosis and phagocytosis, namely the three Rac1 isoforms, RacB, RacC, RacF1, and RacG (8, 11, 13–15). This conclusion is based on localization studies, proteomics, and the analysis of overexpressors, but knockout mutants of these Racs do not show defects in fluid or particle uptake. By contrast, RacH, that associates with membranes of the endoplasmic reticulum and Golgi apparatus, is clearly implicated in sorting of vesicles between compartments along the endocytic pathway (12).

1.3. Assaying Rac-Dependent Processes in D. discoideum

Because it would be impossible to provide detailed methods to investigate all relevant processes in which Rho GTPases have been implicated in *D. discoideum*, here we will restrict ourselves to protocols for monitoring the actin polymerization response and the activation of Rac upon cAMP stimulation of aggregation competent cells. For the investigation of actin localization, chemotaxis, endocytosis, and other processes the reader is referred to other sources (3).

The filamentous (F-)actin content is quantitated by fluorescently labeled phalloidin staining of formaldehyde-fixed pelleted material, an adaptation to *D. discoideum* cells (16) of the method originally devised for neutrophils (17). The method has been applied extensively to monitoring the actin polymerization response induced by cAMP in aggregation competent cells in a variety of circumstances. F-actin exhibits a characteristic biphasic curve upon agonist stimulation. During the first peak, at 5 s, the amount of F-actin approximately doubles. This peak is very short and correlates with the cringe reaction in which the cells round up and produce a uniform cortical accumulation of F-actin. The second peak is significantly broader and lower, shows a maximum at 30–60 s and corresponds to the emergence of pseudopods and cell movement. The actin polymerization response parallels in time and magnitude that of Rac activation, from which it depends (5, 6, 8) and therefore constitutes an excellent tool to monitor alterations in Rac-dependent signaling pathways (see Fig. 1a).

The Rac activation (or pull-down) assay is an adaptation for *D. discoideum* of the method described by Benard et al. (18) for the determination of Rac and Cdc42 activation in chemoattractant-stimulated human neutrophils. The method makes use of a fusion protein of glutathione-S-transferase (GST) and the Rho GTPase-binding domain of an effector molecule to capture active Rac in a cell lysate. In *D. discoideum* this method has been applied to the determination of activated Rac1, RacB, and RacC (5, 6, 8, 19). In principle the method could be adapted to determine the activity of any other Rac, provided a GST fusion with a specific GTPase-binding domain and a specific antibody to detect the Rac by Western blot are available. However Rho GTPase-binding domains of potential use have been identified for a very limited number of *D. discoideum* Racs (Rac1a/1b/1c, RacA, RacB, RacC, and RacF1/2) only, and they include WASP and related proteins, PAK kinases, and IQGAPs ((5, 6, 20) and our unpublished work). We have tested the GTPase binding domains of *D. discoideum* WASP and PAKb as well as human PAK1 for the pull-down assay of activated Rac1 and have obtained better results with WASP (8). WASP has also been used by Han et al. for the determination of RacC (6). Park et al. have used PAKa to pull-down active RacB (5).

Another limitation is posed by the unavailability of good antibodies against *D. discoideum* Racs. At present only endogenous

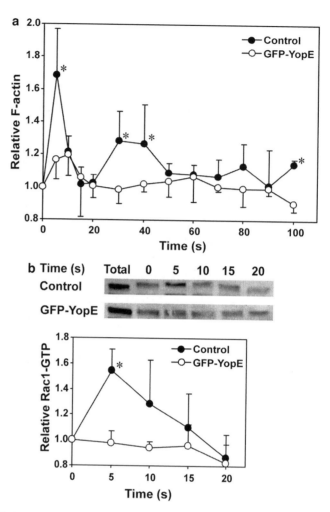

Fig. 1. Reduced actin polymerization response and Rac1 activation upon cAMP stimulation in cells expressing YopE. YopE is an effector protein produced by pathogenic bacteria of the genus *Yersinia*. It displays GTPase activating activity on several Rho GTPases. A green fluorescent protein (GFP) fusion of YopE was expressed in *D. discoideum* cells using a tetracycline-regulatable vector. (a) Relative F-actin content as determined by TRITC-phalloidin staining of aggregation competent cells upon stimulation with 1 μM cAMP. Control cells are non-induced cells carrying the GFP-YopE plasmid. The amount of F-actin was normalized relative to the F-actin level of non-stimulated cells. Data are average ± standard deviation of five independent experiments. For simplicity, error bars are depicted only in one direction. *$P < 0.05$, Student's *t*-test. (b) Activation of Rac1 upon cAMP stimulation in cells expressing GFP-YopE. Rac1-GTP was separated using a pull-down assay with GST-WASP[CRIB]. A representative blot of each strain is shown. Data are average ± standard deviation of four independent pull down experiments. *$P < 0.05$, Student's *t*-test. (Reproduced from ref. 19).

Rac1 can be detected in Western blots after a pull-down assay (8, 11, 19). Two approaches have been taken to overcome this problem. Han et al. have resorted to the use of a strain that expresses a fluorescent protein fusion of RacC and detected the activated protein with an anti-GFP antibody (6). Park et al. replaced endogenous RacB with a gene encoding a myc-tagged RacB under the control of the endogenous promoter (5), an approach that will be very useful if applied in future to the other members of the Rac family.

A limitation of the pull-down assay is that it provides a measurement of the total activity of a given Rac within a population of cells and therefore does not permit the spatial and temporal monitoring of Rac activation within individual living cells. Although fluorescent probes based on fluorescence resonance energy transfer have been developed, they have found very limited use in *D. discoideum* to date (6).

2. Materials

All media and buffers are prepared using deionized water (dH$_2$O), filtered through an ion-exchange unit.

2.1. Culture of Dictyostelium Cells

Dictyostelium cells are cultured in nutrient medium in suspension on a rotary shaker (160 rpm) at 21°C. Cells are harvested and washed at room temperature by centrifugation at $500 \times g$ for 3 min. When analyzing different strains, it is recommended that cells be cultivated under the same conditions for some days preceding the assays. For the assays described here cells should be taken from axenically growing or "log phase" cultures (up to 2×10^6 cells/ml). Dilute cultures the day before in order to have the necessary number of log phase cells.

1. Nutrient medium. AX Medium is available from Formedium™ in powder form. Suspend 40.4 g in 1 L of dH$_2$O and autoclave.

2. Soerensen phosphate buffer: 2 mM Na$_2$HPO$_4$, 14.6 mM KH$_2$PO$_4$, pH 6.0. Autoclave.

2.2. Determination of F-Actin

1. Tetramethylrhodamine isothiocyanate (TRITC)-labeled phalloidin (Sigma P1951). Dissolve to 0.1 mg/mL in methanol to obtain a 76.6 µM stock solution. Store at –20°C (see Note 1).

2. Actin polymerization stop solution: 3.7% formaldehyde, 0.1% Trition X-100, 20 mM potassium phosphate, 10 mM PIPES, 5 mM EGTA, 2 mM MgCl$_2$. Adjust pH to 6.8 with 1 N KOH. Store aliquots at –20°C. TRITC-phalloidin is added to 0.25 µM final concentration fresh before starting the experiment (see Note 2).

3. 0.1 mM cAMP stock solution. Dissolve cAMP in Soerensen buffer. Store aliquots at –20°C.

2.3. Rac Activation Assay

1. Luria broth (LB): dissolve 10 g of Bacto-tryptone, 5 g of yeast extract, 10 g of NaCl in 600 mL of dH$_2$O, adjust pH to 7.5 with 1 N NaOH, complete to 1 L with dH$_2$O, autoclave.

2. 100 mg/mL ampicillin in sterile dH$_2$O, filter sterilize (0.2 μm pore diameter), store aliquots at –20°C.

3. 1 M isopropyl-β-D-thiogalactopyranoside (IPTG) in sterile dH$_2$O, filter sterilize (0.2 μm pore diameter), store aliquots at –20°C.

4. Glutathione Sepharose 4B beads (GE Healthcare).

5. Bacterial lysis buffer: 50 mM Tris–HCl, pH 8.0, 300 mM NaCl, 10 mM MgCl$_2$. Add fresh, prior to use: 4 mM dithiotreitol (DTT), protease-inhibitor cocktail (Roche), 5 mg/mL lysozyme.

6. Washing buffer: 50 mM Tris–HCl, pH 8.0, 1 M NaCl, 10 mM MgCl$_2$, Add fresh, prior to use: 4 mM DTT.

7. Sonicator UP200S (Hielscher Ultrasonics GmbH).

8. 2× SDS sample buffer: 0.125 M Tris–HCl, pH 6.8, 4% SDS, 20% glycerol, 10% 2-mercaptoethanol, 0.004% bromophenol blue.

9. 5× reaction buffer: 50 mM HEPES, pH 7.5, 500 mM NaCl, 100 mM MgCl$_2$, 2.5% Triton X-100. Add fresh, prior to use: 1 mM DTT, protease-inhibitor cocktail (Roche).

10. 1× reaction buffer: prepare from 5× reaction buffer by dilution with sterile dH$_2$O.

11. Spin columns, 0.9-mL capacity with screw caps and press-on bottom plugs (Thermo Scientific Pierce).

3. Methods

3.1. Determination of the Relative F-Actin Content

Using this method cells are permeabilized immediately by Triton X-100, allowing very rapid fixation (within 1 s) by formaldehyde. TRITC-phalloidin binds specifically to F-actin. TRITC-phalloidin is methanol-extracted from the Triton insoluble pellet and quantitated fluorimetrically, providing a measurement of the amount of F-actin present in the cell sample. Here we describe the determination of the basal F-actin content of a cell strain relative to a control strain.

1. Harvest and resuspend cells at 2×10^7 cells/mL in fresh nutrient medium. Allow cells to recuperate for 30 min.

2. Thaw actin polymerization stop solution and transfer the amount needed into a conical tube. Add TRITC-phalloidin to 0.25 μM final concentration (32.6 μL of the 0.1 mg/mL stock

for 10 mL of stop solution). Dispense 450 μL into labeled eppendorf tubes (3 replicates for each strain) (see Note 3).

3. Withdraw three 50-μL samples of cell suspension and transfer into tubes containing 450 μL of stop solution.

4. Incubate samples for 1 h at room temperature in the dark.

5. Spin samples for 5 min at $15,000 \times g$.

6. Carefully remove supernatant without disturbing the pellet. Pellets have a pale pink color.

7. Extract each pellet with 1 mL of methanol for 1 h in the dark prior to fluorimetric measurement (excitation 540 nm, emission 565 nm). Alternatively allow the extraction to proceed overnight at −20°C in the dark.

8. Determination of total protein content. Withdraw three 50-μL samples of the cell suspension. Pellet cells for 10 s at maximum speed in a tabletop centrifuge and remove supernatant. The pellets can be frozen or can be processed immediately for the determination of protein content using a standard method.

9. Normalize the fluorescence values to the protein content of the sample and calculate the F-actin content relative to the control strain.

3.2. Actin Polymerization Response upon cAMP Stimulation

The same principle as in Subheading 3.1 is applied to the determination of the time course of F-actin formation upon agonist stimulation.

1. Harvest cells and wash twice with Soerensen buffer.

2. Resuspend cells at 2×10^7 cells/mL in Soerensen buffer.

3. Starve cells for 6–8 h at 21°C in an Erlenmeyer flask on a rotary shaker (160 rpm).

4. Shortly before starting the experiment thaw actin polymerization stop solution and transfer the amount needed into a conical tube. Add TRITC-phalloidin to 0.25 μM final concentration (32.6 μL of the 0.1 mg/mL stock for 10 mL of stop solution). Dispense 450 μL into labeled eppendorf tubes (see Note 3).

5. Transfer 1 mL of cell suspension into a well of a 24-well plate.

6. Place the 24-well plate on a mini shaker (see Note 4).

7. Collect four 50-μL samples and transfer to tubes containing actin polymerization stop solution (see Note 5). These will be the time point 0 measurements.

8. Stimulate actin polymerization in the remaining 800 μL of cell suspension with 8 μL of 0.1 mM cAMP (1 μM final concentration).

9. Immediately collect 50-μL samples at following time points: 5, 10, 15, 20, 25, 30, 40, 50, 60, 70, 80, 90, 100, and 120 s after

cAMP stimulation (see Note 6). Each sample is dispensed in its corresponding tube containing stop solution.

10. Proceed as in Subheading 3.1, steps 4–7.

11. For the calculations refer the fluorescence values to the value of the 0 time point. In wild type cells a peak of F-actin usually occurs 5 s after cAMP stimulation. This peak is 1.6–2 times the value of unstimulated cells (see Fig. 1a). Alternatively the amount of fluorescence can be normalized to the protein content of the cell suspension. In this case withdraw one sample for protein determination as indicated in Subheading 3.1, step 8.

3.3. Pull-Down Assay for Activated Rac

The method makes use of a fusion protein of GST and the Rho GTPase-binding domain of an effector molecule. The fusion protein is expressed recombinantly in *E. coli* and is purified by attachment to glutathione agarose. The glutathione agarose beads carrying the fusion protein are applied to a cell lysate. The beads are then separated from the lysate followed by washing and elution of the captured active Rac in sample loading buffer. The active Rac is detected by Western blot. The protocol given below is designed for the determination of activated Rac1 using the CRIB (Cdc42 and Rac interactive binding) domain of WASP but can in principle be adapted for the detection of any activated Rho GTPase, provided a specific effector is available and the GTPase can be identified on a Western blot.

3.3.1. Production of GST-WASP^CRIB

The CRIB domain of *D. discoideum* WASP is cloned in frame to GST in a pGEX series vector (GE Healthcare) and the plasmid is introduced in a suitable *E. coli* strain, e.g. XL-1 blue or BL21 (8).

1. Inoculate 50 mL of LB/ampicillin (50 μg/mL) with a single colony of *E. coli* transformed with the pGEX-WASP^CRIB plasmid. Grow overnight at 37°C.

2. Inoculate 1 L of LB/ampicillin (50 μg/mL) with the overnight culture, determine the optical density of the culture at 600 nm (OD_{600}) and grow at 37°C.

3. Induce expression with 0.2 mM IPTG when the culture has reached an OD_{600} around 0.7. Continue growth at 37°C for 2–3 h.

4. Collect bacteria by centrifugation at $2,000 \times g$ for 10 min (see Note 7). Resuspend bacteria in 10 mL of pre-chilled lysis buffer supplemented with protease-inhibitor cocktail and lysozyme.

5. Sonicate the bacterial suspension six to ten times for 30 s at 50–70% duty cycle on ice (see Note 8).

6. Remove cell debris by spinning at $21,000 \times g$ for 10–20 min at 4°C. Collect supernatant and store aliquots at –80°C (see Note 9).

7. Monitor the efficiency of fusion protein expression and purification by SDS-PAGE. Proceed as in steps 4–7 of Subheading 3.3.2 using one spin column and 50 μL of glutathione Sepharose slurry. After the last washing step place the spin column in an open eppendorf cup and remove residual wash buffer by spinning at $10,000 \times g$ for 1 min. Place the spin column into a clean eppendorf cup, apply 50 μL of 2× SDS sample buffer and spin at $10,000 \times g$ for 1 min to collect the affinity purified GST-WASP[CRIB]. Denature samples for 5 min at 95°C. Examine 15 μL of eluate by SDS-PAGE and Coomassie staining (see Note 10).

3.3.2. Pull-Down of Activated Rac upon cAMP Stimulation

Following protocol is devised for the determination of activated Rac1 in one *D. discoideum* strain during the first 20 s upon agonist stimulation (see Fig. 1b). If a longer time course is required, increase the amount of cell suspension and glutathione agarose beads and the number of spin columns accordingly. Start with the preparation of the glutathione agarose beads loaded with GST-WASP[CRIB] approximately 2 h before the end of the starvation period (see Note 9). Except for the Soerensen buffer and the 2× SDS sample buffer, all other buffers should be chilled.

1. Harvest *D. discoideum* cells and wash twice with Soerensen buffer.

2. Resuspend cells at 2×10^7 cells/mL in Soerensen buffer. 6 mL of cell suspension is required.

3. Starve cells for 6–8 h at 21°C in an Erlenmeyer flask on a rotary shaker (160 rpm).

4. Prepare five spin columns, labeled 0, 5, 10, 15, and 20. Prepare a suspension of 250 μL of glutathione Sepharose slurry in 2250 μL of bacterial lysis buffer. Allow equilibration of the beads in lysis buffer for 10 min. Load 500 μL of this suspension into each column. This will ensure that each column receives the same amount of beads. Remove lysis buffer by gravity flow and apply bottom plug to column.

5. Load 700 μl of bacterial lysate containing GST-WASP[CRIB] fusion protein onto each column, cap and allow coupling of fusion protein to the beads on a rotator for 1 h at 4°C.

6. Remove bottom plug and cap from column. Clear bacterial lysate by gravity flow.

7. Wash beads four times with washing buffer by gravity flow. Apply bottom plug to column.

8. Equilibrate beads with 600 μL of 1× reaction buffer for 10 min. Remove reaction buffer by gravity flow and apply bottom plug to column.

9. Label eppendorf tubes with 0, 5, 10, 15 and 20 s. Dispense 175 μL of 5× reaction buffer and keep on ice.

10. Harvest and resuspend cells at 4×10^7 cells/mL in Soerensen buffer.

11. Transfer 700 μL of cells into the time 0 tube containing 5× reaction buffer. Mix well on a Vortex. Stimulate 5 mL of cells with 50 μL of 0.1 mM cAMP (1 μM final concentration). Mix well.

12. Immediately transfer 700-μL samples at 5, 10, 15 and 20 s into the corresponding tubes containing 5× reaction buffer. Mix well on a Vortex and place samples on ice.

13. Transfer 700 μL of cell lysate onto the corresponding spin column containing the fusion protein coupled to glutathione Sepharose beads (see steps 4–8). Cap and incubate for 45 min on a rotator at 4°C.

14. Remove a 50-μL sample from the time point 0 lysate and mix with 50 μL of 2× SDS sample buffer. This sample will be used to monitor total Rac1.

15. Remove bottom plug and cap from columns. Release lysate from column by gravity flow.

16. Wash columns four times with 1× reaction buffer by gravity flow. Place the spin column in an open eppendorf cup and remove residual reaction buffer by spinning at $10,000 \times g$ for 1 min.

17. Place the spin columns into clean labeled eppendorf cups, apply 50 μL of 2× SDS sample buffer and spin at $10,000 \times g$ for 1 min to collect protein complexes.

18. Denature samples for 5 min at 95°C. Load 15 μL of each sample onto 18% polyacrylamide gels, resolve by SDS-PAGE and detect Rac1 by Western blotting with antibody 273-461-3 (11) using standard procedures.

19. Quantitate the intensity of the Rac1 bands by densitometry and refer the values to the value of the 0 time point (see Note 11) (see Fig. 1b).

4. Notes

1. Caution should be exerted while working with phalloidin and methanol. Phalloidin is very toxic, may be fatal if swallowed, inhaled, or absorbed through the skin. For personal protection, wear glasses and gloves. Avoid dispersal of material in the air or on working surfaces. Methanol is toxic, may cause heart and liver damage, blindness, or death if ingested or inhaled. Methanol is highly flammable. The flame above burning methanol is invisible. Always wear safety glasses, clear any source of ignition from the working area, and ensure the working area is well ventilated.

2. The amount of phalloidin present in the stop solution should be such that it saturates the amount of F-actin of the cells. The concentration given here works well for the number of cells indicated in the protocols but may need to be adjusted if substantial changes are introduced (21).

3. TRITC-phalloidin is expensive, thus prepare the volume of stop solution needed for the number of samples in an experiment plus some excess. Leftover stop solution containing TRITC-phalloidin can be stored at −20°C and used in the next experiment by mixing the leftover with the new stop solution, which will make a homogenous stop solution.

4. Use a shaker with low amplitude of displacement, like the ones designed for shaking ELISA plates. Adjust rotation speed so that the cell suspension is visibly shaking to prevent cells from settling while avoiding spilling of the cell suspension.

5. Measurements of four samples at time point 0 are important to obtain an accurate estimate of the basal amount of F-actin prior to cAMP stimulation. Avoid collecting time point 0 samples from the cell suspension before transferring into the 24-wells, as values will differ. All samples of one experiment should be collected from the same well holding a 1-mL aliquot of cell suspension.

6. To achieve accurate sample collection at the recommended time points, it is essential to stay attentive near the shaker, keep tubes containing stop solution open and a pipette ready to withdraw samples and dispense them into the corresponding tubes once cAMP has been added. Avoid contact of the pipette tip with the stop solution, use the same tip used to take the time point 0 samples throughout collection of all time points. Vortexing of samples is dispensable, because lysis occurs immediately upon contact with the stop solution. The sample at time point 5 s is critical, as the first peak of polymerization is expected in this sample.

7. Bacterial pellets obtained after protein expression can be stored at −20°C for processing at later step.

8. Parameters for sonication need to be adjusted to the device available. Bacterial lysis can be achieved by other methods (freezing and thawing, French press).

9. It is more convenient to store bacterial supernatant for attachment to agarose beads freshly prior to the pulldown assay rather than to store a batch of purified fusion protein, as this would need to be dialyzed to remove glutathione and will have to be attached to beads anyway. It is also not advisable to store coupled beads.

10. The amount of fusion protein pulled down using glutathione Sepharose beads can be estimated by comparison to a bovine serum albumin standard resolved on the same SDS-PAGE, followed by Coomassie staining.

11. Several exposure times should be used when documenting the detected Rac1 in the Western blots so that the density of the protein bands varies linearly with the amount of protein on the blot.

References

1. Kessin, R.H. (2001) *Dictyostelium* – Evolution, cell biology, and the development of multicellularity. Cambridge Univ. Press, Cambridge.

2. Steinert, M., and Heuner, K. (2005) *Dictyostelium* as host model for pathogenesis. *Cell. Microbiol* 7, 307–314.

3. Eichinger, L., and Rivero, F. (2006) *Dictyostelium discoideum* protocols. Methods in molecular biology 346. Humana Press Inc, Totowa.

4. Vlahou, G., and Rivero, F. (2006) Rho GTPase signaling in *Dictyostelium discoideum*: insights from the genome. *Eur J Cell Biol* 85, 947–959.

5. Park, K.C., Rivero, F., Meili, R., Lee, S., Apone, F., and Firtel, R.A. (2004) Rac regulation of chemotaxis and morphogenesis in *Dictyostelium*. *EMBO J* 23, 4177–4189.

6. Han, J. W., Leeper, L., Rivero, F., and Chung, C.Y. (2006) Role of RacC for the regulation of WASP and phosphatidylinositol 3-kinase during chemotaxis of *Dictyostelium*. *J Biol Chem* 281, 35224–35234.

7. Rivero, F., Illenberger, D., Somesh, B.P., Dislich, H., Adam, N., and Meyer, A.K. (2002) Defects in cytokinesis, actin reorganization and the contractile vacuole in cells deficient in RhoGDI. *EMBO J*. 21, 4539–4549.

8. Somesh, B.P., Vlahou, G., Iijima, M., Insall, R.H., Devreotes, P., and Rivero, F. (2006) RacG regulates morphology, phagocytosis, and chemotaxis. *Eukaryot Cell* 5, 1648–1663.

9. Chung, C.Y., Lee, S., Briscoe, C., Ellsworth, C., and Firtel, R.A. (2000) Role of Rac in controlling the actin cytoskeleton and chemotaxis in motile cells. *Proc. Natl. Acad. Sci. USA* 97, 5225–5230.

10. Larochelle, D.A., Vithalani, K.K., and De Lozanne, A. (1996) A novel member of the Rho family of small GTP-binding proteins is specifically required for cytokinesis. *J. Cell Biol.* 133, 1321–1329.

11. Dumontier, M., Hocht, P., Mintert, U., and Faix, J. (2000) Rac1 GTPases control filopodia formation, cell motility, endocytosis, cytokinesis and development in *Dictyostelium*. *J. Cell Sci.* 113, 2253–2265.

12. Somesh, B.P., Neffgen, C., Iijima, M., Devreotes, P., and Rivero, F. (2006) *Dictyostelium* RacH regulates endocytic vesicular trafficking and is required for localization of vacuolin. *Traffic* 7, 1194–1212.

13. Lee, E., Seastone, D.J., Harris, E., Cardelli, J.A., and Knecht, D.A. (2003) RacB regulates cytoskeletal function in *Dictyostelium* spp. *Euk. Cell* 2, 474–485.

14. Rivero, F., Albrecht, R., Dislich, H., Bracco, E., Graciotti, L., Bozzaro, S., and Noegel, A.A. (1999) RacF1, a novel member of the Rho protein family in *Dictyostelium discoideum*, associates transiently with cell contact areas, macropinosomes, and phagosomes. *Mol. Biol. Cell* 10, 1205–1219.

15. Seastone, D.J., Lee, E., Bush, J., Knecht, D., and Cardelli, J. (1998) Overexpression of a novel Rho family GTPase, RacC, induces unusual actin-based structures and positively affects phagocytosis in *Dictyostelium discoideum*. *Mol. Biol. Cell* 9, 2891–2904.

16. Hall, A. L., Schlein, A., and Condeelis, J. (1988) Relationship of pseudopod extension to chemotactic hormone-induced actin polymerization in amoeboid cells. *J. Cell. Biochem.* 37, 285–299.

17. Howard, T.H., and Oresajo, C.O. (1985) The kinetics of chemotactic peptide-induced change in F-actin content, F-actin distribution, and the shape of neutrophils. *J Cell Biol* 101, 1078–1085.

18. Benard, V., Bohl, B.P., and Bokoch, G.M. (1999) Characterization of rac and cdc42 activation in chemoattractant-stimulated human neutrophils using a novel assay for active GTPases. *J Biol Chem* 274, 13198–13204.

19. Vlahou, G., Schmidt, O., Wagner, B., Uenlue, H., Dersch, P., Rivero, F., and Weissenmayer, B.A. (2009) *Yersinia* outer protein YopE affects the actin cytoskeleton in *Dictyostelium discoideum* through targeting of multiple Rho family GTPases. *BMC Microbiol* 9, 138.

20. de la Roche, M., Mahasneh, A., Lee, S.F., Rivero, F., and Cote, G.P. (2005) Cellular distribution and functions of wild-type and constitutively activated *Dictyostelium* PakB. *Mol Biol Cell* 16, 238–247.

21. Condeelis, J., and Hall, A.L. (1991) Measurement of actin polymerization and cross-linking in agonist-stimulated cells. *Methods Enzymol* 196, 486–496.

Chapter 26

FRAP-Based Analysis of Rho GTPase-Dependent Polar Exocytosis in Pollen Tubes

An Yan and Zhenbiao Yang

Abstract

The regulation of exocytosis by Rho GTPases is conserved in eukaryotic kingdoms. Given the spatiotemporal dynamics of Rho GTPase activity, a method for visualizing exocytosis with high space and time resolution will be an important tool for the functional analysis of Rho GTPases. During tip growth of pollen tubes, both cell wall material and plasma membrane are renewed via rapid exocytosis. ROP1 GTPase, a plant Rho GTPase, is localized to the apical plasma membrane of pollen tubes and controls polar exocytosis via regulation of F-actin dynamics. Here, we describe a fluorescence recovery after photobleaching (FRAP) method for the analysis of exocytosis dynamics in living pollen tubes. This method has been shown to be a valuable tool for studying the effect of ROP1-dependent F-actin dynamics on polar exocytosis in pollen tubes and may also be applicable to other systems.

Key words: Exocytosis, FRAP analysis, ROP1 GTPase, Pollen tubes

1. Introduction

The regulation of polar growth and morphogenesis by Rho-family GTPases is conserved in all eukaryotic kingdoms, and the underlying mechanisms also seem to be conserved. One shared mechanism is apparently polarized exocytosis/secretion mediated by the exocyst. The active form of Rho GTPases or Rho effector proteins are known to interact with subunits of the exocyst complex in different eukaryotic organisms (1–6). However, direct demonstration of the relationship between Rho GTPases and exocytosis has been scarce because of lack of an easy and efficient method for measuring exocytosis, especially in living cells. We have developed an easy assay for measuring exocytosis in living cells in the ROP1 Rho GTPase regulation of polarized tip growth in pollen tubes (7).

Francisco Rivero (ed.), *Rho GTPases: Methods and Protocols*, Methods in Molecular Biology, vol. 827,
DOI 10.1007/978-1-61779-442-1_26, © Springer Science+Business Media, LLC 2012

Rho-like GTPases from *plants* (ROP) is a plant-specific subfamily of Rho GTPases, which acts as a versatile signal transduction switch regulating various processes, such as polar growth, hormonal responses, disease defense, and nonbiotic stress responses (8). The *Arabidopsis* genome encodes 11 different ROP GTPases, and ROP1 is the major ROP GTPase regulating polar growth in pollen tubes. ROP1 is activated at the apical region of the plasma membrane (PM) in pollen tubes (9–11). Overexpression of a constitutive active ROP1 (CA-ROP1) mutant leads to depolarized growth of pollen tubes, whereas overexpression of a dominant negative ROP1 (DN-ROP1) mutant suppresses pollen tube growth (12, 13). These results suggest that ROP1 controls polarized tip growth in pollen tubes.

ROP1 regulates pollen tip growth through at least two downstream pathways that are dependent on ROP1 effector proteins RIC3 and RIC4. RICs (ROP interactive CRIB motif-containing proteins) belong to a group of ROP1-interacting proteins, which bind active ROP GTPases via a conserved Cdc42 and Rac interactive binding (CRIB) motif (10). In pollen tubes, active ROP1 promotes the accumulation of tip-localized F-actin and calcium through the RIC4- and RIC3-dependent pathways, respectively. Calcium further regulates actin depolymerization, leading to actin dynamics (7, 9). In oscillating tobacco pollen tubes, ROP1 activity also oscillates, slightly before growth rate oscillation, while cytosolic calcium oscillation is behind growth rate oscillation (11, 14).

Tip growth is strictly dependent on polar exocytosis. Given the activation of ROP1 in the apical PM, we speculated that ROP1 controls tip growth via its regulation of polar exocytosis. Transmission electron microscopy analysis reveals the accumulation of high density vesicles in the tip of pollen tubes, leading to the hypothesis that exocytosis occurs in the apex of pollen tubes (15). To test our hypothesis that ROP1 regulates polar exocytosis, we developed a noninvasive method for the analysis of exocytosis, which is based on fluorescence recovery after photobleaching (FRAP). We fused GFP to an *Arabidopsis* receptor-like kinase (RLK), which is targeted to the apical PM of pollen tubes via exocytosis. This RLK, which shares high sequence similarity to a previously reported tobacco pollen PM marker RLK (16), is mainly localized to the apical PM when transiently expressed in tobacco pollen tube (7). The apical PM localization pattern of this RLK-GFP marker is strictly dependent on exocytosis activity because disruption of exocytosis by Brefeldin A treatment completely abolished the PM localization pattern (7). In this FRAP assay, we first photobleached the RLK-GFP signal in the apical PM of pollen tubes, and then conducted a time course analysis of the recovery of the RLK-GFP signal in the bleached region. We reasoned that the rate of the recovery reflects the rate of exocytosis in tobacco pollen tubes.

Using this FRAP method, it has been revealed that ROP1 GTPase regulates polar exocytosis via the control of F-actin dynamics in pollen tubes (7). In this chapter, we introduce the protocol for the FRAP method to analyze exocytosis in growing tobacco pollen tubes.

2. Materials

2.1. Tobacco Plant Growth Conditions

Tobacco (*Nicotiana tabacum*) plants are grown in a growth chamber at 25°C under a light regime of 12 h of light and 12 h of dark.

2.2. Media, Buffers, and Reagents

1. Liquid tobacco pollen germination medium: 18% sucrose, 0.01% H_3BO_3, 1 mM $MgSO_4$, 5 μM $CaCl_2$, 5 μM $Ca(NO_3)_2$, pH 7.0 (see Note 1).

2. 2.5 M $CaCl_2$ solution (autoclaved).

3. 0.1 M spermidine solution.

4. Nitrocellulose membrane (Whatman).

5. Constructs for transient gene expression in tobacco pollen tubes (available from authors' laboratory): pLat52:RLK-GFP, pLat52:CA-ROP1, pLat52:DN-ROP1, pLat52:RIC4, and pLat52:RIC3 (see Note 2).

6. Gold particles (1 μm diameter, Biorad).

2.3. Equipment

1. Gene gun system (Biorad PDS-1000).

2. Confocal microscope: A Leica SP2 confocal microscope equipped with 10× and 63× water immersion objective lenses and 488 nm laser is used for FRAP analysis of RLK-GFP in tobacco pollen tubes (see Note 3).

3. Image analysis tool. Images collected by the confocal microscope are analyzed using Image J software, available for free download at http://rsbweb.nih.gov/ij/download.html.

3. Methods

3.1. Transient Expression in Tobacco Pollen Tubes

To study exocytosis in normal growing control pollen tubes, RLK-GFP plasmid DNA is used. To study the effect of ROP1 signaling on exocytosis, pLat52:CA-ROP1, pLat52:DN-ROP1, pLat52:RIC3, or pLat52:RIC4 plasmid DNA is mixed with pLat52:RLK-GFP plasmid DNA. Constructs are introduced into pollen tubes using a gene gun system (17). This transient gene expression method is quick (less than 1 day from gene transformation to imaging) and efficient (usually more than five transformed pollen tubes can be found per slide of 15 μl liquid medium containing pollen tubes).

1. Aliquot 2 μL of plasmid DNA in a fresh 1.5-mL centrifuge tube. Mix 0.5 μg of pLat52:RLK-GFP with 0.2 μg of pLat52:CA-ROP1, pLat52:DN-ROP1, pLat52:RIC3, or pLat52:RIC4 plasmid DNA.

2. Add 8 μL of well-suspended gold particles. Gold particles are first washed with ethanol and suspended well in 50% glycerol.

3. Add 20 μL of 2.5 M CaCl$_2$ and mix with the gold particles.

4. Add 8 μL of 0.1 M spermidine and mix with gold particles.

5. Vortex the mixture thoroughly for 8 min and let stand for 1 min.

6. Briefly spin down (for about 6 s) the gold particles.

7. Remove the solution and rinse the gold particle pellet with 40 μL of 70% ethanol (see Note 4).

8. Remove the solution and rinse the gold particle pellet with 40 μL of absolute ethanol (see Note 4).

9. Add 15 μL of absolute ethanol and resuspend the gold particles.

10. Put a macrocarrier in the holder and spread the suspended gold particles on the center of the macrocarrier. Let it dry and then put in the macrocarrier launch assembly.

11. Collect fresh mature tobacco pollen grains in 1.5-mL centrifuge tubes. Pollen grains from two tobacco flowers are enough for one bombardment.

12. Suspend pollen grains in 120 μL of liquid pollen germination medium.

13. Place a filter paper on a lid of a 10-cm Petri dish and moisten it with 1.2 mL of liquid pollen germination medium.

14. Place a piece of nitrocellulose membrane (2×2 cm) on the center of filter paper and let it moisten.

15. Spread 100 μL of suspended pollen grains on the nitrocellulose membrane. Let it dry until extra liquid disappears while pollen grains are still wet.

16. Place the pollen sample into the gene gun. Bombard gold particles coated with DNA into tobacco pollen tubes using the gene gun system following the instructions on the operating manual. The rupture pressure for the rupture disk is 1,100 psi.

17. The bombarded pollen grains are washed into 1.6 mL of liquid pollen germination medium and incubated for 4 h at room temperature before observation.

3.2. FRAP Assay

1. Four hours after bombardment, pollen tubes are mounted on a glass slide to be observed under the 10× water immersion lens of the confocal microscope.

2. Choose pollen tubes with moderate RLK-GFP expression levels for FRAP analysis.

3. Adjust the position of the pollen tube to the center of field and set the zoom value to 4. Switch to the 63× water immersion lens. Take one control image of a RLK-GFP expressing pollen tube using 25% of the 488 nm excitation laser power and the emission channel at 500–540 nm.

4. Use the "Zoom in" function of the Leica confocal software to select the region of interest (ROI) for photobleaching. The ROI region is photobleached by 10–15 iterations of the 488 nm laser line at 100% emission laser power (see Note 5).

5. After photobleaching, set the zoom value back to four immediately and begin recording RLK-GFP fluorescence recovery with 25% of 488 nm laser power at 5-s intervals for 3 min. The emission channel for RLK-GFP signal is 500–540 nm. At the end of 3 min, save this time course images as one set of FRAP data for further analysis. Examples of these FRAP images can be found in Figs. 1 and 2. The pollen tube preparation can be maintained for imaging for as long as 30 min before it drying out.

3.3. Image Analysis

1. Start ImageJ, open a time lapse file of a FRAP time course generated with the confocal microscope.

2. Enlarge the image using the "zoom in" tool (displayed in the tool box icon area) of image J. A "segmented line" tool (displayed in the tool box icon area) of Image J is used to select the PM of the ROI.

3. The mean RLK-GFP intensity on the PM of the ROI is measured and recorded by a built in analysis tool ("Measure" function under "Tool" menu) of Image J. The average intensity is displayed as "mean" value on screen. Copy the data of the average intensity and transfer it to an Excel spreadsheet for further analysis.

4. Plot the intensities of RLK-GFP after photobleaching as a function of time using the Excel plot function. Typical examples of the recovery curves of FRAP are shown in Figs. 1 and 2.

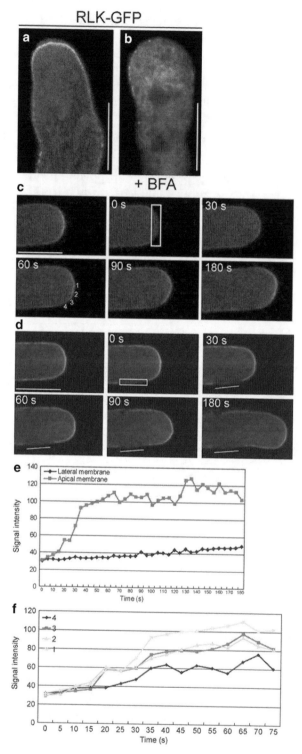

Fig. 1. FRAP of RLK-GFP-stained apical PM in tobacco pollen tubes. (**a**) PM localization of RLK-GFP in a tobacco pollen tube transiently expressing a Lat52:RLK-GFP construct. Confocal images were taken from the midplane. (**b**) BFA effect on the localization of RLK-GFP. Transformed pollen tubes expressing RLK-GFP were treated with 1 μg/mL BFA for 1 h before imaging.

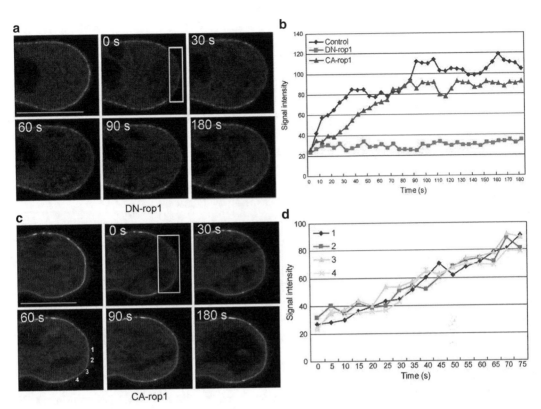

Fig. 2. FRAP analysis of RLK-GFP in pollen tubes expressing CA-rop1 or DN-rop1. (**a, c**) RLK-GFP was coexpressed with DN-ROP1 (**a**) or CA-ROP1 (**c**) in tobacco pollen tubes. After a 4-h incubation, transformed pollen tubes underwent FRAP analysis. The series of images shows before bleaching, immediately after bleaching (0 s), and recovery of fluorescence after bleaching at the indicated time points. The bleached area is marked by the *boxes*. Bars, 10 μm. (**b**) Quantitative FRAP analysis of (**a**) and (**c**). The mean intensity of recovered fluorescence was measured by drawing a line on the photobleached PM regions and plotted as a function of time. (**d**) Quantitative FRAP analysis. The intensity of the recovered GFP signal at four PM regions as indicated by the numbers indicated in (**c**) is plotted as a function of time. Note that fluorescence reappeared simultaneously at the whole bleached membrane. Recovered GFP signal was measured every 5 s for 80 s. Reproduced from ref. 7.

Fig 1. (continued)   BFA completely disrupted PM localization of RLK-GFP. The focus is on the mid-plane. Bar, 10 μm. (**c, d**) FRAP analysis of RLK-GFP in growing pollen tubes. Photobleaching was performed and recovery was analyzed in the apical area (**c**) and subapical area (**d**) of the pollen tube PM. A series of time-lapse images was recorded every 5 s for 3 min. The numbers in each image indicate elapsed time after photobleaching. The bleached area is marked by *boxes*. Note that photobleaching did not affect pollen tube elongation. Bars, 10 μm. (**e**) Quantitative analysis of FRAP for (**c**) and (**d**). Fluorescence recovery was measured by calculating the mean GFP signal intensity on the PM of the region of interest. Fluorescence at the apical PM area recovered completely within 75 s, with a 35 ± 3-s half-time of recovery (*filled square*), whereas little fluorescence recovery occurred in the subapical PM area (*filled diamond*). Similar results were obtained from five individual experiments. (**f**) Quantitative FRAP analysis at four different regions of the membrane. The intensity of recovered GFP signal is plotted as a function of time. Fluorescence recovery was fast at the apical PM. The rate of signal recovery gradually became slower away from the center of the apex. The recovered GFP signal was measured every 5 s for 80 s. The regions (1–4) where FRAP was measured are indicated in (**c**). Reproduced from ref. 7.

4. Notes

1. The pH of the medium is adjusted using 1 N KOH. We usually prepare fresh liquid germination medium at the day of the transient gene expression. Alternatively, the medium may be stored at 4°C for about 1 week.

2. Transient expression of these constructs in tobacco pollen tubes is driven by a pollen-specific Lat52 promoter. For higher transient expression efficiency, all the plasmids are purified using a Qiagen plasmid purification kit.

3. Other confocal systems may also be used for FRAP analysis, such as the Zeiss 510 system. The microscope can be upright or inverted.

4. Be gentle when rinsing, do not disturb the pellet.

5. Avoid over photobleaching due to possible side effects on pollen tube growth.

References

1. Hazak, O., Bloch, D., Poraty, L., Sternberg, H., Zhang, J., Friml, J., and Yalovsky, S. (2010) A rho scaffold integrates the secretory system with feedback mechanisms in regulation of auxin distribution. *PLoS Biol* 8, e1000282.

2. Hala, M., Cole, R., Synek, L., Drdova, E., Pecenkova, T., Nordheim, A., Lamkemeyer, T., Madlung, J., Hochholdinger, F., Fowler, J.E., and Zarsky, V. (2008) An exocyst complex functions in plant cell growth in *Arabidopsis* and tobacco. *Plant Cell* 20, 1330–1345.

3. Lavy, M., Bloch, D., Hazak, O., Gutman, I., Poraty, L., Sorek, N., Sternberg, H., and Yalovsky, S. (2007) A Novel ROP/RAC effector links cell polarity, root-meristem maintenance, and vesicle trafficking. *Curr Biol* 17, 947–952.

4. Lee, Y. J., and Yang, Z. (2008) Tip growth: signaling in the apical dome. *Curr Opin Plant Biol* 11, 662–671.

5. Li, S., Gu, Y., Yan, A., Lord, E., and Yang, Z.B. (2008) RIP1 (ROP Interactive Partner 1)/ICR1 marks pollen germination sites and may act in the ROP1 pathway in the control of polarized pollen growth. *Mol Plant* 1, 1021–1035.

6. Wu, H., Rossi, G., and Brennwald, P. (2008) The ghost in the machine: small GTPases as spatial regulators of exocytosis. *Trends Cell Biol* 18, 397–404.

7. Lee, Y. J., Szumlanski, A., Nielsen, E., and Yang, Z. (2008) Rho-GTPase-dependent filamentous actin dynamics coordinate vesicle targeting and exocytosis during tip growth. *J Cell Biol* 181, 1155–1168.

8. Yang, Z. (2008) Cell polarity signaling in *Arabidopsis*. *Annu Rev Cell Dev Biol* 24, 551–575.

9. Gu, Y., Fu, Y., Dowd, P., Li, S., Vernoud, V., Gilroy, S., and Yang, Z. (2005) A Rho family GTPase controls actin dynamics and tip growth via two counteracting downstream pathways in pollen tubes. *J Cell Biol* 169, 127–138.

10. Wu, G., Gu, Y., Li, S., and Yang, Z. (2001) A genome-wide analysis of *Arabidopsis* Rop-interactive CRIB motif-containing proteins that act as Rop GTPase targets. *Plant Cell* 13, 2841–2856.

11. Hwang, J. U., Gu, Y., Lee, Y. J., and Yang, Z. (2005) Oscillatory ROP GTPase activation leads the oscillatory polarized growth of pollen tubes. *Mol Biol Cell* 16, 5385–5399.

12. Kost, B., Lemichez, E., Spielhofer, P., Hong, Y., Tolias, K., Carpenter, C., and Chua, N. H. (1999) Rac homologues and compartmentalized phosphatidylinositol 4, 5-bisphosphate act in a common pathway to regulate polar pollen tube growth. *J Cell Biol* 145, 317–330.

13. Li, H., Lin, Y., Heath, R.M., Zhu, M.X., and Yang, Z. (1999) Control of pollen tube tip

growth by a Rop GTPase-dependent pathway that leads to tip-localized calcium influx. *Plant Cell* 11, 1731–1742.

14. McKenna, S.T., Kunkel, J.G., Bosch, M., Rounds, C.M., Vidali, L., Winship, L.J., and Hepler, P.K. (2009) Exocytosis precedes and predicts the increase in growth in oscillating pollen tubes. *Plant Cell* 21, 3026–3040.

15. Hepler, P.K., Vidali, L., and Cheung, A.Y. (2001) Polarized cell growth in higher plants. *Annu Rev Cell Dev Biol* 17, 159–187.

16. Cheung, A.Y., Chen, C.Y., Glaven, R.H., de Graaf, B.H., Vidali, L., Hepler, P.K., and Wu, H.M. (2002) Rab2 GTPase regulates vesicle trafficking between the endoplasmic reticulum and the Golgi bodies and is important to pollen tube growth. *Plant Cell* 14, 945–962.

17. Fu, Y., Wu, G., and Yang, Z. (2001) Rop GTPase-dependent dynamics of tip-localized F-actin controls tip growth in pollen tubes. *J Cell Biol* 152, 1019–1032.

INDEX

Francisco Rivero (ed.), *Rho GTPases: Methods and Protocols*, Methods in Molecular Biology, vol. 827,
DOI 10.1007/978-1-61779-442-1, © Springer Science+Business Media, LLC 2012